26th Technical and Scientific Conference 'Transport, Ecology - Sustainable Development' (EKO Varna 2020)

IOP Conference Series: Materials Science and Engineering
Volume 977

Varna, Bulgaria
8 - 10 October 2020

Editors:

R. Hristov
R. Dimitrov

ISBN: 978-1-7138-3205-8
ISSN: 1757-8981

Printed from e-media with permission by:

Curran Associates, Inc.
57 Morehouse Lane
Red Hook, NY 12571

CURRAN ASSOCIATES INC. proceedings.com

Some format issues inherent in the e-media version may also appear in this print version.

Institute of Physics
Dirac House, Temple Back
Bristol BS1 6BE UK

Phone: 44 1 17 929 7481
Fax: 44 1 17 920 0979

techtracking@iop.org

Additional copies of this publication are available from:

Curran Associates, Inc.
57 Morehouse Lane
Red Hook, NY 12571 USA
Phone: 845-758-0400
Fax: 845-758-2633
Email: curran@proceedings.com
Web: www.proceedings.com

TABLE OF CONTENTS

IOP Conf. Series: Materials Science and Engineering **977** (2020) 011001 doi:10.1088/1757-899X/977/1/011001

PREFACE

We are pleased to introduce the Proceedings of the 26th TECHNICAL AND SCIENTIFIC CONFERENCE "TRANSPORT, ECOLOGY – SUSTAINABLE DEVELOPMENT" EKO Varna 2020. The conference was held from October 8th to October 10th, 2020 in the city of Varna, Bulgaria. It was organised by Transport Engineering and Technologies Department at Technical University of Varna.

The conference was attended by scientists and researchers from Bulgaria, Russia, Ukraine and Poland. The conference was inaugurated by the Rector of Technical University of Varna and was attended by representatives of Varna Regional Administration and the City Hall of Varna.

The main conference topics were:

- Vehicle design and technology
- Ecology of transport engines
- Bio fuels for internal combustion engines
- Road safety
- Electric vehicles and charging stations

The conference was held for the first time in a hybrid format - attendance in the conference hall of Rosllyn Dimyat Hotel Varna and online on the Zoom platform. Despite the difficulties associated with the global Covid 19 pandemic and the postponement to October, the conference was held in compliance with all safety measures. The whole conference could be watched by participants and spectators live through the ZOOM platform. The presentations of the online participants were watched on the main screen in the conference hall. Each participant had 15 minutes to present their work. There were lengthy discussions for some articles. As a result, the program of the conference had to be changed and two of the presentations from first day were postponed to the second day.

The editors would like to thank all authors for their high-quality contributions and the members of the international programme committee for their commitment. We thank also to Academic Management of Technical University of Varna and to our commercial partners and exhibitors.

Official conference website: http://ecovarna.org

Editors
R. Hristov
R. Dimitrov

IOP Conf. Series: Materials Science and Engineering **977** (2020) 011002 doi:10.1088/1757-899X/977/1/011002

Peer review declaration of EKO Varna 2020

All papers published in this volume of ***IOP Conference Series: Materials Science and Engineering*** have been peer reviewed through processes administered by the Editors. Reviews of the professional and scientific standards required by the proceedings journal published by IOP Publishing have been carried out by expert referees.

- **Type of peer review:** Double-blind
- **Conference submission management system:** Microsoft CMT, E-mail: ecovarna@tu-varna.bg
- **Number of submissions received:** 57
- **Number of submissions sent for review:** 30
- **Number of submissions accepted:**28
- **Acceptance Rate (Number of Submissions Accepted / Number of Submissions Received X 100):** 49,12 %
- **Average number of reviews per paper:** 2
- **Total number of reviewers involved:** 14
- **Any additional info on review process:**
  After preliminary review by the editors of the received applications, 30 articles were approved for review. People with extensive experience were selected as reviewers, who with their reviews helped the participants to present high-quality papers.
- **Contact person for queries:**
  Assoc. Prof. Rosen Hristov, PhD
  E-mail: rosen.hristov@tu-varna.bg

IOP Conf. Series: Materials Science and Engineering **977** (2020) 012001 doi:10.1088/1757-899X/977/1/012001

The investigation of stability of wheeled vehicle model in curved sections of overpass

A Yefymenko[1*], V Verbitskii[2], V Khrebet[3]

[1]Department of Automobiles, National Transport University, 1 Mykhaila Omelianovycha-Pavlenka Street, Kyiv, 01010, Ukraine
[2]Department of Automation System Maintenance, Zaporizhzhia State Engineering Academy, 226 Soborny Avenue, Zaporizhzhia, 69006, Ukraine
[3]Department of basic and special disciplines, Educational and scientific Institute of continuing education of the National aviation University, 1 Cosmonaut Komarova Avenue, Kyiv, 103058, Ukraine

*E-mail: alla.yefimenko@yandex.ru

Abstract. It deals with the reassignment of vertical and transverse reactions of wheel assembly of monorail vehicle. Characteristics of force interaction of monorail wheels with overhead road surface at passage in curved sections of overpass were investigated. Also it was determined the intervals of vertical stiffness of bearing wheels and transverse stiffness of guide wheels with the safety movement of monorail vehicle.

1. Introduction

Metropolitans are crowded with vehicles that create traffic jams. The high concentration of cars reduces not only negative effects, such as traffic noise, air pollution and traffic accidents. All this has a negative environmental impact on the quality of life. The appearance of monorail transport in metropolitan solves a number of the above-mentioned problems. This mode of transport carries passengers without delay during rush hours, compared to common urban transport, thanks to the placement of monorail overpasses at some distance above the ground. The main advantage of the monorail road is that it, like the metro, does not occupy a place on the crowded highways of the city, but unlike the metro, is cheaper to build. The monorail stricter can overcome steeper vertical slopes compared to double rail transport and is much quieter.

The monorail with the "Alweg" support system [1,2] kindles practical interest. The structural features of the monorail car were considered in V.P. Sahno and A.N. Yefymenko's works [3,4,5]. The mathematical model of the wheeled vehicle is described in N.L. Korotenko [6], A.N. Yefymenko and V.G. Verbitsky's works [7].

The monorail vehicle moves on pneumatic tyres (bearing wheels). Stability of monorail in longitudinal and transverse direction is provided by guide wheels. Bearing and guide wheels are structurally fixed on truck (Figure 1).

The previous article presents the results of the study of vehicle stability in straight movement [8]. The purpose of this investigation is to check the possibility of implementing the least loaded mode of movement of the monorail car in curves of rounding of the specified radius R on the basis of the proposed spatial model of the monorail vehicle.

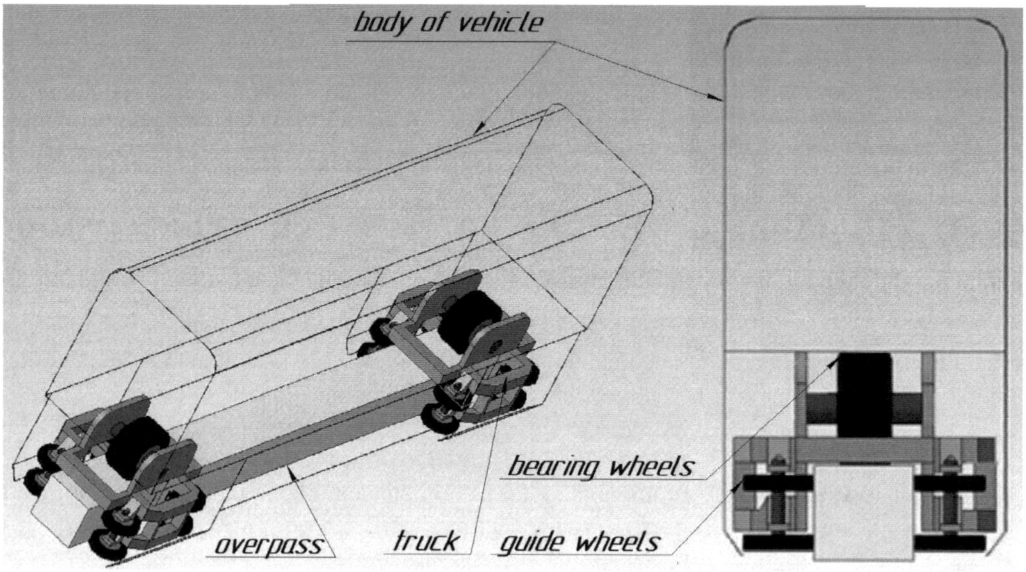

Figure 1. Wheeled vehicle

2. Mathematical model of spatial motion

During the passing of a section with a certain radius of curvature, inertia force is added to all active forces and reactions of connections. The value of inertia force is determined by the product of mass and the value of normal acceleration of the car mass center (Figure 2).

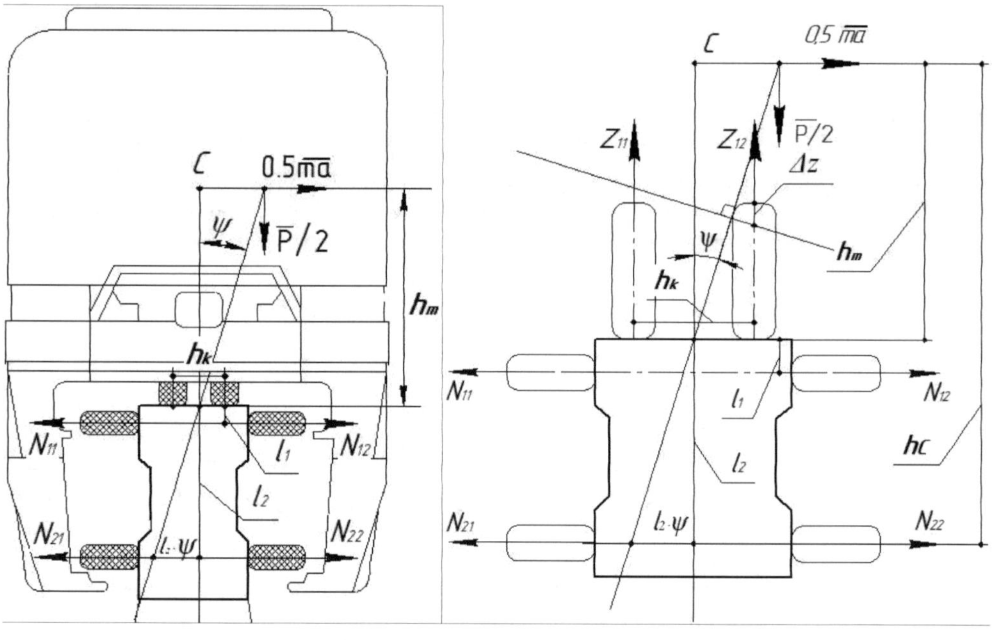

Figure 2. Scheme of monorail vehicle in vertical plane

The magnitude of normal acceleration is determined by the radius of curvature in the trajectory that the mass center is passing now.

$$a^n = V^2/R; \tag{1}$$

V is longitudinal velocity of the mass center, m/s;
R is radius of curvature of mass center trajectory, m.

Following is a system of differential motion equations (2) linking the plane-parallel motion equations and the vertical dynamics equations (jump z(t), pitch φ(t), roll ψ(t)).

$$\begin{aligned}
&z(t):\ \mathrm{m}\cdot\ddot{z}(t)+\mathrm{k}_{11}\left(-\mathrm{a}\cdot\varphi(t)-\mathrm{h}\cdot\psi(t)+z(t)\right)+\mathrm{k}_{12}\left(-\mathrm{a}\cdot\varphi(t)+\mathrm{h}\cdot\psi(t)+z(t)\right)+\\
&\quad+\mathrm{k}_{21}\cdot\left(\mathrm{b}\cdot\varphi(t)-\mathrm{h}\cdot\psi(t)+z(t)\right)+\mathrm{k}_{22}\left(\mathrm{b}\cdot\varphi(t)+\mathrm{h}\cdot\psi(t)+z(t)\right)+\mathrm{kd}_{11}\times\\
&\quad\times\left(-\mathrm{a}\cdot\dot{\varphi}(t)-\mathrm{h}\cdot\dot{\psi}(t)+\dot{z}(t)\right)+\mathrm{kd}_{12}\left(-\mathrm{a}\cdot\dot{\varphi}(t)+\mathrm{h}\cdot\dot{\psi}(t)+\dot{z}(t)\right)+\\
&\quad+\mathrm{kd}_{21}\left(\mathrm{b}\cdot\dot{\varphi}(t)-\mathrm{h}\cdot\dot{\psi}(t)+\dot{z}(t)\right)+\mathrm{kd}_{22}\left(\mathrm{b}\cdot\dot{\varphi}(t)+\mathrm{h}\cdot\dot{\psi}(t)+\dot{z}(t)\right)=0;\\
&\varphi(t):\ \mathrm{Jy}\cdot\ddot{\varphi}(t)+\left(z(t)+\mathrm{hm}\right)\mathrm{m}\cdot\mathrm{u}(t)\cdot\omega(t)-\mathrm{k}_{11}\left(-\mathrm{a}\cdot\varphi(t)-\mathrm{h}\cdot\psi(t)+z(t)\right)\cdot\mathrm{a}-\mathrm{k}_{12}\times\\
&\quad\times\left(-\mathrm{a}\cdot\varphi(t)+\mathrm{h}\cdot\psi(t)+z(t)\right)\mathrm{a}+\mathrm{k}_{21}\left(\mathrm{b}\cdot\varphi(t)-\mathrm{h}\cdot\psi(t)+z(t)\right)\cdot\mathrm{b}+\mathrm{k}_{22}\left(\mathrm{b}\cdot\varphi(t)+\mathrm{h}\cdot\psi(t)+z(t)\right)\times\\
&\quad\times\mathrm{b}-\mathrm{kd}_{11}\cdot\left(-\mathrm{a}\cdot\dot{\varphi}(t)-\mathrm{h}\cdot\dot{\psi}(t)+\dot{z}(t)\right)\mathrm{a}-\mathrm{kd}_{12}\cdot\left(-\mathrm{a}\cdot\dot{\varphi}(t)+\mathrm{h}\cdot\dot{\psi}(t)+\dot{z}(t)\right)\cdot\mathrm{a}+\mathrm{kd}_{21}\times\\
&\quad\times\left(\mathrm{b}\cdot\dot{\varphi}(t)-\mathrm{h}\cdot\dot{\psi}(t)+\dot{z}(t)\right)\cdot\mathrm{b}+\mathrm{kd}_{22}\times\left(\mathrm{b}\cdot\dot{\varphi}(t)+\mathrm{h}\cdot\dot{\psi}(t)+\dot{z}(t)\right)\cdot\mathrm{b}=0;\\
&\psi(t):\ \mathrm{Jx}\cdot\ddot{\psi}(t)+\left(z(t)+\mathrm{hm}\right)\cdot\mathrm{m}v\omega(t)+2\mathrm{KK}\cdot\psi(t)-\mathrm{k}_{11}\left(-\mathrm{a}\cdot\varphi(t)-\mathrm{h}\cdot\psi(t)+z(t)\right)\mathrm{h}+\mathrm{k}_{12}\times\\
&\quad\times\left(-\mathrm{a}\cdot\varphi(t)+\mathrm{h}\cdot\psi(t)+z(t)\right)\cdot\mathrm{h}-\mathrm{k}_{21}\left(\mathrm{b}\cdot\varphi(t)-\mathrm{h}\cdot\psi(t)+z(t)\right)\mathrm{h}+\mathrm{k}_{22}\left(\mathrm{b}\cdot\varphi(t)+\mathrm{h}\cdot\psi(t)+z(t)\right)\times\\
&\quad\times\mathrm{h}-\mathrm{kd}_{11}\cdot\left(-\mathrm{a}\cdot\dot{\varphi}(t)-\mathrm{h}\cdot\dot{\psi}(t)+\dot{z}(t)\right)\mathrm{h}+\mathrm{kd}_{12}\left(-\mathrm{a}\cdot\dot{\varphi}(t)+\mathrm{h}\cdot\dot{\psi}(t)+\dot{z}(t)\right)\mathrm{h}-\mathrm{kd}_{21}\times\\
&\quad\times\left(\mathrm{b}\cdot\dot{\varphi}(t)-\mathrm{h}\cdot\dot{\psi}(t)+\dot{z}(t)\right)\mathrm{h}+\mathrm{kd}_{22}(\mathrm{b}\cdot\dot{\varphi}(t)+\mathrm{h}\cdot\dot{\psi}(t)+\dot{z}(t))\cdot\mathrm{h}=0.
\end{aligned} \tag{2}$$

Initial equations system of "two-wheeled" vehicle motion along straight section of overpass is transferred toward inertial coordinate system (3), that moves at constant speed (parameter v) along center line of overpass.

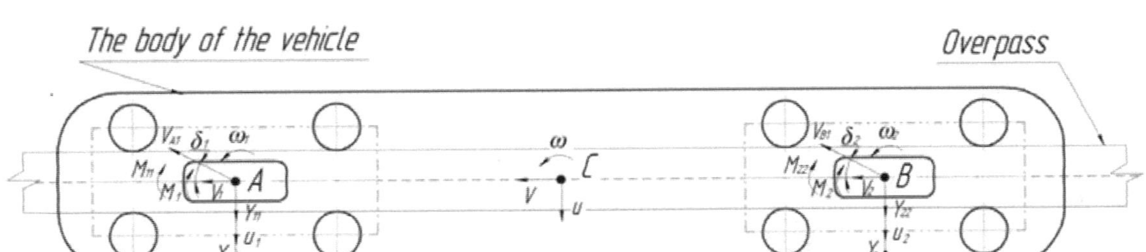

Figure 3. Calculated bicycle scheme of a wheeled vehicle

Equations of plane-parallel motion take into account rescheduling of vertical loads on bearing wheels, with conditional absence of guide wheels. Vehicle motion is described by two half degrees of freedom: transverse component of mass center velocity u and angular velocity ω relative to vertical axis.

Forces of bearing wheels slip Y_{ii} in contact spot are considered. During the movement of monorail radially of rounding R, front and rear trucks turn through angle $\theta = l/(2R)$ and $\theta = -l/(2R)$, accordingly, that is taken into account in determination of slip angles δ_{ii} (5).

$$m(\omega v + \dot{u}) = cos\theta(Y_{11} + Y_{12}) + cos\theta(Y_{21} + Y_{22});$$
$$J_Z\dot{\omega} = -b \cdot cos\theta(Y_{21} + Y_{22}) + a \cdot cos\theta(Y_{11} + Y_{12}). \tag{3}$$

The pull factor depends on the vertical load of the bearing wheel that is determined by the following ratios for each bearing wheel:

$$ky_{11} = (-\alpha Z_{11} + kky) \cdot Z_{11}; \quad ky_{12} = (-\alpha Z_{12} + kky) \cdot Z_{12};$$
$$ky_{21} = (-\alpha Z_{21} + kky) \cdot Z_{21}; \quad ky_{22} = (-\alpha Z_{22} + kky) \cdot Z_{22}; \tag{4}$$

kky and α - coefficients defined empirical for bearing wheels;
Z_{11}, Z_{12}, Z_{21}, Z_{22} - vertical load of bearing wheel of *i*-th truck;

$$Z_{11} = \frac{mgb}{2l} + k_{11}(-a\varphi(t) - h\psi(t) + z(t)); \quad Z_{12} = \frac{mgb}{2l} + k_{12}(-a\varphi(t) + h\psi(t) + z(t));$$

$$Z_{21} = \frac{mga}{2l} + k_{21}(b\varphi(t) - h\psi(t) + z(t)); \quad Z_{22} = \frac{mga}{2l} + k_{22}(b\varphi(t) + h\psi(t) + z(t));$$

k_{11}, k_{12}, k_{21}, k_{22} - radial stiffness of bearing wheels of *i*-th truck/

The total vertical load of the twin bearing wheels Z_1 and Z_2 was defined according to the structural arrangement of the bearing wheels and their rigidity.

$$Z_1 = Z_{11} + Z_{12}; \; Z_2 = Z_{21} + Z_{22}$$

Quiescent value of vertical reactions on bearing axes:

$$Z_{10} = \frac{mgb}{l}; \; Z_{20} = \frac{mga}{l}.$$

Pull angels of bearing wheels of *i*-th truck:

$$\delta_{11} = \theta - \arctan\left(\frac{a\omega + u}{-h\omega + v}\right); \quad \delta_{12} = \theta - \arctan\left(\frac{a\omega + u}{h\omega + v}\right);$$
$$\delta_{21} = -\theta + \arctan\left(\frac{b\omega - u}{-h\omega + v}\right); \quad \delta_{22} = -\theta + \arctan\left(\frac{b\omega - u}{h\omega + v}\right); \tag{5}$$

h is ½ distance between longitudinal symmetry planes of bearing wheels.

The approximation of the nonlinear dependencies of pull forces can be featured as (6).

$$Y_{11}=\frac{0.5\cdot ky_{11}\cdot\delta_{11}}{\sqrt{1+\left(0.5\cdot ky_{11}\cdot\delta_{11}/(\kappa_1 Z_{11})\right)^2}};\ Y_{12}=\frac{0.5\cdot ky_{12}\cdot\delta_{12}}{\sqrt{1+\left(0.5\cdot ky_{12}\cdot\delta_{12}/(\kappa_1 Z_{12})\right)^2}}$$
$$Y_{21}=\frac{0.5\cdot ky_{21}\cdot\delta_{21}}{\sqrt{1+\left(0.5\cdot ky_{21}\cdot\delta_{21}/(\kappa_2 Z_{21})\right)^2}};\ Y_{22}=\frac{0.5\cdot ky_{22}\cdot\delta_{22}}{\sqrt{1+\left(0.5\cdot ky_{22}\cdot\delta_{22}/(\kappa_2 Z_{22})\right)^2}} \quad (6)$$

3. Results of mathematical simulation of wheel vehicle spatial model

The following numerical values of the design parameters of the wheeled vehicle were used for the research:

$m_1 = m_2 = 1320$ kg и $m_B = 16000$ kg – the mass of the truck and the car accordingly;
m = 18640 kg – total mass of wheeled vehicle;
Jy=10000 kg·m^2 – the central cross moment of inertia;
Jx= 10000 kg·m^2 – the central longitudinal moment of inertia;
Jz= 234864 kg·m^2 – the central vertical moment of inertia;
L= 12 m – base of wheeled vehicle;
$a = b = 6$ m – the distance from car mass center to hinge pivots "A" and "B";
h_m = 1,2 m – the distance between the mass center of the car and the bearing overpass area;
h_k= 0,43 m – the distance between longitudinal symmetry planes of bearing wheels;
l_1 =0,204 m – the distance between longitudinal symmetry plane of upper row guide wheels and bearing overpass area;
l_2=1,296 m – the distance between longitudinal plane of symmetry of lower row guide wheels and bearing overpass area;
KK = 640000 N/m - vehicle stiffness factor assigned to one truck;
$k_{11} = k_{12} = k_{21} = k_{22}$ = 477700 N/m – radial stiffness of bearing wheels;
$kd_{11} = kd_{12} = kd_{21} = kd_{22}$ = 7500 N/m – total coefficient of damping effect suspension over each bearing wheel;
h = 0,215 m – ½ distances between longitudinal symmetry planes of bearing wheels;
R = 300 m – trajectory curve radius;
$\kappa_1 = \kappa_2$ = 0,8 – wheel friction coefficient in transverse direction;
V = 10..20 m/s – motion speed of wheeled vehicle;
kky = 1.222692064 – empirical coefficient;
$\alpha = 1.19130658\cdot 10^{-6}$ 1/N – empirical coefficient.

In response to more rational selection of stiffness characteristics, it is possible to reach of significant reduction of reassignment vertical loads and comfort criteria on the speed interval up to 20 m/s when moving along curve with radius of curvature 300 m. The graph shows the corresponding characteristics of reassignment of vertical loads on double bearing wheels of front and rear trucks.

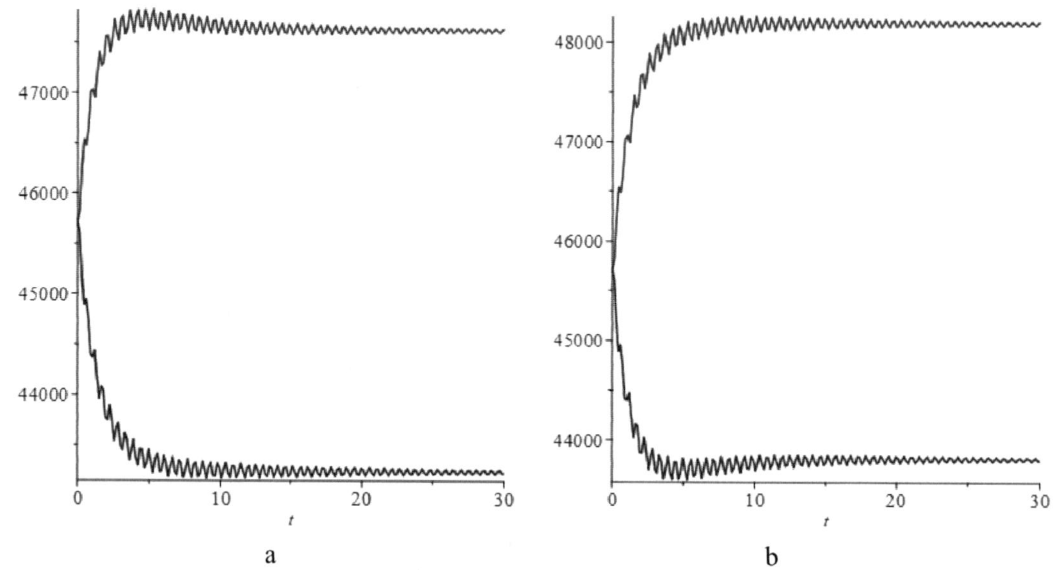

Figure 4. Vertical reactions of double bearing wheels of front (a) and rear (b) truck

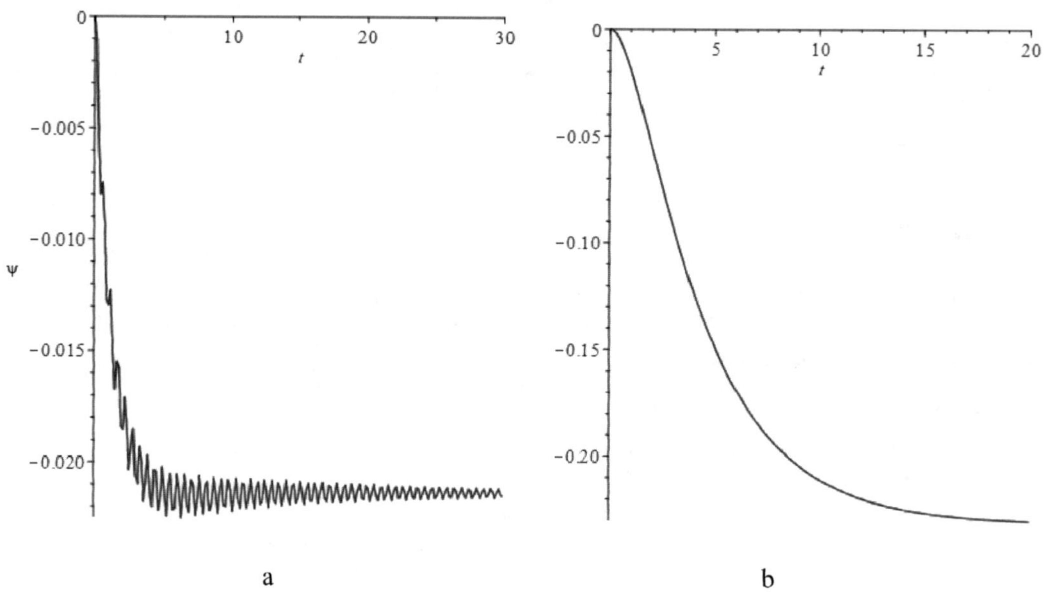

Figure 5. Time dependence of roll angle (a) and sliding angle (b)

4. Conclusions
It was determined a speed range by numerical analysis of vertical dynamics monorail vehicle model in curves of constant bend radius, where the motion of the monorail circular curve is realized with the least pressure on the side of guide wheels (without them), that will affect only realized roll angles. So for radius R = 300 m the possible interval of travel speed is up to 20 m/s. Simulating the spatial model of the monorail vehicle in circular curves, it was taking into account the effects of load reassignment on the factors of run out of bearing wheels' model, the given results illustrating the rearrangement of vertical support pressure and the measure of the roll and pitch angle in the system. Proposed model of

spatial motion allows at design stage to make rational selection of elastic and damping characteristics of wheel vehicle suspension providing required level of safety and comfort during exploitation.

References

[1] Kennedy R R 2004 *Considering Monorail Rapid Transit for North American Cities* (Narrative report) p 43
[2] www.STC-IN.com & www.skytraincorp.com 2004 *Monorail History and Technology of Successful Technology* (Narrative report) p 31
[3] Sakhno V P and Yefimenko A M 2016 *Bulletin of the National Transport University* **2** 204-210
[4] Yefymenko A M 2014 *Transport system and technologies* **25** 51-55
[5] Sakhno V P and Yefymenko A M 2016 *The Journal of ZSTU / Engineering* **2(77)** 259–265
[6] Korotenko M L and Dontsova N V 1984 *Problems of dynamics and strength of railway rolling stock* **232** 53–58
[7] Yefimenko A, Verbitskii V and Demchenko V 2015 *Transport, Ecology - Sustainable Development* **21** 341–346
[8] Gerlici J, Sakhno V, Yefymenko A and Verbitskii V 2018 *MMS 2017 MATEC Web of Conferences* **157** (01007 (2018))

IOP Conf. Series: Materials Science and Engineering **977** (2020) 012002 doi:10.1088/1757-899X/977/1/012002

Determining the criteria and the degree of the stratification of the air-fuel charge in a cylinder of a spark-ignition engine during injecting fuel

V Korohodskyi[1,4], A Voronkov[1], V Migal[2], I Nikitchenko[1], E Zenkin[1], V Rublov[3], N Rudenko[1]

[1]Kharkiv National Automobile and Highway University, Department of Internal Combustion Engines, 61002 Kharkiv, Ukraine
[2]Kharkiv Petro Vasylenko National Technical University of Agriculture, Department of Tractors and Cars, 61002 Kharkiv, Ukraine
[3]Ukrainian State University of Railway Transport, Department of Heat Engineering, Heat Engines and Energy Management, 61002 Kharkiv, Ukraine

[4]Corresponding author: korohodskiy@ukr.net

Abstract. In internal combustion engines with spark ignition and the direct fuel injection, the indicators are offered to evaluate the organization of the working process in the form of a criterion for the stratification of the air-fuel charge and the degree of stratification of the air-fuel mixture. Using the experimental data of a two-stroke engine with spark ignition and the direct injection of fuel with load characteristics, the values of the excess air coefficient in a cylinder and in the volume of the air-fuel mixture are determined. A comparative assessment of the suggested indicators in the organization of the workflow with the stratification of depleted fuel-air charge is presented. It has been defined that a more promising way of organizing the workflow to improve the fuel and economic performance of a spark ignition engine is organizing the fuel injection with the stratification of a depleted air-fuel charge.

1. Introduction

One of the promising areas for improving the technical, economic and environmental operation of spark ignition (SI) engines is the use of a power system for the direct fuel injection (DFI) into a cylinder. The formation of internal mixture contributes to the creation of conditions for the engine operation on depleted air-fuel mixtures (AFM). The air excess in the AFM allows to efficiently burn fuel with a low level of emissions of harmful substances with exhaust gases.

However, the features of organizing the working process in the internal mixture formation related to the shaping of a homogeneous or layered composition of the AFM depending on the engine operating conditions, affect the efficiency of the combustion processes and the fuel consumption respectively.

Therefore, to evaluate the effectiveness of the organization of internal mixture formation in SI engines, it is proposed to use indicators characterizing the features of the formation of the AFM composition by the head-end volume.

2. Determining the stratification of the air-fuel charge in a cylinder of an SI engine

For a qualitative assessment of the processes of internal mixture formation it is offered to use *L* criterion, which characterizes the stratification of the air-fuel charge (SAFC) in a cylinder of an SI engines during DFI [1].

$$L = \frac{\lambda_{cyl} - \lambda_{e.f.f.}}{\lambda_{e.f.f.}} = \frac{(\lambda_\Sigma - \lambda_\Sigma \cdot \upsilon) - \lambda_{e.f.f.}}{\lambda_{e.f.f.}}. \tag{1}$$

where $\lambda_{e.f.f.}$ normalized air-fuel ratio (AFR / AFR$_{stoichiometric}$) is the coefficient of excess air which takes into account the lower limit of the emission of the flame front in the petrol and air mixtures of SI engines; λ_Σ is the total coefficient of excess air (AFR / AFR$_{stoichiometric}$); λ_{cyl} is the coefficient of excess air in an engine cylinder (AFR / AFR$_{stoichiometric}$); υ is the coefficient which takes into account the leakage of purge air or AFM from an engine cylinder.

The value of the coefficient $\lambda_{e.f.f.}$ depends on the parameters (pressure and temperature) in an engine cylinder, the composition of the gasoline used [2] and the intensity of the oxidant supply to the combustion zone.

To estimate the lower limit of the emission of the flame front in petrol and air mixtures, to a first approximation $\lambda_{e.f.f.} = 1.35$ is used [2]. Values of $\lambda_{e.f.f.}$ can be specified for a particular case after systematizing the statistical experimental data which takes into account the conditions and characteristics of the fuel used.

At the same time, the use of the stratification *L* criterion allows to evaluate the influence of the organization of internal mixture formation processes on the technical, economic and environmental operation of an SI engines and to characterize the excess of the lower limit of the flame front emission in the AFM.

For a qualitative assessment of the features of the working process organization during the internal mixture formation with SAFC in SI engines, it is suggested to use the SAFC degree indicator:

$$\varepsilon_L = \frac{\lambda_{cyl}}{\lambda_{AFM}}, \tag{2}$$

where λ_{cyl} is the coefficient of excess air in a cylinder:

$$\lambda_{cyl} = \frac{M_{inc.ch.}}{B_{cycl} \cdot M_0} \cdot (1 - \upsilon), \tag{3}$$

where $M_{inc.ch.}$ is the mass of incoming charge, kg; $B_{cycl.}$ is the cyclic fuel supply, kg; M_0 is the mass of air theoretically necessary for the complete fuel combustion in an engine cylinder, kg; υ is the leakage coefficient of purge air or AFM; λ_{AFM} is the average coefficient of excess air in the AFM volume at the moment of its ignition, corresponding to the ignition dwell angle $\theta_{ign.}$ when a spark discharge is applied to the electrodes of a spark plug:

$$\lambda_{AFM} = \frac{M_{airAFM}}{B_{cycl} \cdot M_{0_{AFM}}}, \tag{4}$$

where M_{airAFM} is the air mass in the AFM volume, kg; M_{0AFM} is the air mass theoretically necessary for the complete fuel combustion in the AFM volume, kg.

3. Experimental methodology

At values of the SAFC power coefficient $\varepsilon_L > 1$ the composition of the AFM or AFM and air layers is distributed in the engine head-end volume, and at the values $\lambda_{cyl} = \lambda_{AFM}$ and accordingly $\varepsilon_L = 1$ there is no SAFC, which is characterized by a uniform distribution of fuel particles over the entire head-end volume.

The AFM volume in the head-end volume at the DFI and due to feeding the AFM into a divided combustion chamber can be determined by its size taking into account the possible flow of a part of AFM into an engine cylinder. At DFI, the AFM volume can be determined in a semi-separated and non-separated combustion chamber, for example, as a result of experimental studies using high-speed multi-channel video recording [3] or laser diagnostics [4] when registering any luminescent radiation in the ultraviolet wavelength range [5].

The obtained experimental images taken by a high-speed video camera synchronized with the laser radiation can also be used for the comparison in modeling the mixture formation processes and the SAFC combustion [6].

The AFM volume in a semi-separated and non-separated engine combustion chamber at the time of ignition can be determined by *3-D* modeling [7].

To organise the conditions for the internal mixture formation in a two-stroke SI engines at DFI on the surface of a hemispherical combustion chamber with the off-centre exhaust port and a hemispherical symmetric combustion chamber [8], the AFM volume is formed inside the combustion chamber until the voltage is applied to the electrodes of a spark plug. Due to the fact that the air charge flows during the compression stroke are directed inside the combustion chamber [9], the AFM do not travel from the volume of the combustion chamber to an engine cylinder. For that reason, the λ_{AFM} value at the moment of the AFM ignition is determined by the amount of the air M_{airAFM} that is in the volume of the hemisphere of the combustion chamber according to its design dimensions, the amount of injected fuel cycle supply B_{cyl} and the values of the theoretical air mass M_{0AFM} required for the complete combustion of the fuel used.

4. Experimental data and results

On a two-stroke SI engines 1D 8.7 / 8.2 with DFI, two work processes are organized: with the separation of the air-fuel charge (SAFC) and the separation of the depleted air-fuel charge (SDAFC).

In general, the term SAFC should be understood as such an organization of the mixture formation process in which the different AFM concentration with the head-end volume or the air excess coefficient α is provided.

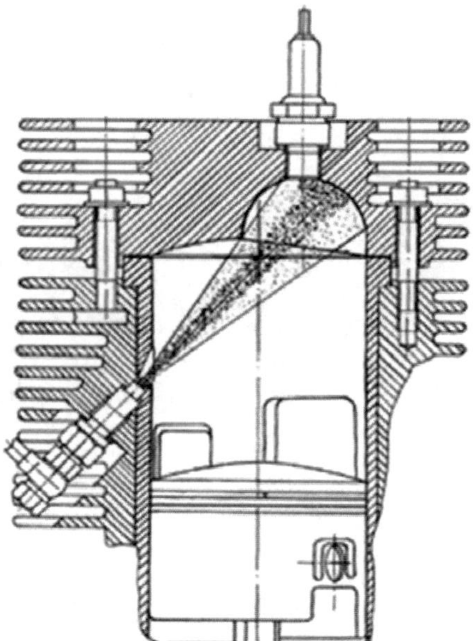

Figure 1. Organization of the work process with SAFC.

In the first workflow, the SAFC was organized with the formation of an enriched AFM spark plugs ($1.0 > \lambda > 0.8$) in the electrode zone and the air ($\lambda = \infty$) at the walls of the head-end volume.

The working process with SAFC is organized at DFI into an engine cylinder on the surface of a hemispherical combustion chamber, displaced to the exhaust port and the movement of the air charge inside the combustion chamber during the compression stroke (Figure 1) [8].

The peculiarity of the SAFC organization is that the direction and intensity of the movement of the air charge in the combustion chamber throat during the volume and film-type mixture formation does not allow to transfer fuel particles into the cylinder volume. As a result, at the time of the ignition a fuel-air charge is formed in layers in the head-end volume, in the volume of the hemisphere of the combustion chamber - AFM, and in the cylinder volume - the air.

By the amount of the air in the hemisphere of the combustion chamber, the cyclic supply of injected fuel, and the position of a piston in the cylinder at the moment of the AFM ignition, at a first approximation, the values of λ_{AFM} at the load characteristic modes have been determined (Figure 2).

The values of the total coefficient of the excess air α_{Σ} are determined due to the experimental studies obtained on the basis of measuring the air output at the inlet and the mass fuel consumption by the load characteristic at n=3000 min^{-1}.

The values of the excess air coefficient in a cylinder λ_{cyl} are determined by the amount of air entering the cylinder, the air leakage during the purge of a cylinder and the cyclic supply of the fuel directly injected into the engine head-end volume.

With increasing the load from $b_{mep} = 0.15$ *MPa* to $b_{mep} = 0.29$ *MPa* and the opening of a choke valve, the AFM composition $\lambda_{AFM} = 1.21$, respectively the coefficients of the excess air in a cylinder $\lambda_{cyl} = 1.46$ and the total coefficient of the excess air $\lambda_{\Sigma} = 1.7$ are combined up to $\lambda_{AFM} = 1.35$, $\lambda_{cyl} = 1.64$, $\lambda_{\Sigma} = 2.03$.

In the second working process, the SDAFC is organized with the formation of a depleted AFM spark plug in the electrode area ($1.5 > \lambda > 1.0$), and at the walls of the combustion chamber there was almost only the air ($\lambda = \infty$) at the value of the excess air coefficient in a cylinder $\lambda_{cyl} < 5$.

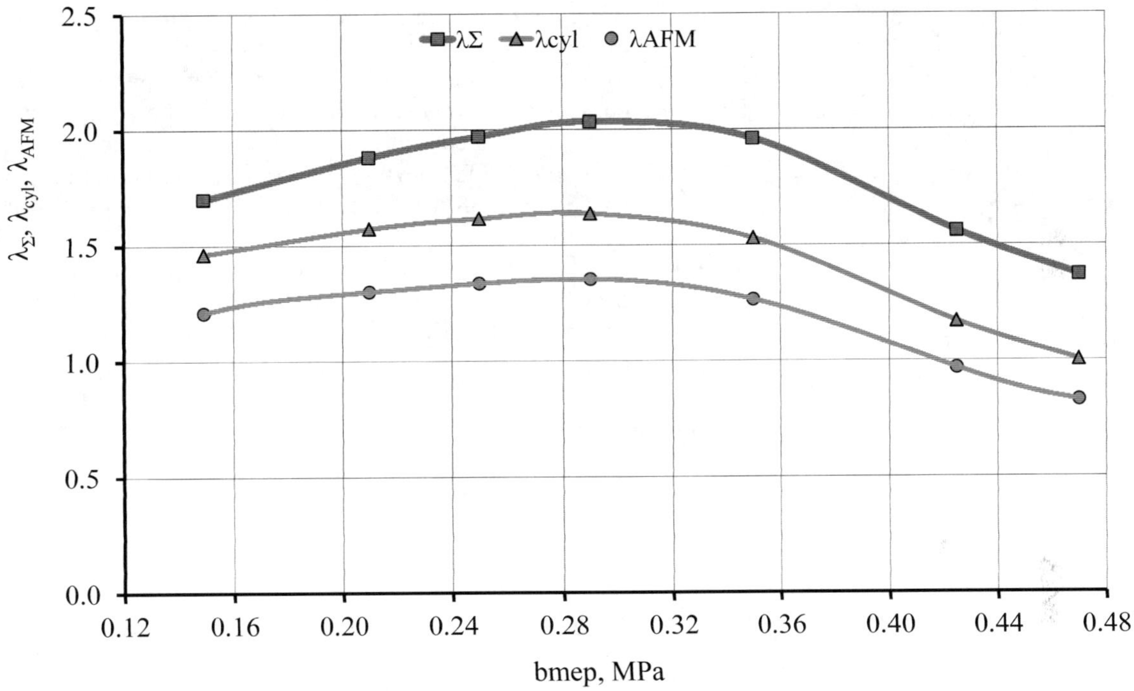

Figure 2. Values of the coefficients of the excess air: the total coefficient (λ_{Σ}), in the cylinder (λ_{cyl}) and in the volume of the air-fuel mixture (λ_{AFM}) during the organization of the working process with SAFC during different load characteristic conditions at $n = 3000$ min^{-1}.

The workflow with SDAFC is organized by installing a symmetrical hemispherical combustion chamber and using an outward-opening pintle-type nozzle that forms the peripheral distribution of fuel particles in the fuel jet (Figure 3).

A specific feature of the SDAFC organization is that the AFM is formed at the compression stroke during the fuel injection on the surface of a symmetrical hemispherical combustion chamber and the air movement from a cylinder into the combustion chamber volume, which is directed to the electrodes of a spark plug [8]. The nature of the "pressing" of the AFM volume from a cylinder from all sides by the air contributes to the organization of the layered distribution of AFM and the air in the head-end volume.

The organized directed layer-by-layer distribution of AFM and the air in the head-end volume at the time of ignition and the movement of air into the mixture volume contributes to the formation of a uniform AFM composition within the flammability range with the subsequent feeding of the air into the combustion zone.

To evaluate the AFM composition, the experimental data were used at the load characteristic conditions at $n = 3000\ min^{-1}$ and the functional features of an engine at the time of ignition.

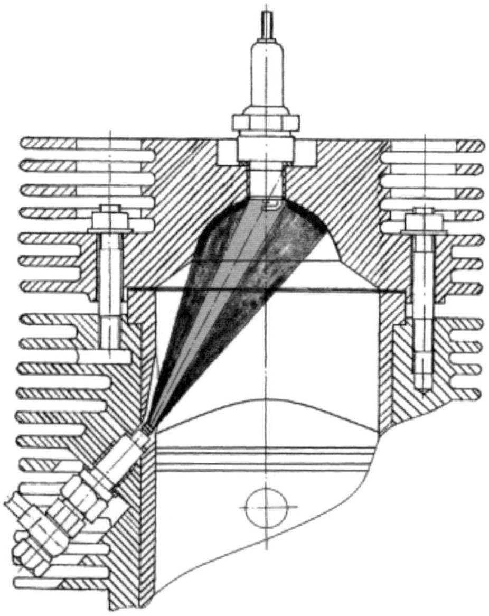

Figure 3. Organization of the workflow with SDAFC.

For this purpose, the total values of the excess air coefficient (λ_Σ) are determined at the investigated operating modes of the 1D 8.7 / 8.2 two-stroke engine taking into account the incoming air quantity at the inlet and the cyclic fuel supply.

Based upon the defined values of the working fluid leakage during the gas exchange, the amount of the air in a cylinder after closing the units of the gas distribution is determined and, taking into account the cyclic fuel supply the coefficient of the excess air in the engine cylinder (λ_{cyl}) is calculated. Considering that AFM are formed in the volume of the hemisphere of the combustion chamber so the amount of the air contained in it is defined with regard for its size and by the evaporated fuel amount, in a first approximation the coefficients of the excess air in the AFM (λ_{AFM}) are determined.

The results of certain values λ_{AFM} during the load characteristic conditions at $n = 3000\ min^{-1}$ (Figure 4) show that when the load decreases from $b_{mep} = 0.428\ MPa$ to $b_{mep} = 0.192\ MPa$ the AFM composition $\lambda_{AFM} = 1.00$, respectively the coefficient of the excess air in a cylinder $\lambda_{cyl} = 1.31$ and the total coefficient of the excess air $\lambda_\Sigma = 1.74$ are depleted up to $\lambda_{AFM} = 1.49$, $\lambda_{cyl} = 1.94$, $\lambda_\Sigma = 2.30$.

With a further decrease to a minimum load ($b_{mep} = 0.144\ MPa$), the AFM composition is enriched up to $\lambda_{AFM} = 1.30$, $\lambda_{cyl} = 1.70$, $\lambda_\Sigma = 1.97$ (Figure 4).

The expansion of the AFM depletion limit during the SDAFC organization during the partial load conditions up to $\lambda_{AFM} = 1.49$ ($b_{mep} = 0.192\ MPa$) is contributed by the increased compression ratio $\varepsilon = 11.3$, the intensive AFM turbulization and the air supply to the combustion zone.

However, the increase in the content of combustion products in the working fluid and, accordingly the increase in the values of the coefficient of residual gases while reducing up to a minimum load ($b_{mep} = 0.144\ MPa$) and increasing up to a maximum load ($b_{mep} = 0.428\ MPa$) facilitate the AFM enrichment.

5. Analysis of the results according to the indicators of the stratification of the air-fuel charge in an SI engine cylinder

According to the experimental studies of a two-stroke SI engines 1D 8.7 / 8.2 with DFI and the SAFC organization considering the load characteristic ($n = 3000\ min^{-1}$), the maximum value of the SAFC criterion $L_{SAFC} = 0.212$ is determined by the formula (1) [1] taking into account the distribution limit of the flame front in petrol-air mixtures ($\lambda_{e..f..f} = 1.35$) and the coefficient of the excess air in a cylinder λ_{cyl}, which corresponds to the maximum SAFC at $b_{mep} = 0.29\ MPa$ (Figure 5).

With the decrease in load up to $b_{mep} = 0.15\ MPa$, the level of values L_{SAFC} decreases up to 0.08, and with the increase up to the maximum loads in the range from $b_{mep} = 0.39\ MPa$ to $b_{mep} = 0.47\ MPa$, the L level has negative values, which characterizes the SAFC absence. To ensure the increased pressure in an engine cylinder during the maximum load conditions, the fuel assembly composition is enriched (Figure 2), and the AFM volume tends to increase up to the size of the head-end volume, which is consistent with the results of L_{SAFC} values obtained in the experimental studies (Figure 5).

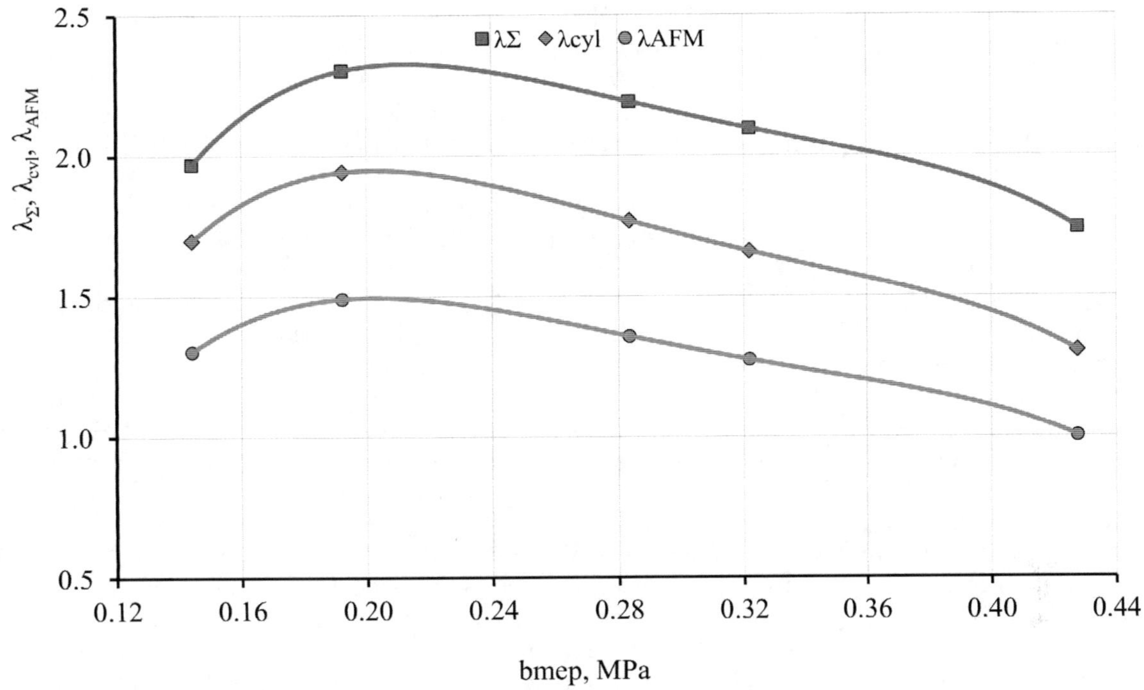

Figure 4. Values of the coefficients of the excess air: the total coefficient (λ_Σ), in the cylinder (λ_{cyl}) and in the volume of the air-fuel mixture (λ_{AFM}) during the organization of the working process with SDAFC during different load characteristic conditions at $n = 3000\ min^{-1}$.

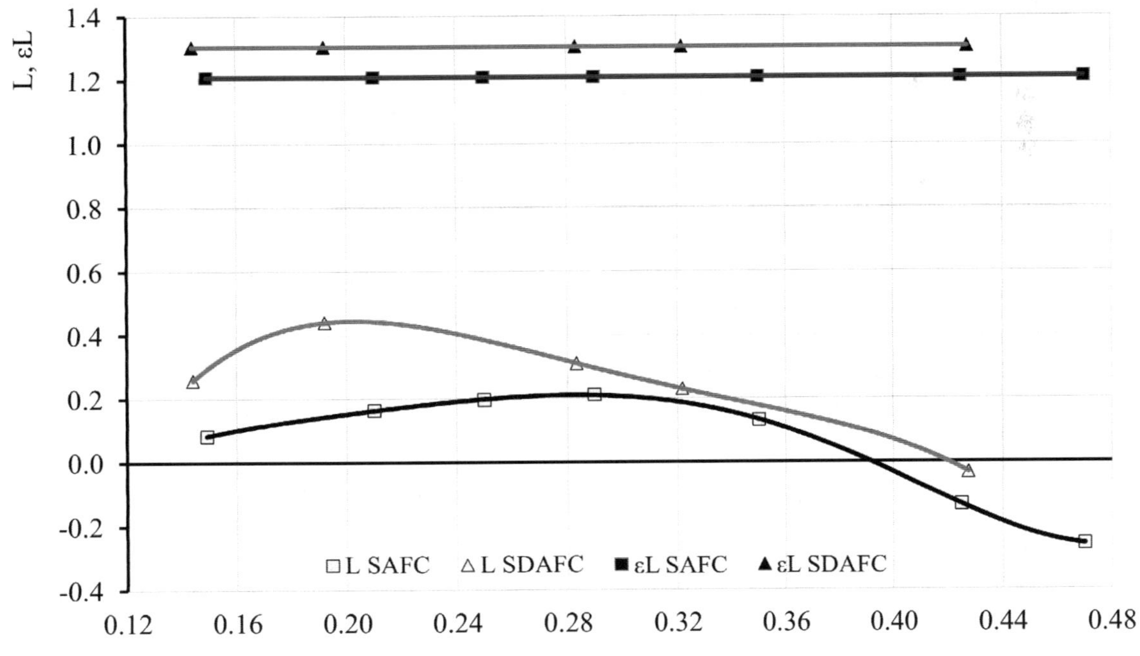

Figure 5. Values of the stratification of the air-fuel charge L and the degree of stratification ε_L when organizing the working process with SAFC and SDAFC according to the load characteristic conditions at $n = 3000\ min^{-1}$.

The values of SAFC criterion L_{SAFC} by the load characteristic ($n = 3000\ min^{-1}$) for the SDAFC working process and the increase in load from $b_{mep} = 0.144\ MPa$ to $b_{mep} = 0.192\ MPa$ increase from 0.26 to 0.44. With a further increase in load, the level of the stratification criterion decreases and $L_{SAFC} = 0$ is at $b_{mep} = 0.42\ MPa$, which is characterized by the SAFC absence. In the load range over $b_{mep} = 0.42\ MPa$, it corresponds to negative values of the stratification criterion and the SAFC absence (Figure 5).

The change in the L_{SAFC} values is related to the nature and data of the change in the coefficient of the excess air in a cylinder (λ_{cyl}) and the value adopted in the first approximation ($\lambda_{e.f.f.} = 1.35$).

The level of the degree of the SAFC separation when organizing the work process with SDAFC ($\varepsilon_L = 1.3$) is higher than when organizing SAFC ($\varepsilon_L = 1.2$) (Figure 5), considering the higher values of λ_{cyl} and λ_{AFM} (Figure 2 and Figure 4). Due to the fact that an engine with large values of the SAFC degree (ε_L) has a higher indicator efficiency (η_i), it can be stated that increasing the ε_L values contributes to a more rational organization of the working process for the efficient fuel combustion and improving the fuel and economic operation of SI engine. The degree of separation (ε_L) characterizes the perfection of the organization of the working process with SAFC in an SI engine.

6. Conclusion

The proposed criterion for the separation of the air-fuel charge (L) characterizes the presence of the separation of the air-fuel charge and allows a qualitative assessment of the influence of the internal mixture formation and combustion processes on the characteristics of an engine as a whole.

The proposed indicator of the degree of the separation of the air-fuel charge (ε_L) characterizes the perfection of the organization of the working process with the separation of the air-fuel charge in an SI engine.

It has been proved that a more promising way of organizing a workflow to increase the fuel and economic operation of an SI engine is to organize the internal mixture formation with the stratification of a depleted air-fuel charge.

References

[1] Eroshhenkov S, Korogodskij V, Voropaev E and Vasilenko O 2013 Raschetno-jeksperimental'nye issledovanija dvuhtaktnogo dvigatelja vnutrennego sgoranija s neposredstvennym vpryskivaniem topliva i iskrovym zazhiganiem *Polzunovsky Vestnik* (Barnaul: FGBOU VO AltGTU im. I.I. Polzunova) vol 4/3 pp 106–109

[2] Semenov I and Utkin P 2011 Chislennoe modelirovanie detonacionnyh processov v gazah (Moscow: IAP RAN) p 69

[3] Merker G, Schwarz C and Teichmann R 2011 Verbrennungsmotoren. Funktionsweise, Simulation, Messtechnik *Vieweg+Teubner Verlag* (Wiesbaden) p 795

[4] Fissenewert U 2005 Laserspektroskopische Untersuchungen zur Stickoxid - Bildung an einem optisch zugänglichen Viertakt-Ottomotor mit Direkteinspritzung (Berlin: Technischen Universität - Diploma Thesis) p 107

[5] Kazunari K 2003 In-Cylinder Phenomena Diagnostics for Gasoline Engine Development *Technical Review* vol 15 (Tokyo: Mitsubishi Motors) pp 21-31

[6] Rotondi R 2004 Modeling mixture formation in a GDI engine *Mec'anica Computacional* vol. XXIII ed G Buscaglia, E Dari and O Zamonsky (Bariloche) pp 2249-70

[7] Sendyka B and Cygnar M 2013 Stratified Charge Combustion in a Spark-Ignition Engine with Direct Injection System *Advances in Internal Combustion Engines and Fuel Technologies* ed H Kiat Ng (Rijeka: InTech) pp 85-117

[8] Korogodskij V 2013 Organizacija racional'nogo sposoba regulirovanija moshhnosti v dvigatele s iskrovym zazhiganiem *Dvigateli vnutrennego sgoranija* (Har'kov: NTU «HPI») 1 pp 11–16

[9] Korohodskyi V, Khandrymailov A and Stetsenko O 2016 Dependence of the coefficients of residual gases on the type of mixture formation and the shape of a combustion chamber *Eastern-European Journal of Enterprise Technologies* vol 1 no. 5(79) pp 4-12

IOP Conf. Series: Materials Science and Engineering **977** (2020) 012003 doi:10.1088/1757-899X/977/1/012003

A modelling approach to the transport support for the harvesting and transportation complex under uncertain conditions

Ievgen Medvediev[1*], Peter Eliseyev[2], Iryna Lebid[3], Olha Sakno[4]

[1]Department of Logistics Management and Traffic Safety in Transport, Volodymyr Dahl East Ukrainian National University, 59-a Ave. Central, Sievierodonetsk, 93406, Ukraine
[2]Department of Machine Science and Industrial Enterprises Equipment, Volodymyr Dahl East Ukrainian National University, 59-a Ave. Central, Sievierodonetsk, 93406, Ukraine
[3]Department of International Transportation and Customs Control, National Transport University, 1 Mykhaila Omelianovycha-Pavlenka Street, Kiev, 01010, Ukraine
[4]Department of Operation and Maintenance of Machines, Prydniprovs'ka State Academy of Civil Engineering and Architecture, 24A Chernyshevsky Street, Dnipro, 49600, Ukraine

*E-mail: medvedev.ep@gmail.com

Abstract. The article proposes a modelling approach based on structural and parametric identification of the transport support of the harvesting and transportation complex. The efficiency and effectiveness of the proposed methods of structural and parametric identification for the development of a system for harvesting and transportation complex operation has been proved. A mathematical model based on fuzzy logic has been developed. It reflects the interaction and interference of the weather-climatic, technical and technological parameters of transport support and its elements. A simulation method of transport support for harvesting and transport complex has been developed. The basis of the method is an algorithm for constructing a flow chart for the transport system operation consisting of structural and parametric identification levels and a mathematical model for the winter wheat harvesting transportation. A special feature of the method is the consideration of the weather-climatic, technical and technological parameters of the harvesting-transportation complex.

1. Introduction

The large number of parameters of agricultural production, that influence the transport support for the harvest and transportation complex, their compound interrelationships and the lack of certainty of data make it very difficult to construct an adequate mathematical model. Moreover, the very rare repetition of events makes statistical methods of data processing almost unacceptable. First and foremost, it is impossible to obtain a representative sample of data to produce statistically reliable estimates. All this has led us to turn to the mathematical apparatus of the theory of fuzzy sets. Less stringent conditions to a fuzzy measure, versus probabilistic, allow a wide range of subjective data to be added in addition to objective observations. This has made it possible to draw on the experience and expert judgement accumulated over many years of work.

Methods for coordinating harvesting and transport operations in the process of their operational planning and event models in projects at different stages of planning for early grains harvesting are considered in [1,2]. The agrometeorological group of factors is uncontrollable, the disclosure of its impact on the grain harvesting projects intensity is the basis for improving the harvest management efficiency [3]. The need to develop methods and models to take into account the essentials of the agrometeorological conditions influence on the state of the subject of the field work and the trends of its change, as well as the formation of a naturally determined time fund for the relevant work implementation, is an important stage in the development of statistical simulation models of virtual projects and the assessment on their basis of the managerial decisions effectiveness as of reactions to technological risk [4,5]. The stochastic nature of agro-meteorological conditions and their impact on the soil condition determine the variability of the naturally permitted time fund and the methods of cultivation of the soil and crops. Development of methods and models that take into account the agrometeorological conditions influence allows to study the features and tendencies of variability of the efficiency parameters of the adaptive machines technological complexes [6]. The harvesting and transportation process is considered as a nonlinear dynamic deterministic-stochastic model [7] and pays attention to the distribution of harvesting equipment across the fields, based on artificial intelligence methods [8].

However, the available analytical-statistical and statistical-simulation models do not take into account the weather and climate conditions influence, as well as technical and technological parameters, on the transportation process organization for harvesting and solve the problem in mostly deterministic setting.

The implementation of technical and technological solutions alone to improve the transport operations efficiency has not been shown to lead to a fundamental change in the situation with regard to the improvement of the transportations quality and volume [9].

The article is aimed at developing an approach to modelling the transport support of the harvesting and transportation complex in conditions of uncertainty, in particular weather and climatic parameters, on the basis of fuzzy logic with the development of the specific key tasks implementation algorithms.

The need to apply and use the fuzzy logic theory apparatus in transport processes is based on: a large number of unknowns; process complexity; the difficulty to establish quantitative relationships between parameters that affect the control process; the possibilities of formulating expert knowledge about an object (process) only in verbal expression; operational accounting of many conflicting parameters; creating a simpler mathematical model, consistent with practical application. The use of fuzzy logic apparatus in transport systems and processes is one of the modern and promising methods for levelling uncertainty, as evidenced by [10-13]

2. Justifying the use of the weather and climate parameters in modelling transportation support for the harvesting and transport complex

The agricultural production's feature is that it is carried out under the influence of a natural factor, under the conditions of a divergence of the working period and the production period, and the tangible dependence of agricultural production on natural, transport and other features [14].

Significant variability of weather conditions is the main cause of uncertainty in obtaining the results of agricultural activities, in particular with the transport operations.

In general, there is such a pattern: the less defined the external environment, the more difficult it is to make effective decisions. However, in all circumstances, for making informed decisions, it is important to forecast the external environment, that is, to anticipate possible changes and the occurrence of possible events under the influence of such a change [15].

Apart from differences in personal assessments, the environment in which decisions are made is a typical difficulty in determining the best alternatives. For example, the uncertainty of the technology system will be significantly reduced if additional information is provided in the form of weather forecasts for the forthcoming period, or, say, yield forecasts before harvest [16].

According to Handozhko L.A. the development of modern transport systems and technologies is impossible without taking into account the climate factor and is included in the development of technical projects. Productivity, reliability and economy are directly (reliable and economical) and indirectly (productivity) dependent on weather and climate conditions [17].

Hrybyniuk O.M. in his work [18] indicates that the basis for modelling and justifying objective management decisions in the harvesting of early grains is the disclosure of cause-effect relationships resulting from the conditions in which, the grain-stem mass, harvesters and vehicles are.

The difficulty of designing and managing production processes is also that most production and technological factors are a function of the weather factors. Weather factors dictate the choice of work organization, determine its course, qualitative and technical-economic indicators. Failure to take into account weather factors leads to impossibility of objective design and use of technical means, correct organization of works, to low quality indicators, high material and labour costs [19].

Based on processed sources, it is established that weather-climatic factors affect:

– the need to manoeuvre the equipment to minimize the costs of supporting technological processes [20];

– on the correct justification of the annual need for oil products [21]. The climate has an impact on energy conservation in agriculture. Thus, [22] states that scientifically based standards for fuel and energy consumption in climatic zones, levels of electrification and mechanization of agriculture should be adhered to. Efficient organization of use of vehicles is one of the basic principles of measures aimed at energy saving in transport.

– on the variability of the composition of the mutually agreed and interacting machines during the working period and day [23]. According to work [24] in accordance with the existing or anticipated weather conditions, differential application of agricultural machinery is required. In addition, poor logistics are enhanced by the effects of weather conditions on grain yields [25]. The material and technical base of agriculture is largely influenced by natural and climatic conditions. In addition, agriculture located in different natural zones requires different technical means, machine systems, and costs of production per unit of land [26].

According to the author [27], due to insufficient attention to technological issues depending on climatic and soil conditions, the flexible response to changes in weather conditions led to an unjustified increase in costs of production during harvesting.

M.E. Mayboroda in his work [28] emphasizes that the natural and climatic conditions in which the transport is carried out are one of the factors on which the level of technical and operational indicators of the transport process depends. At the same time, one of the factors on which the coefficient of release of the rolling stock on the line depends on the natural and climatic conditions.

Mokhnenko A.S. believes that the efficiency of agriculture depends on the environment (soil, climate, weather conditions) [29].

The implementation of scientific methods for the use of the meteorological information helps to reduce losses in the economy of the state in terms of taking into account the weather conditions, get the greatest effect at the lowest cost [30].

Eric E. Massey emphasizes the relevance of taking into account the effects of climate change into long-term transport planning in [31] while pointing to its relative novelty.

Due to the complex formalization of weather-climatic conditions in the transport process of harvesting, they are usually idealized. For example, in work [32] the author idealizes the downtime of the combine harvester because of the bad weather, and the climatic conditions at a certain point in time for all fields with ripe crops consider the same.

In the work [33] when considering weather factors affecting field work, the authors in the model assume that the delay in field operation due to 1 mm of precipitation in the form of rainfall is equal to one hour.

3. General provisions for the structural and parametric identification of the system

The algorithm for constructing the scheme of transport support for the harvesting-transport complex operation consists of two main levels - structural identification of the system and parametric identification of the system.

In our study, the level of structural identification of the system includes the following steps (Figure 1):

1. Definition of the structure of the model and its elements; and involves development of formal analytical qualities of the model elements, experts' theoretical and heuristic considerations, a priori information about the research object.

2. Defining the dependant and independent parameters in the organization and planning the transport process based on the general analysis. The purpose of this stage is to analyse and explain the components of the system, as well as determine dependant and independent parameters.

3. The next step is the theoretical justification for the influence of independent parameters. For this purpose our research uses the method of Saati analytic hierarchy process [34]. This method allows theoretically to justify the influence of independent parameters of the transport process as ascending for the stage of simulation with obtaining quantitative estimates.

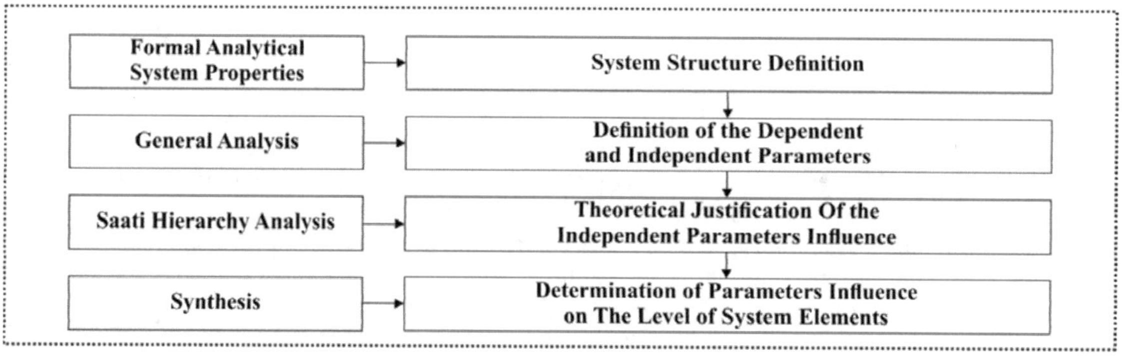

Figure 1. Algorithm for Structural Identification of the System.

4. The next step is to determine the influence of the parameters dependence based on the synthesis method. The purpose of this stage is to establish the relationship between the system parameters. It will occur both in the context of "dependent parameters - independent parameters" and in the context of "dependent parameters - dependent parameters". This step is transitional to the parametric identification of the system.

The level of parametric identification of the system consists of the following steps (Figure 2):

1. The assignment of terms to each of the dependent and independent parameters of the system based on the method of expert estimates. After the terms are assigned for all parameters, go to the next step.

2. Fuzzification of system parameters, i.e. the presentation of fuzzy values of dependent and independent parameters in the form of linguistic truth values.

3. Fuzzy inference is realized by mapping the input data vector to a scalar initial value [35]. To date, the most common models of fuzzy logical inference of Mamdani and Sugeno [36]. The advantages of the Mamdani model are model transparency, comprehensibility to customers - farms, high accuracy of identification with small training samples, and the ability to explain and justify the decision. Each such fuzzy inference forms a separate model, which is responsible for calculating a specific parameter.

4. Defuzzification, that is, the procedure for obtaining a clear result with a fuzzy one. The best method of defuzzification in the construction of applied fuzzy systems is the centre of gravity method [37]. The advantages of the centroid method include a reduction in the amount of computation and an increase in speed [38].

5. Getting the result of the calculation.

6. Validation of the result. When establishing that the calculation result is correct, the final goal is achieved.

If, as a result of modelling, it turns out that the system gives an incorrect result as assessed by practitioners (experts), then it will be necessary to reconfigure the system, for example, set more detailed terms, choose a different fuzzification method, etc.

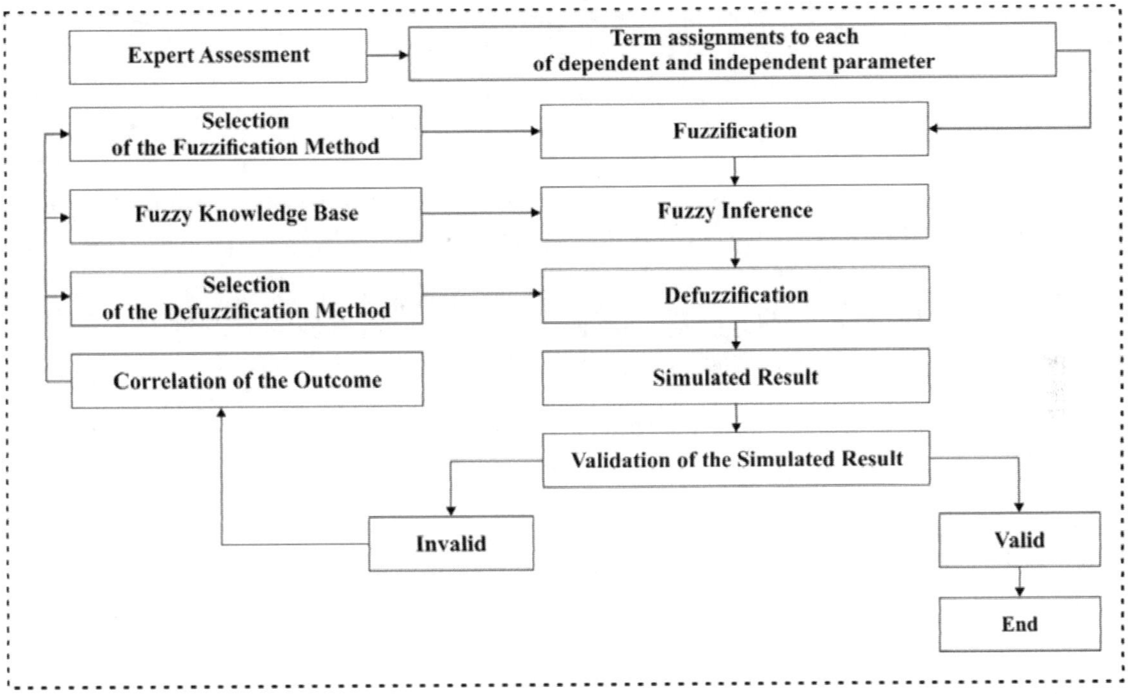

Figure 2. System Parametric Identification Algorithm.

System identification is a cyclic process, which consists in repeatedly choosing the model structure, calculating, according to the selected criterion, the optimal model parameters of the selected structure, and evaluating the properties of the resulting model [39].

When developing the system, the following specialists are involved: analysts, agronomists, and transport engineers, transport operation engineers, managers of agricultural enterprises that are involved in early wheat harvesting, programmers.

4. The implementation of structural and parametric identification of the transportation support system for the harvesting and transport complex

When studying the system of organizing the transport process, the main elements were identified: “Weather and Climate Conditions”, characterizing the influence of the external environment, “Field”, “Culture”, “Harvesting Method”, “Harvester”, “Road”, “Vehicle”.

Each of the aforementioned elements of the system for organizing the transport process has been given the appropriate parameters for it, according to which the model was identified (Figure 3).

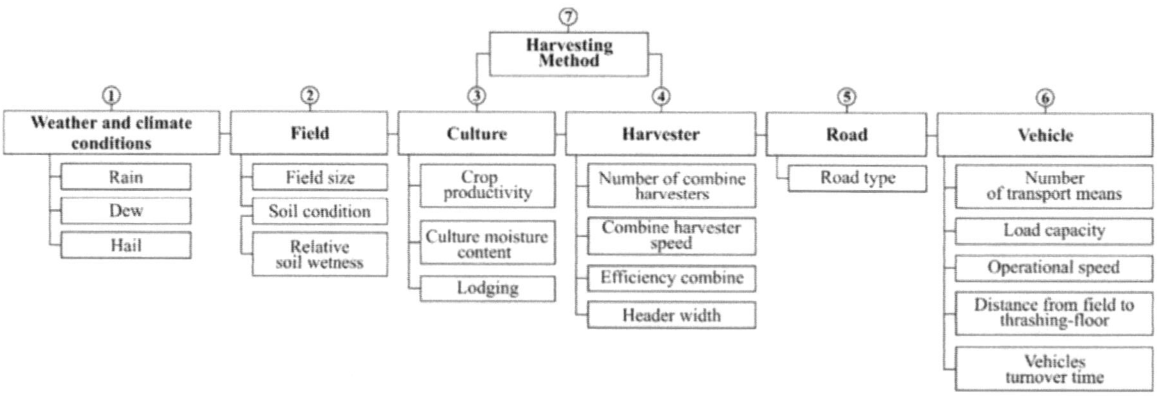

Figure 3. Determining the Structure of the System and its Elements with Parameters in the Organization of the Transport Process of Wheat Early Harvesting.

The second step in accordance with the above methodology of structural identification is the determination of dependent and independent parameters.

Dependent and independent parameters are presented in the form of a diagram (Figure 4).

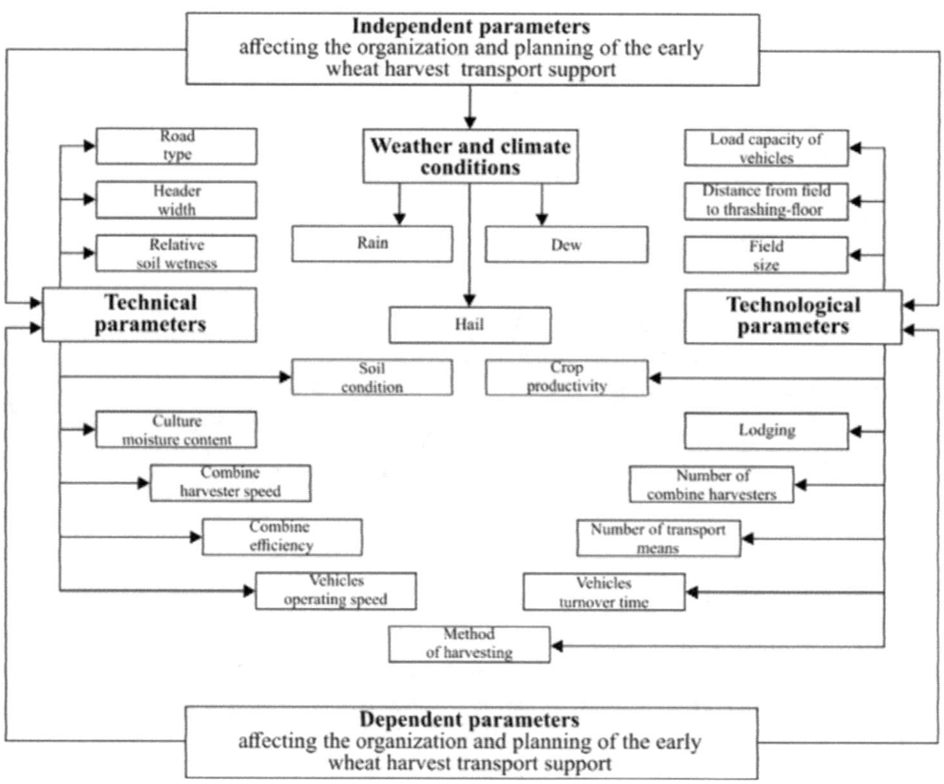

Figure 4. Dependent and Independent Parameters of the Elements of the Transport Process Organization for the Early Wheat Harvesting.

The third step according to the structural identification algorithm of the system (Figure 1) is the theoretical justification of the independent parameters based on the method of Saati analytic hierarchy process. This approach is described in detail in [40].

IOP Conf. Series: Materials Science and Engineering **977** (2020) 012003 doi:10.1088/1757-899X/977/1/012003

The next step is to determine the influence of the dependence of the parameters based on the synthesis method. The purpose of this stage is to establish the relationship between the system parameters. It will occur both in the context of "dependent parameters - independent parameters" and in the context of "dependent parameters - dependent parameters".

As an example, here are a few dependencies. Let us illustrate the influence of the independent parameter "Type of Road" on the dependent parameter "Operational Speed of Vehicles (automobiles)" (Figure 5) and the influence of the independent parameter "Rain" on the dependent parameters "Lodging", "Soil Condition", "Culture Moisture Content" and "Operational Speed of Vehicles (automobiles)" (Figure 6).

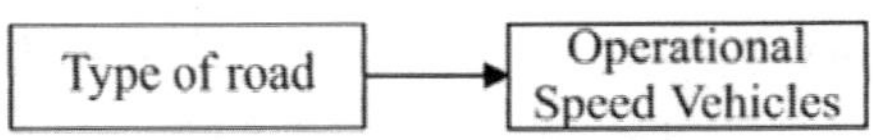

Figure 5. Influence of the independent parameter "Type of Road" on the dependent parameter "Operational Speed of Vehicles (Automobiles)".

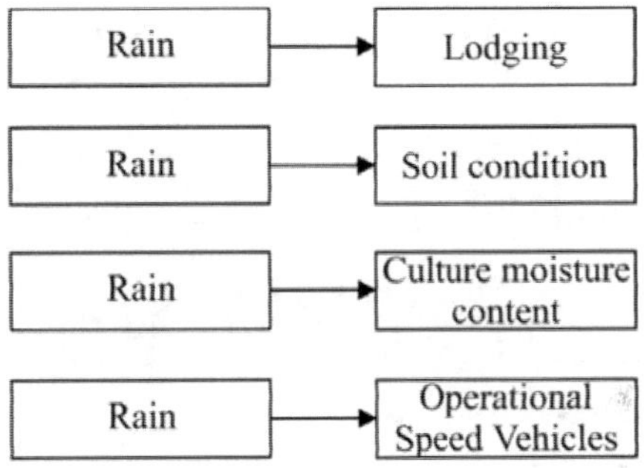

Figure 6. Influence of the independent parameter "Rain" on the dependent parameters "Lodging", "Soil Condition", "Culture Moisture Content" and "Operational Speed of Vehicles (Automobiles)".

The same procedure has been applied to each of 17 parameters. The step of determining the influence of the dependence of the parameters on the basis of the synthesis method is transitional to the system parametric.

The first step of system parametric identification (Figure 2) is the assignment of terms to each of the system parameters based on the expert assessment method.

The research experts [41] in amount of 63 respondents, having more than 10 years practical experience in harvesting wheat, served as resource persons.

The terms are assigned to each of the dependent and independent parameters, using Figure 4. Thus, the terms for the dependent parameters are presented in Table 1.

Table 1. Terms for Parameters.

Parameter	Term	Index
Rain, mm	Lingering	< 20
	Heavy	10
	Short-term	3-5
Operational Speed of Vehicles (Automobiles), km/h	High	51-60
	Higher than average	35-50
	Average	13-35
	Low	12-18
Distance from field to thrashing-floor, km	Long	40 and more
	Higher than average	20-35
	Medium	10-19
	Short	2-9

After the terms are assigned to all parameters, we proceed to the next step - fuzzification of parameters [42,43].

Fuzzy systems can have membership functions of an arbitrary structure, but from a practical point of view, triangular and trapezoidal functions are most popular [44].

We give examples of fuzzification. So, for instance, the analytical form of fuzzification for the "Rain" will be as follows:

Short-term rain

$$\mu(u)=\begin{cases}0, & u\le 0 \text{ or } u\ge 7{,}5\\ 1, & 0\le u\le 5\\ \dfrac{7{,}5-u}{7{,}5-5}=\dfrac{7{,}5-u}{2{,}5}, & 5\le u\le 7{,}5\end{cases} \tag{1}$$

Heavy rain

$$\mu(u)=\begin{cases}0, & u\le 6{,}25 \text{ or } u\ge 17{,}5\\ \dfrac{u-6{,}25}{10-6{,}25}=\dfrac{u-6{,}25}{3{,}75}, & 6{,}25\le u\le 10\\ \dfrac{17{,}5-u}{17{,}5-10}=\dfrac{17{,}5-u}{7{,}5}, & 10\le u\le 17{,}5\end{cases} \tag{2}$$

Lingering rain

$$\mu(u)=\begin{cases}1, & 20\le u\le 23{,}75\\ \dfrac{u-15}{20-15}=\dfrac{u-15}{5}, & 15\le u\le 20\end{cases} \tag{3}$$

The graphic view of fuzzification for the "Rain" parameter will have the following form (Figure 7).

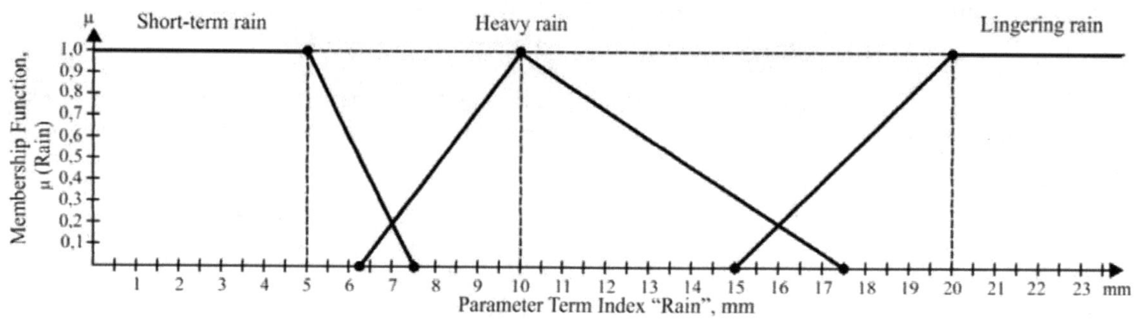

Figure 7. Fuzzification Graph for the "Rain" Parameter

In Figure 7 for the Rain parameter there is a trapezoidal membership function in the initial and final parts of the term, a triangular membership function for the middle part of the term. Let us calculate the membership function μ (u) for the Rain parameter.

At the next stage, after all the parameters are fuzzified, we proceed to the step of fuzzy inference.

A Mamdani model fuzzy inference for the parameters of the transport support system is developed, as illustrated in Table 2.

Table 2. Mamdani model fuzzy inference for the parameters of the transport support system.

IF	Parameter	Term	THEN	Parameter	Term
If	Rain	“Lingering”	Then	Lodging	“Heavy”
If	Rain	“Heavy”	Then	Lodging	“Middle”
If	Rain	“Short-term”	Then	Lodging	“Light”

Thus, based on the implementation of structural and parametric identification of the system, a general approach to modelling transport support for harvesting winter wheat has been developed.

5. Development of an original mathematical model of transport support for the harvesting and transport complex

The mathematical model reflects the interconnection and mutual influence of weather, climate, technical and technological parameters on the elements “Field”, “Culture”, “Combine Harvester”, “Road”, “Vehicle”.

Based on the decomposition of the system, we obtain individual mathematical models, which simplifies the construction and identification of the model, while maintaining the overall efficiency of the system. Using the model, it becomes possible to conduct a detailed study of transport support during the winter wheat harvesting.

The mathematical model of transport support for winter wheat harvesting is a graph with 9 sources (B), 5 levels of impact stratification (I, II, III, IV, V) and one runoff (C) presented in Figure 8.

In the model, the input parameters act as sources (B): weather and climate - rain, hail, dew; technological parameters - field size, distance from field to threshing floor, load capacity; technical parameters - type of road, header width, and relative soil wetness.

Impact stratification levels break systems down into subsystems in more detail and indicate the interplay and interdependence between system parameters.

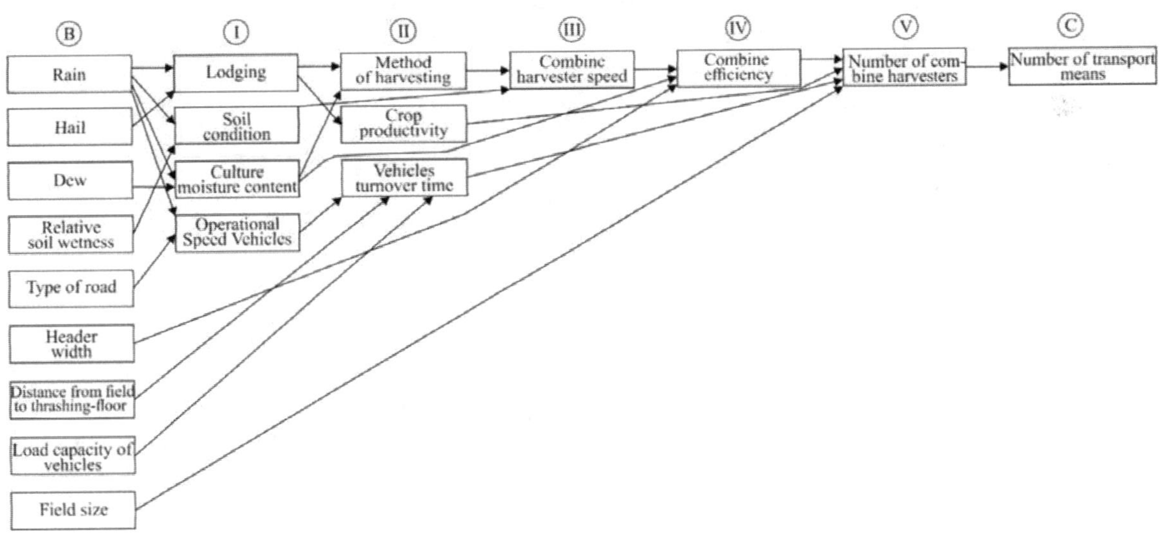

Figure 8. Mathematical Model of Transport Support for Winter Wheat Harvesting.

To better understand and reveal the essence of the mathematical model of transport support in harvesting winter wheat, which is presented on Figure 8, fuzzy mathematical models in the form of linguistic models “Lodging”, “Soil Condition”, “Culture Moisture Content”, “Vehicles (Automobile) Operational Speed”, “Harvesting Method”, “Crop productivity”, “Vehicles (Automobile) Time of Turnover”, “Combine Harvester Speed”, “Combine efficiency”, “Number of Combine Harvesters”.

So, according to [45], the linguistic model is a rule base that reflects the general semantic statement of the problem.

As an example, we show a fuzzy mathematical model of "Lodging" in the form of a linguistic model, which will look as follows:

$$\begin{cases} \text{IF "Rain Short term"} \cap \text{"Hail Small"} \rightarrow \text{"Lodging Slight"} \\ \text{IF "Rain Heavy"} \cap \text{"Hail Moderate"} \rightarrow \text{"Lodging Middle"} \\ \text{IF "Rain Lingering"} \cap \text{"Hail Heavy"} \rightarrow \text{"Lodging Heavy"} \\ \text{IF "Rain Short term"} \rightarrow \text{"Lodging Slight"} \\ \text{IF "Rain Heavy"} \rightarrow \text{"Lodging Middle"} \\ \text{IF "Rain Lingering"} \rightarrow \text{"Lodging Heavy"} \\ \text{IF "Hail Small"} \rightarrow \text{"Lodging Slight"} \\ \text{IF "Hail Moderate"} \rightarrow \text{"Lodging Middle"} \\ \text{IF "Hail Heavy"} \rightarrow \text{"Lodging Heavy"} \end{cases} \tag{4}$$

The fuzzy mathematical model of "Lodging" consists of 9 function rules.

A graphical representation of the software implementation of the linguistic model "lodging" will be implemented using the Fuzzy Logic Toolbox application. The "Lodging" model depends on the "Rain" and "Hail" parameters, and when they are taken into account, it is responsible for determining the level of lodging when harvesting winter wheat. Figure 9 presents the view of the Fuzzy Logic Toolbox FIS-editor window after setting the structure of the model "Lodging".

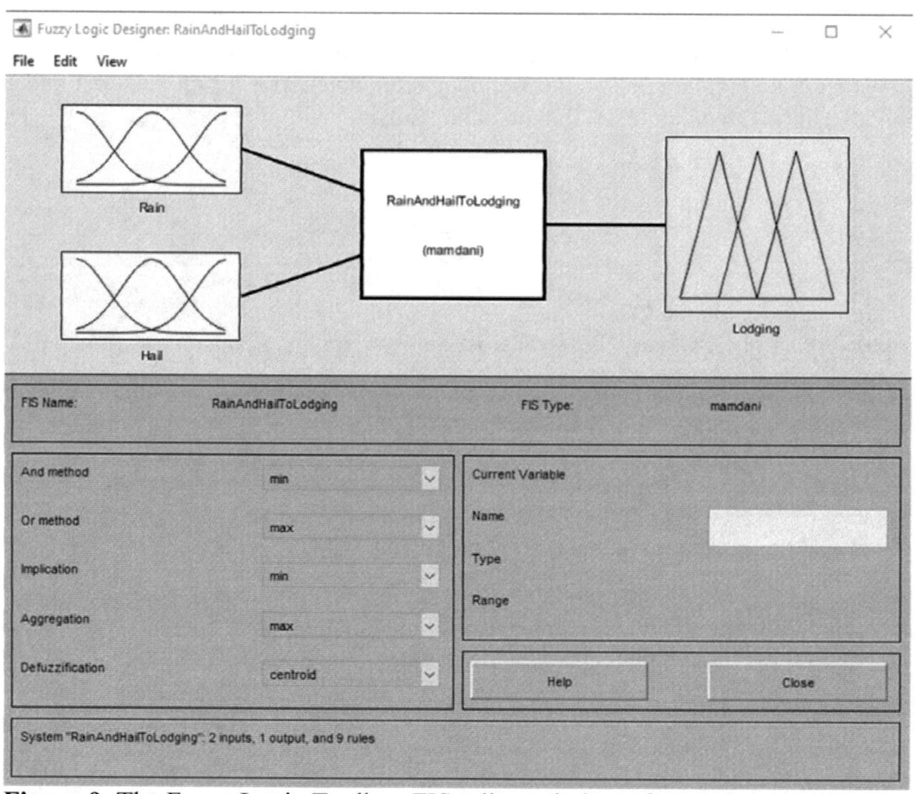

Figure 9. The Fuzzy Logic Toolbox FIS-editor window after setting the structure of the model "Lodging".

The visualization of the fuzzy inference of the "Lodging" model Figure 10 and its graphic view from the input variables "Rain" and "Hail" are shown in Figure 11.

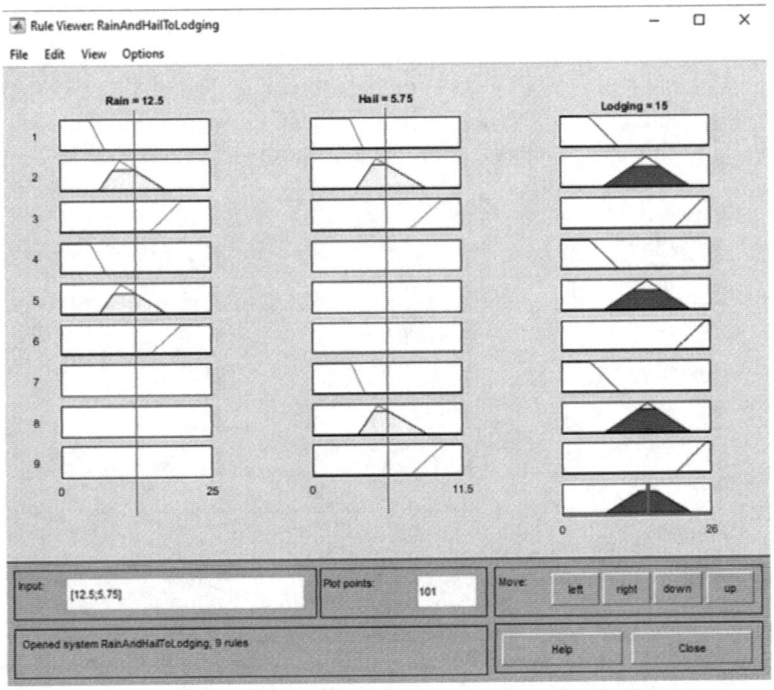

Figure 10. The visualization of the fuzzy inference of the "Lodging" model

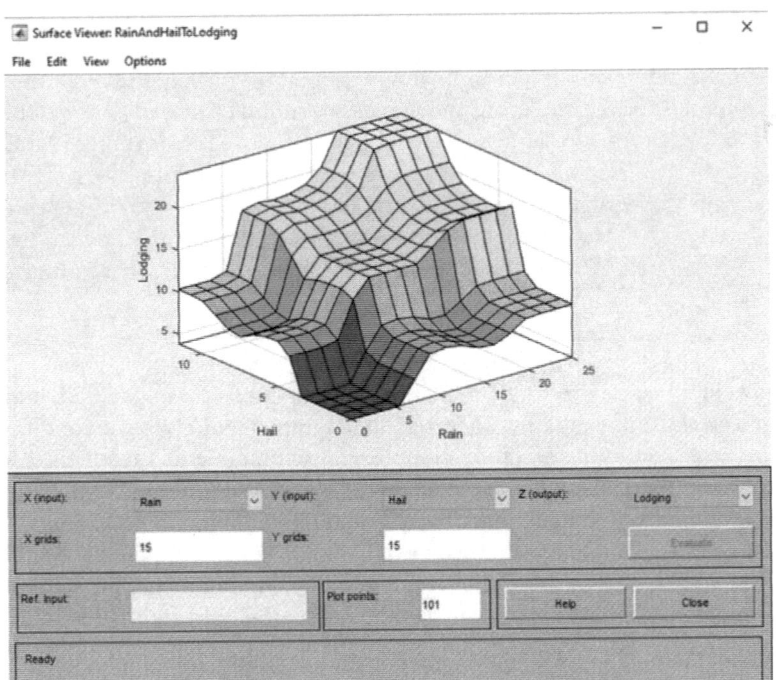

Figure 11. Graphic view from the input variables "Rain" and "Hail"

A mathematical model of transport support built in the Simulink graphical simulation environment is presented in the works [46, 47].

Based on the proposed general mathematical model of transport support for harvesting winter wheat and the algorithms for its implementation, the software was developed in the form of a computer program "Weather Climatic Condition Calculation Agriculture Transport (WCC CAT)" [48] for planning a rational number of vehicles when harvesting winter wheat. Test simulation was conducted on the basis of the peasant (farmer) economy – hereinafter PFE SASHA.

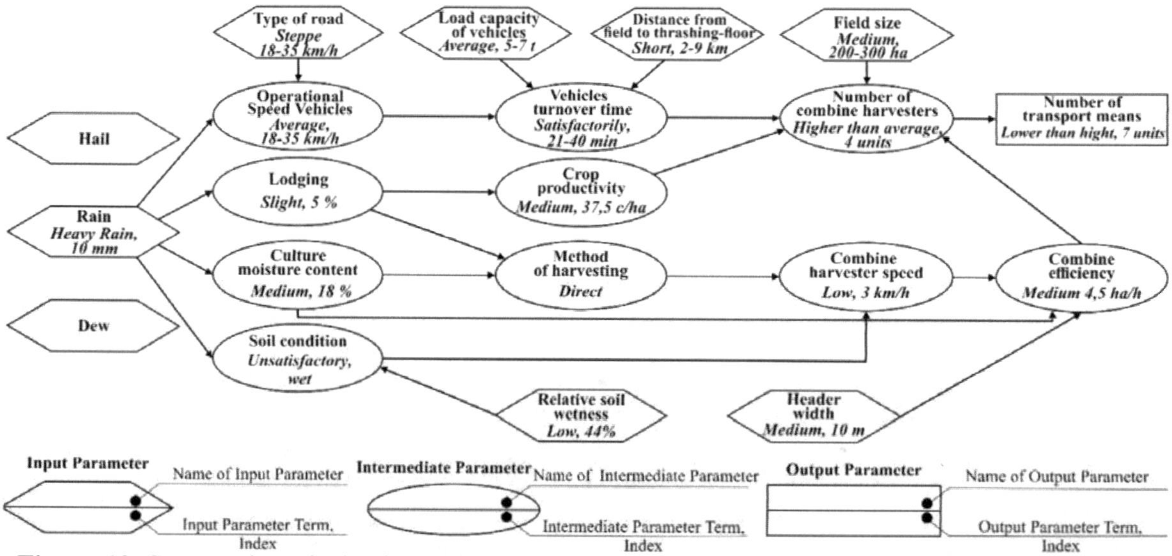

Figure 12. Structural-Logical Scheme for Determining the Number of Vehicles by Day in a Situation When There Was Rain

Using the WCC CAT program, based on the implementation of the structural and logical diagrams of the mathematical model for the situations, the necessary number of cars was calculated that provide the transport process for harvesting winter wheat at the PFE SASHA (Figure 12). Expert analysis of the obtained results shows: weather and climatic conditions have the most significant impact on transport support during the winter wheat harvest; taking into account the influence of the parameters of weather conditions WCC CAT allows to obtain a rational number of vehicles, thereby allowing to reduce the PFE SASHA's total planned need in vehicles by one unit, i.e. 8 vehicles used instead of 9, without violating the harvesting time – 12 days

6. Conclusions

Based on the research, it was found that the existing analytical-statistical and statistical-simulation transport support models do not take into account the weather and climatic conditions and solve the problems of the harvesting and transportation complex functioning in a deterministic setting.

A mathematical model of transport support is developed, which is a graph with 9 inputs, 5 levels of the influence hierarchy and one output. This model, unlike the existing ones, allows to formalize the relationship of the weather, climate, technical and technological parameters of the harvesting and transportation complex.

The simulation method of transport support for the harvesting and transportation complex has been developed. The peculiarity of the method is the accounting of weather-climatic, technical and technological parameters of the harvesting and transportation complex.

The developed method can be used as the basis for the creation of an automated complex for calculating the number of vehicles during harvesting.

References

[1] Komarnitskyi S. P., Sydorchuk O. V., Duhanets V. I. Method of coordination of harvesting and

transport operations in the process of their operational planning. *Eastern European Journal of Advanced Technology*. 2012. #1/10(55). P. 35-38.
[2] Sydorchuk O. V., Dnes V. I., Skibchyk V. I., Zhul S. H., Komarnitskyi S. P. Analysis of research methods and event models in projects at different stages of early grain harvest planning. *Computer-integrated technologies: education, science, production*. Lutsk: LNTU, 2011. #7. P. 141-144.
[3] Sydorchuk O. V., Dnes V. I., Komarnitskyi S. P. The intensity of work in grain harvesting projects. *Scientific Bulletin of the International Humanities University*. 2010. #1. P. 30-33.
[4] Sydorchuk O. V. The impact of the agrometeorological component on the risk of crop production projects. *East European Journal of Advanced Technology*. 2012. #1/10 (55). P. 49-51.
[5] Sydorchuk O. V., Lub P. M. Ivasiuk I. P. Impact of the subject conditions on the terms of performing soil tillage and sowing operations of summer-autumn period. *MOTROL. Commission of motorization and energetics in agriculture*. Lublin. 2012. Vol. 14, №4. P. 16–20.
[6] Sydorchuk O., Lub P., Malanchuk O. Impact of meteorogical conditions on the need in adaptive perfoming of technological operations of soil tillage and crop sowing. *Econtechmod*. 2014. Vol. 3. № 4. P. 35-38.
[7] Seliakova S. M. Development of a submodel of the lower level of a three-level model of the harvesting and transport process. *Artificial Intelligence*. 2008. # 1. P. 117-121.
[8] Seliakova S. M. Solving the problem of dividing harvesting equipment by fields using artificial intelligence methods. *Electronic and computer systems*. 2010. # 1(42). P. 91-95.
[9] Nefedova Ya. Y. Modeling the process of situational management in the system "industrial enterprise - port". Scientific works SWorld. 2015. # 4 (41). Vol. 1. P. 16-23.
[10] Sarkar A., Sahoo G., Sahoo U.C. Application of Fuzzy Logic in Transportation Planning. International Journal on Soft Computing (IJSC). Vol. 3, № 2. 2012. P. 1-21.
[11] Medvediev Ye. P., Lebid I. H., Yelisieiev P. Y. Justification of the appropriateness of using the fuzzy logic apparatus in transport processes of harvesting wheat. Technical service of agricultural, forestry and transport complexes. 2018. # 11. P. 269-279. (in Ukrainian)
[12] Fuzzy flow control methods in geoinformation systems / C.L. Belyakov, A.V. Bozhenyuk, L.A. Ginis, Ye.M. Gerasimenko. 2013. 176 c. (in Russian)
[13] Shramenko N., Muzylyov D. (2020) Forecasting of Overloading Volumes in Transport Systems Based on the Fuzzy-Neural Model. In: Ivanov V. et al. (eds) Advances in Design, Simulation and Manufacturing II. DSMIE 2019. Lecture Notes in Mechanical Engineering. Springer, Cham, pp 311-320, https://doi.org/10.1007/978-3-030-22365-6_31
[14] Plechko A.G. Uncertainty and risk as determinants of diversification changes in the agricultural sector. Agromir. 2012. # 9. P. 57-59.
[15] Andriichuk V. H. Economics of agricultural enterprises: a textbook. K .: KNEU, 2013. 79 p.
[16] Pyskarev A. V. Reliability of technological systems of machine-using in crop production: improvement of methods of design and operation on the basis of system approach. 2011. 385 p. (in Russian)
[17] Khandozhko L. A. Economic meteorology: a textbook. 2005. 489 p
[18] Grybyniuk O.M. Gribinyuk A.N. Research of the operating conditions and development of a method for optimizing the park of combine harvesters of an agricultural enterprise: Abstract. dis. ... cand. tech. Sciences: 05.20.01 / Grybyniuk Oleksandr Mykolaiovych. National Scientific Center “Institute of Mechanization and Electrification of Agriculture”. 1994. – 16 p. (in Ukrainian)
[19] Aryutov B. A., Vazhenin A. N., Pasin A. V., Novozhilov A. I. Methods for improving the efficiency of mechanized production processes in terms of their functioning in crop production: a textbook. 2010. 335 p. (in Russian)
[20] Popov L.A. Operation of the machinery and tractor fleet in the agro-industrial complex. 2004. 152 p. (in Russian)
[21] Kiiashko V. M., Tikhonov O. V., Rovnyi Ye. V. Operation of machinery and equipment Term project. 2015. 174 p. (in Ukrainian)

[22] Energy saving. Methods for determining the economic efficiency of energy conservation measures: national standard DSTU 2155-93. [Effective as of 01.01.95]. National Standart of Ukraine. (in Ukrainian)
[23] Nechaev V.I., Paramonov P.F. Organization of production and business activities in agroindustrial complex. 2007. 466 p. (in Russian)
[24] Tkachenko T.H. Agrometeorology. 2015. 268 p. (in Ukrainian)
[25] Nechayev V. I., Moiseyev V. V., Bondarenko V. V., Chernykh G. V., Blinnikova Ye. A., Kovtunenko B. V. The main directions of increasing the stability and efficiency of grain production. 2006. 402 p. (in Russian)
[26] Rusnak P. P., Zhebka V. V., Rudyi M. M., Chalyi A. A. Agricultural Economics. 1998. 320 p. (in Ukrainian)
[27] Romanenko G. A. To ensure the environmentally sound development of the agro-industrial complex. Agricultural science. 2002. #8. P. 2-4. (in Russian)
[28] Mayboroda M. Ye., Bednarskiy V. V. Freight road transportation. 2008. 422 p.
[29] Mokhnenko A. S. Analysis of the efficiency of the farms in the region. Busines-navigator. 2012. # 1. P. 71-77.
[30] Polovyi A. M., Bozhko L. Yu., Volvach O. V. Fundamentals of Agrometeorology. 2012. 250 p. (in Ukrainian)
[31] Eric E. Massey. Experience of the European Union in Adaptation to Climate Change and its Application to Ukraine. 2012. OSCE. 40 p.
[32] Senchuk S.P. Justification of many models for the study of grain harvesting and transport complexes. Agricultural machinery. 2005. #13. P. 153-163.
[33] Shahab Sokhansanja, Amit Kumarc, Anthony F. Turhollowa. Development and implementation of integrated biomass supply analysis and logistics model (IBSAL). 2006. *Biomass and Bioenergy*. P. 1-10.
[34] Saaty T. L. The Analytic Hierarchy Process. New York: McGraw-Hill. 1980. 288 p.
[35] Mamdani E.H. Applications of fuzzy logic to approximate reasoning using linguistic synthesis. IEEE Transanctions on Computers. 1977. Vol. 26, № 12. P. 1182-1191.
[36] Dubovoi V. M., Kovaliuk O. O. Models of decision making in the distributed dynamic systems management. 2008. 185 p.
[37] Rotshtein A. P., Shtovba S. D. The influence of defuzzification methods on the tuning speed of a fuzzy model. Cybernetics and systems analysis. 2002. #5. P.169–176.
[38] Khizhnyakov Yu. N. Fuzzy, neural and neuro-fuzzy control algorithms in real-time systems. 2013. 160 p.
[39] Korzhyk M. V. Modeling of Objects and Asset Management Systems MatLab. 2016. 174 p.
[40] Medvediev, Ie. Organization and planning of harvest transport process on the basis of saati's hierarchy analysis method. PODILIAN BULLETIN: AGRICULTURE, ENGINEERING, ECONOMICS. 2018. No. 29. P. 92-103. (in Ukrainian)
[41] Medvediev, Ie. Assessment of the weather and climate conditions impact on the organization and planning of transport support for wheat harvesting / Ie. Medvediev, I. Lebid, M. Bragin // TEKA of the Commission of Motorization and Energetics in Agriculture. – 2017. Vol. 17, No. 2. – P. 45-54.
[42] Shtovba S. D. Fuzzy systems design using MATLAB. 2007. 288 p.
[43] Subbotin, S.O., Representation and processing of knowledge in the artificial intelligence and decision support. 2008. 341 p. (in Ukrainian)
[44] Dranyshnykov L. V. Intellectual methods in management. 2018. 416 p. (in Russian)
[45] Khizhnyakov Yu.N. Algorithms of fuzzy, neural and neuro-fuzzy control in real-time systems. 2013. 160 p.
[46] Barbaruk L. V., Zubarev D. V., Medvediev Ye. P., Barbaruk V. M. The use of fuzzy models for planning harvesting in various climatic conditions. Bulletin of the Volodymyr Dahl East Ukrainian National University. 2018. # 6 (247). P. 175-179. (in Ukrainian)
[47] Medvediev Ye. P., Lebid I. H., Yelisieiev P. Y., Zubarev D. V. Computer program development

for organizing and planning the transport process for early wheat harvesting. Naukovi Visti Dalivskoho Universitetu 2018. #15. – URL: http://nvdu.pp.ua.
[48] Medvediev Ye. P., Yelisieiev P. Y., V. M. Barbaruk, Zubarev D. V. Copyright Registration of Computer Programs Certificate №4073 "Weather Climatic Condition Calculation Agriculture Transport" (WCC CAT) - № 85087; appl. 06.11.2018; registered 02.01.2019. (in Ukrainian)

Acknowledgments
The composite authors express their gratitude to Sergiy Khliakin, Director, Department of Agro-Industrial Development Luhansk Regional State Administration-Regional Civil Military Administration and Vasyl Hromyak, PhD (Agricultural Sciences), Associate Professor Head of Laboratory of Protection and Rational Use of Lands National Scientific Center «Institute for Soil Science and Agrochemistry Research named after O.N. Sokolovsky.

IOP Conf. Series: Materials Science and Engineering **977** (2020) 012004 doi:10.1088/1757-899X/977/1/012004

Calculation methods for determining of fuel consumption per hour by transport vehicles

S I Krivoshapov[1], **A I Nazarov**[1], **M I Mysiura**[1], **I A Marmut**[1], **V A Zuyev**[1], **V V Bezridnyi**[1], **V N Pavlenko**[1]

[1]Department of Technical operation and service of cars, Kharkov National Automobile and Highway University, 25 Yroslava Mudrogo Street, Kharkov, Ukraine

tesa@khadi.kharkov.ua

Abstract. The paper proposes a mathematical model by which it is possible to calculate the hourly fuel consumption for an internal combustion engine operating at idle mode or with a constant load. Simplification of the technique consisted in the calculation of the coefficients of filling the engine cylinders and the excess air through the approximation by polynomials of the first and second degree. It is established that when the engine is running without load, the hourly fuel consumption increases in proportion to the rotational speed of the crankshaft and the engine displacement. Examples of calculating the standard fuel consumption at idle mode with the engine off without a load are given for a Skoda Octavia 1.6 MPI with a gasoline engine and a Skoda Octavia 2.0 TDI with a diesel engine. Graphic dependencies of the hourly fuel consumption change on the crankshaft rotation speed for the petrol and diesel Skoda Octavia were obtained. The calculation was carried out and graphical dependencies of fuel consumption were obtained when special equipment was operated using the example of the KrAZ-6322 car. It is proposed to use deterministic methods for determining fuel consumption in road transport.

1. Introduction

Fuel efficiency is one of the quality indicators of transport vehicles. Fuel consumption is a significant part of the operating costs of a car. In the structure of the cost of transport operations, the consumption of fuels and lubricants accounts for 40-50% of all costs [1].

In all countries coming from the Soviet Union, fuel consumption in road transport is controlled at the legislative level. In Ukraine, this is an order of the Ministry of Transport No. 43 of 1998 [2], which fixes the normative value of the basic consumption rate for each car, stipulates the conditions for additional fuel consumption and introduces corrective factors for changing the flow rate depending on operating conditions. Similar principles are laid down in the standards: Russia [3], Belarus [4], Moldova [5], Kazakhstan [6], etc.

The disadvantage of the existing methodology is the limited number of car brands for which a basic fuel consumption rate has been established, and the lack of a mechanism for independently expanding the range of standards for new rolling stock.

This problem can be solved by creating a mathematical model that relates fuel consumption to the technical data of the car, taking into account the conditions of its operation.

The aim of this work is to develop a mathematical model for calculating the hourly fuel consumption during engine operation without load and with a fixed constant load on the internal combustion engine.

2. Exposition

Rationing of fuel consumption according to the method [2] is determined through the flow rate in l/100 km.

Hourly consumption is used when the car engine is running without movement in the following cases:

- in case of forced downtime of the car during loading and unloading of goods or passengers, production necessity and downtime in traffic jams during rush hour;
- when starting and warming up gas-filled cars;
- during the operation of specialized equipment with a stationary vehicle;
- the operation of autonomous engines used to drive specialized equipment while driving and stopping the vehicle.

For transport vehicles, hourly consumption is determined through the basic rate of fuel consumption [2]. So the hour of forced downtime is equivalent to 5 km of the car. The start-up norm for a gas-balloon vehicle is (0.5 ... 5.0) % of the base vehicle's liquid fuel consumption per day of operation, depending on the air temperature.

For specialized mobile and autonomous engines, the hourly norm should be numerically presented in the normative [2] for a specific vehicle and equipment.

On the other hand, the hourly fuel consumption can be determined analytically through the speed and load characteristics of the engine, using the dependences from the theory of engines [7].

Hourly consumption determines fuel consumption in kilograms or liters per hour of operation of a power plant. For transport vehicles, it is more common to record fuels and lubricants in liters, and for autonomous engines and construction vehicles in kg / h.

Hourly fuel consumption in l / h:

$$G_t = \frac{q_e \cdot N_e}{1000 \cdot \rho_t} \qquad (1)$$

where: q_e - specific effective fuel consumption, g / (kW · h);

N_e - effective power, kW;

ρ_t - fuel density, g / cm^3 (kg / l).

Effective specific fuel consumption:

$$q_e = \frac{3600 \cdot 10^3}{\eta_e \cdot H_n}, \qquad (2)$$

where: η_e - effective engine efficiency;

H_n - fuel calorific value, kJ / kg.

Effective engine efficiency:

$$\eta_e = \eta_i \cdot \eta_m, \qquad (3)$$

where: η_i - engine efficiency indicator;

η_m - mechanical engine efficiency.

Indicator engine efficiency:

$$\eta_i = \frac{P_i \cdot L_0 \cdot R \cdot T}{H_n \cdot \eta_v \cdot P} \cdot \alpha, \qquad (4)$$

where: P_i - average indicator pressure, kPa;

L_0 - stoichiometric amount of fuel-air mixture, kmol / kg;

R - universal gas constant, J / (mol · K);

T - air temperature, K;

η_v - engine cylinder fill ratio;

P - air pressure, kPa;

α - excess air ratio.

The coefficient of excess air depends on the load on the engine. In figure 1 shows a graphical dependence of the coefficient of excess fuel on the percentage of used power [8].

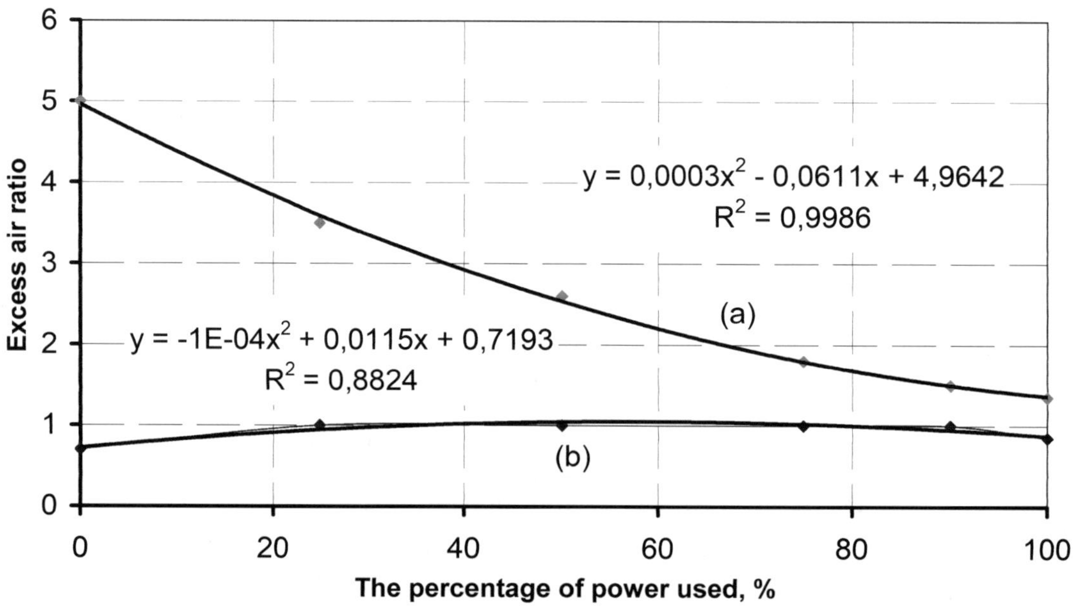

Figure 1. Change in the coefficient of excess air from the percentage of power use: (a) - diesel ICE, (b) - gasoline ICE

This relationship can be approximated by a polynomial of the second degree of the form:

$$\alpha = A_\alpha \cdot N_1^2 + B_\alpha \cdot N_1 + C_\alpha, \tag{5}$$

where: N_1 - power utilization percentage, %;

A_α, B_α, C_α - empirical coefficients depending on engine type.

The values of these coefficients can be taken:

- for diesel engine:

$$A_\alpha = 3 \cdot 10^{-4},\ B_\alpha = -0.06,\ C_\alpha = 5.0;$$

- for gasoline ICE:

$$A_\alpha = -1 \cdot 10^{-4},\ B_\alpha = 0.012,\ C_\alpha = 0.85.$$

The engine fill factor also depends on the load. In figure 2 shows this dependence [8].
The result of approximating a linear function of the form:

$$\eta_v = B_\eta \cdot N_1 + C_\eta, \tag{6}$$

where: A_η, B_η - empirical coefficients depending on engine type;

The values of these coefficients can be taken:

- for diesel: $A_\eta = 9 \cdot 10^{-4}$, $B_\eta = 1.0$;
- for gasoline: $A_\eta = 7 \cdot 10^{-3}$, $B_\eta = 0.17$.

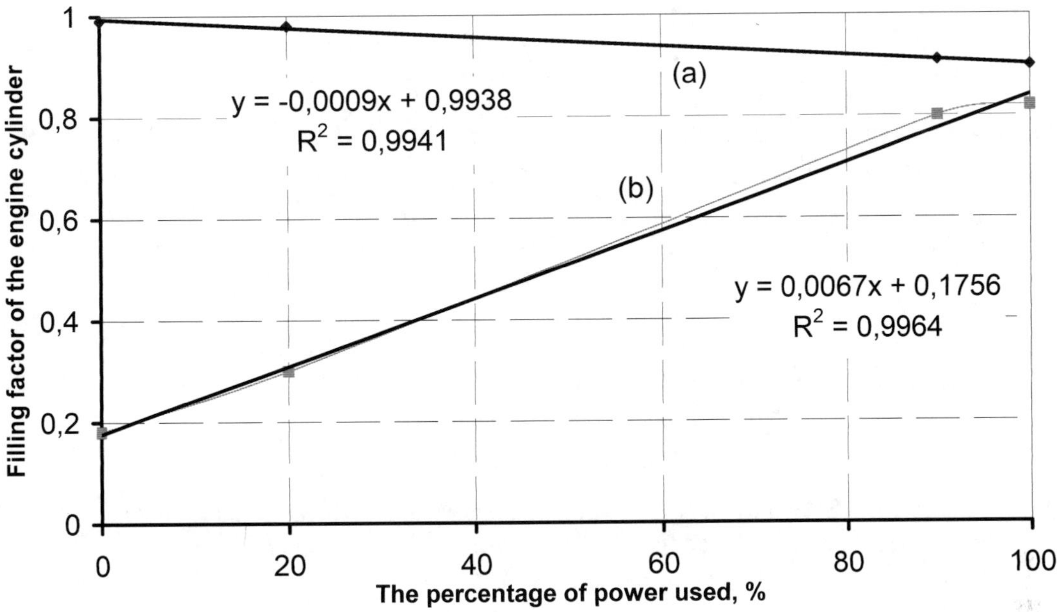

Figure 2. The change in the filling ratio of the percentage of power use: (a) - diesel ICE, (b) - gasoline ICE

Power utilization percentage:

$$N_1 = \frac{N_e \cdot 100}{N_{e\max}}, \tag{7}$$

where: $N_{e\max}$ - maximum effective engine power, kW;
Average indicator pressure:

$$P_i = \frac{P_e}{\eta_m}, \tag{8}$$

where: P_e - average effective pressure, kPa;
Mechanical efficiency:

$$\eta_m = \frac{P_e}{P_e + P_m}, \tag{9}$$

where: P_m - pressure of mechanical losses, kPa;
Average effective pressure:

$$P_e = \frac{N_e \cdot 30 \cdot \tau \cdot 10^3}{V_h \cdot n}, \tag{10}$$

where: τ - engine cycle;
V_h - engine displacement (all cylinders), l;
n - engine rotation speed, rpm.
The average pressure of mechanical losses:

$$P_m = a_m + b_m \cdot W_n$$

where: W_n - average piston speed, m / s;

a_m, b_m - mechanical loss factors in the engine.

$$W_n = \frac{30 \cdot S_n}{n}, \tag{11}$$

where: S_n - cylinder height (distance from TDC to BDC), m.

After substitution, we obtain the expression for determining the hourly fuel consumption in l / h:

$$G_t = \frac{0.12 \cdot P \cdot V_h \cdot n}{L_0 \cdot R \cdot T \cdot \tau \cdot \rho_t} \cdot \frac{B_\eta + \dfrac{10^2 \cdot A_\eta \cdot N_e}{N_{e\max}}}{C_\alpha + \dfrac{10^2 \cdot B_\alpha \cdot N_e}{N_{e\max}} + \dfrac{10^4 \cdot A_\alpha \cdot N_e^2}{N_{e\max}}}. \tag{12}$$

If the effective power is excluded from the equation, i.e. $N_e = 0$, then equations (12) will be:

$$G_t = \frac{0.12 \cdot P \cdot V_h \cdot n \cdot B_\eta}{L_0 \cdot R \cdot T \cdot \tau \cdot C_\alpha \cdot \rho_t}. \tag{13}$$

Equation (13) can be used to calculate fuel consumption when the engine is running without load, and equation (12) to determine the additional flow to the auxiliary equipment drive.

In the formula (13) as constant values are used: $R = 8.31$ J / (mol · K) and $\tau = 4$.

For vehicles with gasoline engine: $\rho_t = 0.76$ g / cm³, $L_0 = 0.5119$ kmol / kg, $B_\eta = 0.17$, $C_\alpha = 0.85$. Then the formula (13) will be simplified:

$$G_t = 0.00185 \cdot V_h \cdot P / T \cdot n. \tag{14}$$

For diesel cars, the following indicator values can be accepted: $\rho_t = 0.84$ g / cm³, $L_0 = 0.495$ kmol / kg, $B_\eta = 1.0$, $C_\alpha = 5.0$. Then formula (13) takes the form:

$$G_t = 0.00174 \cdot V_h \cdot P / T \cdot n. \tag{15}$$

The temperature and air pressure during the intake P and T compression process in the engine cylinders rises. We assume that their change is proportional. Then under and accept the parameters of the atmospheric air. Under normal conditions: $P = 101.325$ kPa and $T = 293$ K. Then: $P/T = 0.346$.

Let's calculate the hourly fuel consumption using the example of a Skoda Octavia 1.6 MPI car with a gasoline engine. For this car we will accept the following initial data: $n_{\min} = 800$ rpm. and $V_h = 1.6$ l. Then the hourly fuel consumption at idle will be:

$$G_t = 0.00185 \cdot 1.6 \cdot 0.346 \cdot 800 = 0.808 \text{ l / h}.$$

Let's calculate the hourly fuel consumption of a Skoda Octavia 1.9 TDI diesel car with the following initial data: $V_h = 1.9$ l and $n_{\min} = 600$ rpm. Then the hourly fuel consumption at idle will be:

$$G_t = 0.00175 \cdot 2.0 \cdot 0.346 \cdot 600 = 0.673 \text{ l / h}.$$

The consumption of a diesel car at idle is less than gasoline due to lower crankshaft revolutions and a more economical engine.

In figure 3 shows a graph of changes in hourly fuel consumption for Skoda Octavia 1.6 MPI and 1.9 TDI vehicles. With an increase in engine speed, fuel consumption increases proportionally.

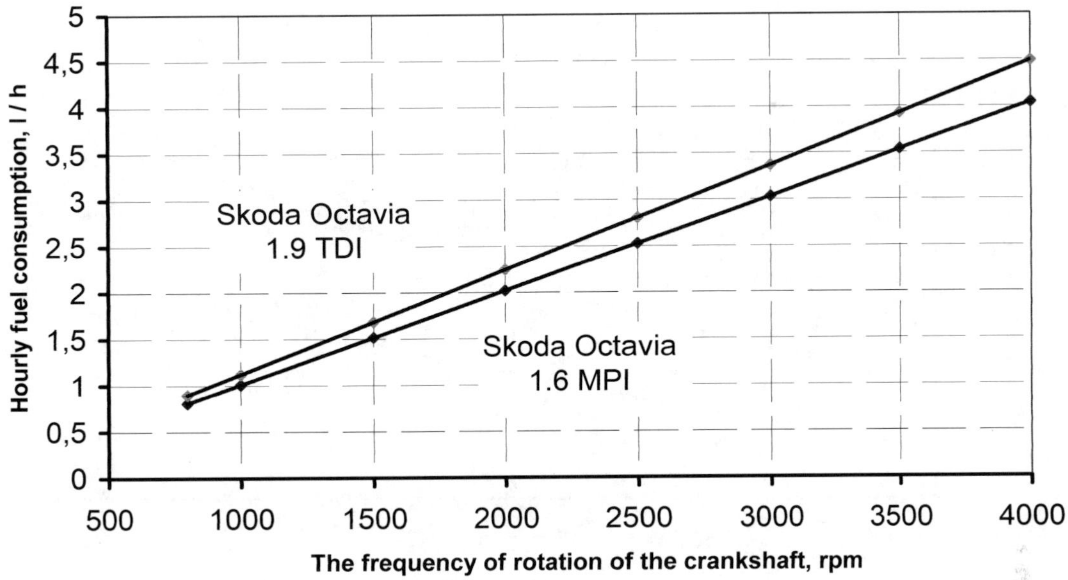

Figure 3. Change in hourly fuel consumption for a car with a diesel and gasoline engine

We calculate the standard fuel consumption for Skoda Octavia cars according to the method of rationing fuel consumption [2]. For a Skoda Octavia car with an AVU engine with a displacement of 1.6 l and a power of 75 kW, the basic rate of gasoline consumption is set at 7.5 l / 100 km. Then at 5 km the car will spend: $7.5 \cdot 5/100 = 0.375$ l. Therefore, according to the method [2], the fuel consumption of a Skoda Octavia will be 0.375 l / h. A Skoda Octavia car with an AGR diesel engine with a working volume of 1.9 l and a power of 66 kW, for which the basic rate of gasoline consumption is set at 5.2 l / 100 km, according to the method [2], the hourly fuel consumption will be $5.2 \cdot 5/100 = 0.26$ l / h. The hourly fuel consumption obtained by the method [2] is significantly lower than the calculated values obtained by the formula (13). Experimental studies have shown idle fuel consumption for Skoda Octavia 1.6 MPI – 0.7 ... 0.8 l / h, and for Skoda Octavia 1.9 TDI – 0.6 ... 0.7 l / h. These data are consistent with our calculations.

We will calculate the fuel consumption for specialized equipment installed on the chassis of the KrAZ-6322 car. The following indicators are accepted for this car: $V_h = 14.86$ l, $N_{e\max} = 243$ kW, $\rho_t = 0.84$ g / cm^3, $A_\eta = 9 \cdot 10^{-4}$, $B_\eta = 1.0$, $A_\alpha = 3 \cdot 10^{-4}$, $B_\alpha = -0.06$, $L_0 = 0.495$ kmol / kg, $C_\alpha = 5.0$. After substituting in the formula (12) the hourly fuel consumption for this car can be calculated as follows:

$$. \ G_t = \frac{864 \cdot n \cdot (N_e / 2700 + 1)}{N_e^2 - 486 \cdot N_e + 9.8 \cdot 10^4}$$

In figure 4 shows a graph of the change in hourly fuel consumption for the KrAZ-6322 car at different engine speeds of the YaMZ-238 diesel engine.

If specialized equipment that works at a frequency of 1,500 rpm and consumes 25 kW of power is connected to the car, the hourly consumption of the KrAZ-6322 will be:

$$G_t = \frac{864 \cdot 1500 \cdot (25/2700 + 1)}{25^2 - 486 \cdot 25 + 9.8 \cdot 10^4} = 15.05 \text{ l/h}.$$

With an increase in average effective power, hourly fuel consumption increases.

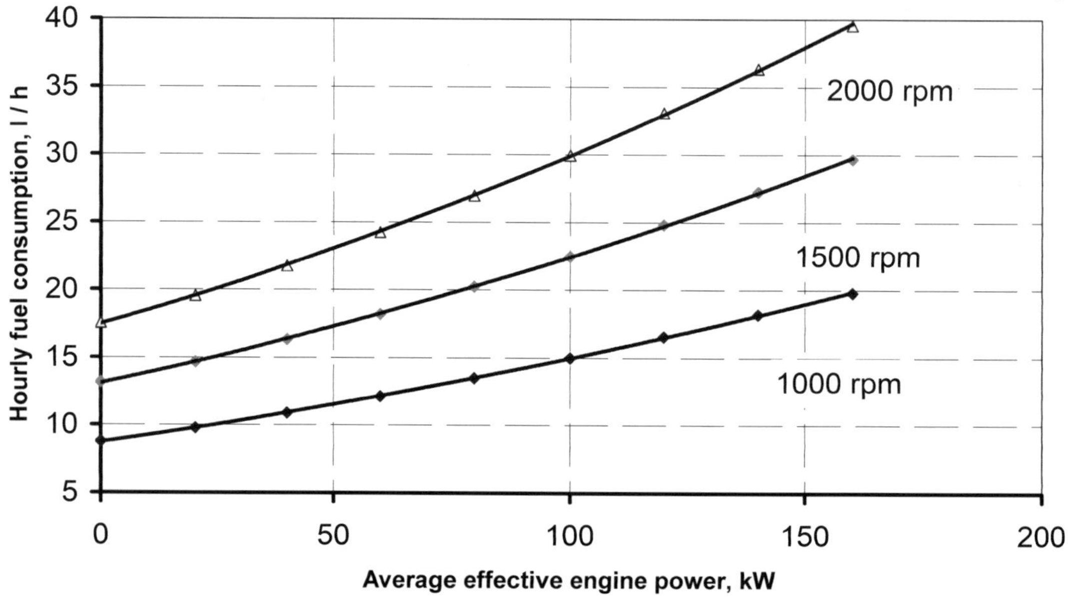

Figure 4. Change in hourly fuel consumption for a KrAZ-6322 truck with a YaMZ-238 diesel engine

3. Conclusion

The current fuel consumption rating system in Ukraine is not perfect. By modeling the processes of the functioning of the system, taking into account the design features and operating conditions, it is possible to get by calculation the standard values of the fuel consumption of a particular car according to its technical and operational characteristics.

References

[1] Govorushhenko N Ja and Turenko A N 1998 Sistemotehnika transporta (na primere avtomobil'nogo transporta) *RIO HGADTU* 474.

[2] Pro zatverdzhennja zmin do Norm vitrat paliva i mastil'nih materialiv na avtomobil'nomu transporti. Nakaz Minis-terstvo Infrastrukturi Ukraïni № 36 vid 24.01.2012. *Verhovna Rada Ukraïni.* URI: https://zakon.rada.gov.ua/rada/show/v0036733-12#n2.

[3] Rasporjazhenie Mintransa Rossii ot 14.03.2008 g. № AM-23-r. Glavbuh. *Bumazhnyj i jelektronnyj zhurnal, soob-shhestvo, spravochnaja sistema i onlajn-servisy.* URI: http://www.glavbukh.ru/doc/2126.

[4] Ob ustanovlenii norm rashoda topliva v oblas-ti transportnoj dejatel'nosti i priznanii utrativ-shimi silu nekotoryh normativnyh pravovyh aktov Ministerstva transporta i kommunikacij Respubliki Belarus'. Postanovlenie Mini-sterstva transporta i kommunikacii Respubliki Belarus', 6 janvarja 2012 g. № 3. *Nacional'nyj pravovoj Internet-portal Respubliki Belarus',*. URI: http://www.pravo.by/main.aspx?guid=3871&p0=W21226157 &p1=1

[5] Prikaz ob utverzhdenii Norm rashoda topliva i smazochnyh materialov na avtomobil'nom trans-porte N 172 ot 09.12.2005. *Contabil Set.* URI: http://www.contabilsef.md/libview.php?l=ru &idc=241&id=1347

[6] Normy rashodov gorjuche-smazochnyh materia-lov i rashodov na soderzhanie avtotransporta. *Plast.* URI: http://plast.com.kz/support/zakon/gsm.php

[7] Borisov G V, Leliovskij K Ja and Pachurin G V 2015 K voprosu o normirovanii rashoda zhidkih topliv na avtomobil'nom transporte. *Fundamental'nye issledovanija.* **3** 28-35.

[8] Krivoshapov S I 1999 Razrabotka metodiki i algo-ritma obshhego diagnostirovanija avtomobilej po izmeneniju kojefficienta poleznogo dejstvija. *HGADTU* 225.

IOP Conf. Series: Materials Science and Engineering **977** (2020) 012005 doi:10.1088/1757-899X/977/1/012005

Simulation with using of functional resonance analysis method the traffic safety control of vehicles, which carrying out the transportation process

Olha Sakno[1*], Svitlana Sharai[2], Tatiana Kolesnikova[1], Viktor Poliakov[3], Danylo Hirman[3]

[1]Department of Operation and Maintenance of Machines, Prydniprovs'ka State Academy of Civil Engineering and Architecture, 24A Chernyshevsky Street, Dnipro, 49600, Ukraine
[2]Faculty of Transport and Information Technologies, Department of International Transportation and Customs Control, National Transport University, 1 Mykhaila Omelianovycha-Pavlenka Street, Kiev, 01010, Ukraine
[3]Faculty of Automotive and Mechanical Engineering, Department of Motor Vehicles, National Transport University, 1 Mykhaila Omelianovycha-Pavlenka Street, Kiev, 01010, Ukraine

*E-mail: sakno-olga@ukr.net

Abstract. In the article, using the method of functional resonance analysis, the process of traffic safety control of vehicles that carry out cargo delivery and transportation of passengers, when their technical condition changes during operation, is simulated. Constant growth in transportation of cargo and passenger by rolling stock of automobile transport requires the involvement in their implementation of a large number of vehicles and increase the efficiency of their operation. In the last time, the geography of the movement of material flows has been expanding, which is manifested, in particular, in an increase of cargo and passenger traffic in international and intercity transportation by road. These trends entail the need to ensure the serviceable technical condition of modern vehicles designs. On the basis of statistical data, the reasons of the change of the technical condition of the chassis of the vehicles and their influence on its operational properties were analyzed. Resilience Decision Support (Resilience DS) software is used to solve this problem. Such a tool made it possible to simulate the socio-technical system "Person-Car-Organization of maintenance and repair system" and to form the formal models for constant evaluation of the effective functioning of the system of maintenance of vehicles and conditions of increasing their operation. To develop a maintenance model using a functional resonance analysis method, data was collected at a proprietary vehicle maintenance station. The model consists of a set of functions that determine the actions executed during maintenance of vehicles.

1. Introduction

Successful management of the transport process can be implemented by identifying the influence of factors on the effectiveness of its implementation and determining the patterns of occurrence of causes that can negatively affect it. The issues of monitoring the safety of the movement of vehicles during the implementation of the processes of cargo delivery and passenger transportation are relevant and require their solution using modern scientific methods. The functional resonance analysis method (FRAM) is used to study complex sociotechnical systems. It was created and described by Eric Hollnagel, who noted that complex technical systems contain a large number of subsystems and components, the variability of which is usually absorbed by the system with minimal impact on the overall system. The main sources of this variability are people, technology and organization, which ensures the serviceable state of the technique.

Hollnagel stated that, since these elements are not linearly connected to each other, they can lead to the accidents. When the variations of the components become too large for the system to absorb, then the result becomes unfavorable or random. This concerns the functional resonance effect arising from a situation where the system is not able to function in normal operation due to changes in its daily productivity.

A model of the FRAM describes the ability of the functions of the system components of a vehicle to resonate and create hazards during its operation that can get out of control and cause accidents or accidents due to the low level of functioning of the vehicle maintenance and repair system.

The FRAM [1] supports a system analysis process aimed at identifying interdependencies and system behaviors that are potentially important for a tool which focuses on the interdependence of the process and its dynamics. This is a method of modeling complex organizational and technical systems obtained by the theory of sustainable health [2], which relates to how to achieve success of the technical system by adapting it to difficult conditions [3]. Recent documents have shown the use of FRAM to understand the legal implementation of guidelines of management the efforts to achieving security [4]. The method of functional resonance analysis involves the identification of functions (technological, human or organizational activities) in the in the day-to-day operation of a technical system [5-6].

The purpose of the paper is to simulate, on the basis of the FRAM, the process of control over the safety of the vehicle movement in a result of changes of its technical condition, which affects the deterioration of its steerability, stability and smoother running characteristics during operation.

2. General vehicular principles of the functional resonance analysis method

The activity of the FRAM is to record the influence of the technical condition of the chassis of the vehicle on the change in its operational properties and traffic safety from a system point of view.

Using the FRAM takes a systematic perspective, which means that the analysis cannot be limited to a certain part of the socio-technical system, but it should consider a broader picture when the organization of the control process is considered as a whole, and not as maintenance and assembly of components of the vehicle's chassis [7-8]. The method analyzes the organization of the traffic safety control process as a socio-technical system, where the technology is embedded in the social context and develops, tests, launches and maintains a serviceable technical condition of the vehicle.

Method is based on four basic principles:

I. Equivalence principle: operability and inoperability are equivalent to the fact that they both stem from shift work and the functioning of the system as a whole. There are no special reasons that only work for failures.

II. Approximate adjustment principle: variability, as a way to adapt a person to the management of the maintenance and repair system and to the operating conditions of the vehicle. People (specialists) always know how to adjust what they do to fit the situation (traffic safety). Such variability of efficiency is inevitable, ubiquitous, and necessary.

III. Principle of surprise: the appearance of operability and disability is not a direct result of variability within a particular task or function, but rather is a combination of the variability of many functions. The variability of the functioning of the system may combine unexpectedly, which leads to disproportionate large results (non-linear effects). The result arises if it cannot be attributed to the functions of the system or explained by them.

IV. Functional resonance principle: unexpected "enhanced" effects of interactions between different sources of variability underlie the phenomenon, which is described by functional resonance.

Functional resonance is a detection signal (it can be detected) that comes out from an unintended combination of the variability of many signals. Functional resonance is an alternative to linear causality.

Models of FRAM can be used to study potential sources of variability by simulating and identifying context-sensitive aspects of the functioning of the system as a whole, such as human, technological, and organizational. This approach supports the assessment of the potential of the system for solving variability, taking into account the expected and unexpected variability, and follows from the work of the technical system (vehicle as a whole).

The basic unit of analysis is the functional hexagon (Fig. 1). Using six aspects of each function (time, control, output, resources, preconditions and input), system interactions aimed at identifying potential sources of resonance are studied.

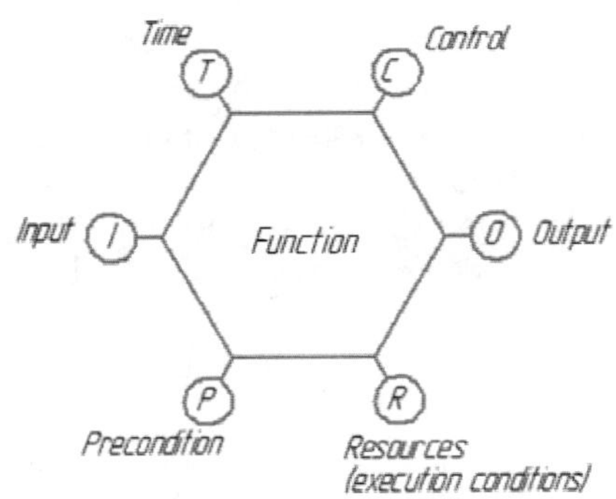

Figure 1. FRAM Function

Thus, the output of the function can be an input, a prerequisite, or even a forced control aspect of one or another function of the system (Table 1). This process can lead to the identification of possible sources of attenuation for undesirable variability. For example, if the resources for a certain function are evaluated as "more than necessary", this may indicate the existence of a "spare system", which can act as a damper barrier to ensure the safety of the vehicle. Research process of possible connections between functions in order to identify both potential undesirable sources of variability and barriers is called an instance of FRAM (i.e., direct proximity).

Table 1. - Characteristics of FRAM functional units

Parameter	Symbol	Description
Input	*I*	Input information (or input data) that characterizes a function in the system's functioning. These are links to previous functions. Input data runs the execution or action of a function.
Output	*O*	The result of the function (output). Provides links to further functions. Represents the result of changing the technical condition of the system elements (vehicle).
Time	*T*	The time, which is required for processing (preparation, diagnosis, maintenance of system elements) by functional subsection or subsystem.
Control	*C*	Limitations, methods and procedures of control. They determine how the function transmits technical data and/or is monitored.
Preconditions	*P*	System conditions which determine the conditions of functioning and which must be provided before function execution.
Resources	*R*	Resources that are needed or used during the processing of a function or operation of system elements (vehicle).

IOP Conf. Series: Materials Science and Engineering **977** (2020) 012005 doi:10.1088/1757-899X/977/1/012005

3. Simulation of a vehicle maintenance system

To achieve this goal, the technical support of Resilience DS software is used [9]. Such a tool allows you to simulate the socio-technical system "Person-Car-Organization of maintenance and repair system" and form formal models to constantly evaluate the effectiveness of the functioning of the system of technical maintenance and repair of vehicles and improve their operating conditions.

Data collection for simulate by FRAM was carried out in three different stages. At the first stage, in the spring of 2019, a two-day visit to the technical service station in the city Dnipro was organized, where technical maintenance of branded vehicles, in particular passenger cars, is carried out. During the visit, there was made the familiarization of the process of organization the maintenance of vehicles in specific conditions. The second stage of data collection was to conduct a seminar, which was attended by researchers, a contact person and employees involved in the process of preparation and execution vehicles maintenance. The purpose of the seminar was to discuss, revise and update the previous FRAM model, developed on the basis of information collected during a visit to the branded technical service station and official documentation. The seminar helped researchers identify some of the functions of the system and determine their aspects (input, output, control, preconditions, time and resources to ensure the serviceable technical condition of vehicles).

The collected data was used to develop by FRAM of a vehicles maintenance model, consisting of a set of functions that describe the actions taken to their service and to ensure their serviceable technical condition.

In our study, to ensure the safe movement of a vehicle during its operation, it is necessary, without fail, to conduct periodic technical impact on its chassis, which consists in: diagnostics; checking the condition of shock absorbers, springs, support cups, brake blocks, discs, hoses, backlash in ball bearings and steering tips, silent blocks; adjustment and replacement of wheel bearings; checking the wheel alignment; checking the condition of the car tire and monitoring its resource (Fig. 2). All this affects the operational properties of vehicles during their activity: stability, controllability, fuel efficiency. So, when the resource of a car tire approaches its limit state, the probability of a traffic accident increases [10]. In fig. 3 shows function of vehicle for FRAM. In fig. 4 shows a diagram of the chassis and the impact of its technical condition on the operational properties of a vehicle.

4. Simulation system of a qualitatively new technology for vehicles maintenance

Thus, it is necessary to form a new system [11-12] of vehicles maintenance, which defines modern principles and methods (Fig. 5). Its functioning to a greater extent will be able to ensure traffic safety when executing various transport-technological schemes for the delivery of cargo and transportation of passenger. This simulation system of a qualitatively new technology for technical maintenance of vehicles is implemented in the shell of a system of a qualitatively new technology for designing vehicles of a new generation.

The system of principles and methods for providing the qualitatively new properties of the maintenance technologies of new generation vehicles can be formed within the limits of a recurrence approach (repeatability) based on the following types of principles:

a) Well-known principles that can be represented by a set $R_s = \{R_{s1}, R_{s2}, R_{s3}, ..., R_{si}, ...R_{sp}\}$;

b) New principles $R_n = \{R_{Sn1}, R_{n2}, R_{n3}, ..., R_{ni}, ...R_{nj}\}$;

c) Qualitatively new principles $R_h = \{R_{h1}, R_{h2}, R_{h3}, ..., R_{hi}, ...R_{hg}\}$;

d) Fundamental new principles $R_m = \{R_{m1}, R_{m2}, R_{m3}, ..., R_{mi}, ...R_{md}\}$;

e) Principles at the level of new solutions $R_x = \{R_{x1}, R_{x2}, R_{x3}, ..., R_{xi}, ...R_{xv}\}$;

f) …;

g) Principles at the level of discovery $R_w = \{R_{w1}, R_{w2}, R_{w3}, ..., R_{wi}, ...R_{wy}\}$.

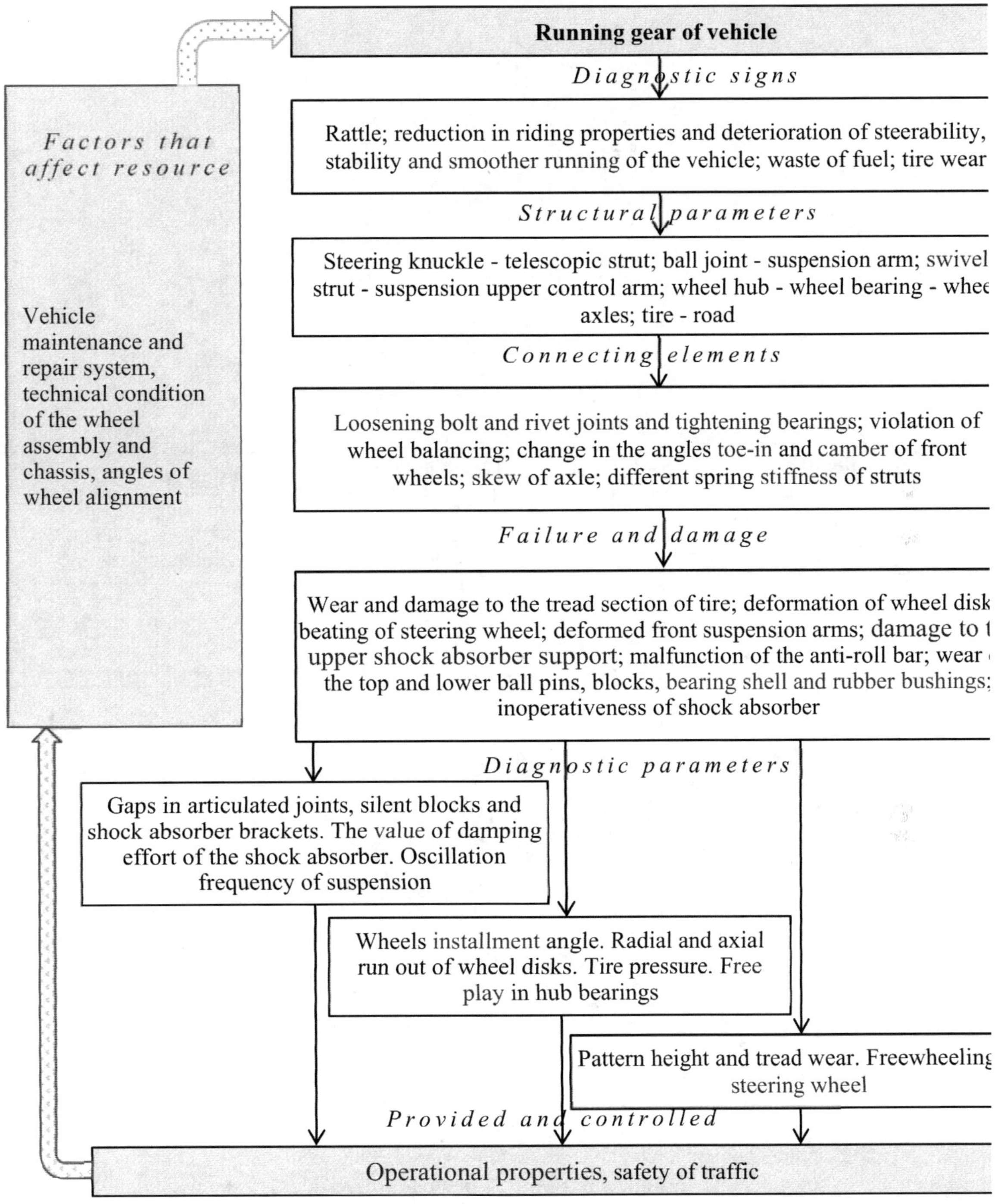

Figure 2. Structural and investigative model of the running gear of a car

The system of principles and methods for providing qualitatively new properties and capabilities of the maintenance technology of new generation vehicles can also be formed within the limits of the generation of a set of variants of compositions of such principles on the basis of the morphological box:

$$R = \begin{Vmatrix} R_{s1} & R_{s2} & R_{s3} & \dots & R_{si} & \dots & R_{sp} \\ R_{n1} & R_{n2} & R_{n3} & \dots & R_{ni} & \dots & R_{nj} \\ R_{h1} & R_{h2} & R_{h3} & \dots & R_{hi} & \dots & R_{hg} \\ R_{m1} & R_{m2} & R_{m3} & \dots & R_{mi} & \dots & R_{md} \\ R_{x1} & R_{x2} & R_{x3} & \dots & R_{xi} & \dots & R_{xv} \\ \dots & \dots & \dots & \dots & \dots & \dots & \dots \\ R_{w1} & R_{w2} & R_{w3} & \dots & R_{wi} & \dots & R_{wy} \end{Vmatrix} \tag{1}$$

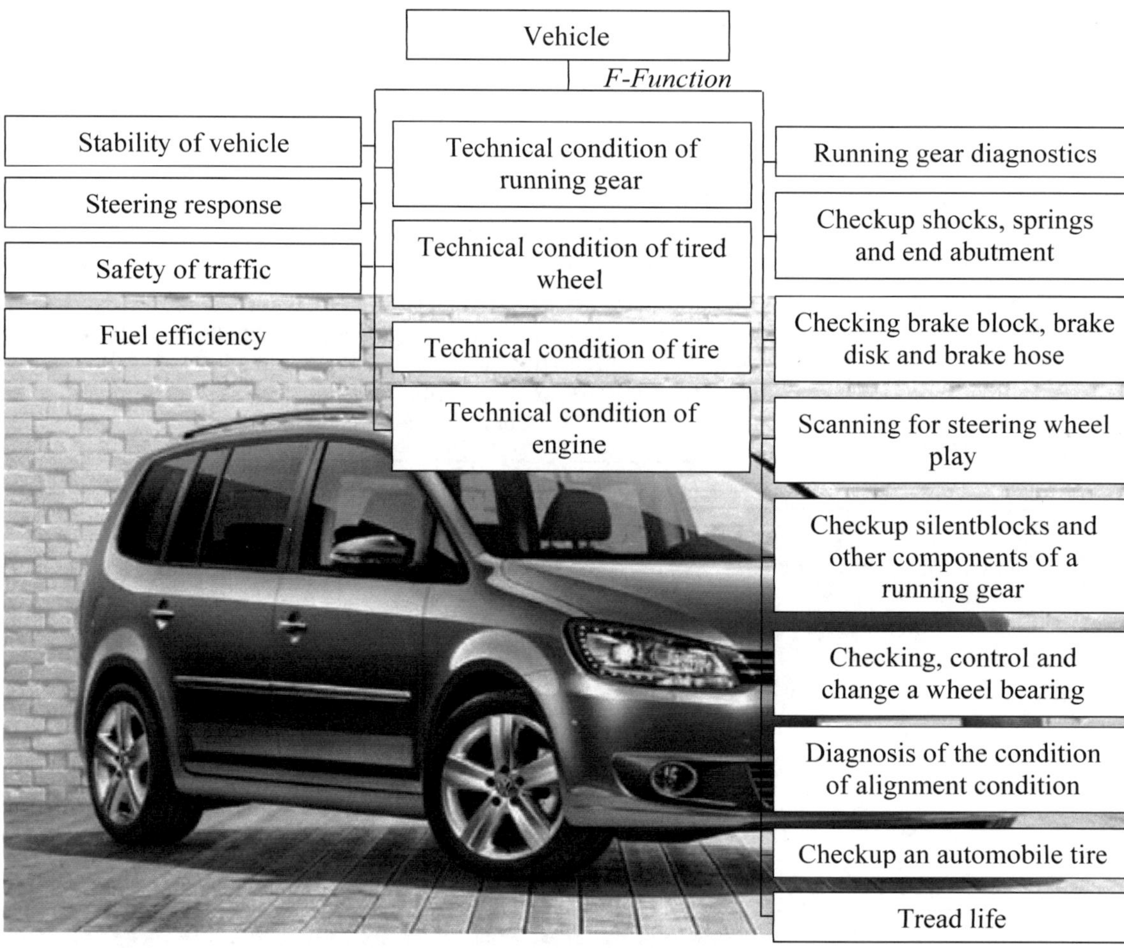

Figure 3. Function of a car for FRAM

Where R is the general system of principles and methods for providing qualitatively new properties and capabilities of the new generation maintenance technology; R_{ij} is the i-th principle of the j-th type of principles for providing qualitatively new properties and capabilities of the new generation maintenance technology.

IOP Conf. Series: Materials Science and Engineering **977** (2020) 012005 doi:10.1088/1757-899X/977/1/012005

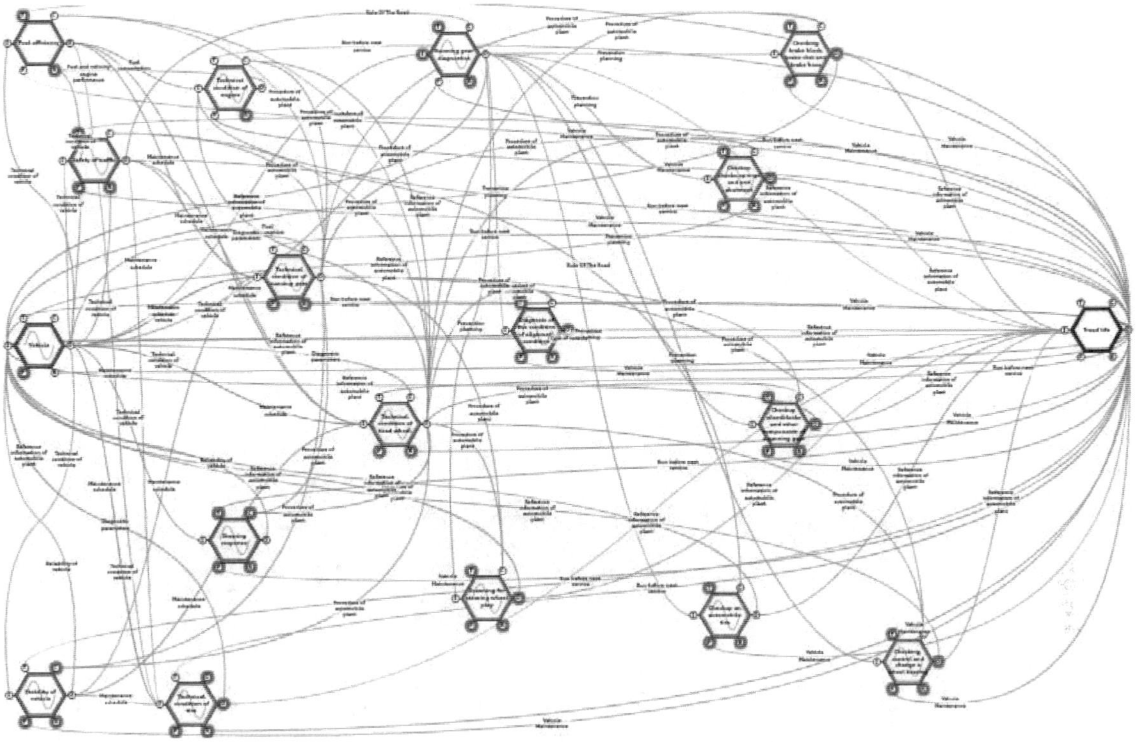

Figure 4. Functional Resonance Analysis Method simulation of the traffic safety control of the vehicle due to changes in its technical condition during operation

It should be noted that any principle of creating new generation technologies R_{nj} can be formed from the intersection of a certain subset of known p_i principles:

$$R_{nj} = \bigcap_{i=1}^{p_j} p_i \, . \tag{2}$$

Where R_{nj} is a field that forms the new principles of technology; R_{si} is many well-known principles of simulation of technology; p_i is power of subset of known principles.

5. Conclusions

Using the method of functional resonance analysis to simulate the influence of the technical condition of the chassis of a vehicle on its operational properties showed that one of the important elements for the effective implementation of the transport-technological transportation process is the a serviceable technical condition of vehicles, which carries out delivery of cargo and transportation of passenger. High-quality maintenance of vehicles, taking into account modern approaches to new generation technologies, ensures the safety of the transportation process. Simulation the influence of the technical condition of the chassis on the service properties of a vehicles using method of functional resonance analysis clearly revealed that its maintenance is a complex type of human activity and requires new approaches to new generation technologies (new generation vehicles are hybrids, electric, autonomous). The FRAM model gave an idea of how successful and high-quality maintenance of vehicles depends on the complex interaction of several of its components working in the technical system. The interconnections are so complex that they are almost impossible to apprehend. Thus, the

important socio-technical mechanisms of the "Person-Car-Organization of a maintenance and repair system" system regulate the communication between the functions of this system. They can be:
a) The informal control mechanisms;
b) The human control mechanisms;
c) A new generation vehicle maintenance technologies or specific tools such as monitoring the technical condition of the main units of vehicles;
d) The technological process of vehicle maintenance and diagnostics (both general and element-wise) and
e) The instructions and regulations for the repair team workers and similar documents.

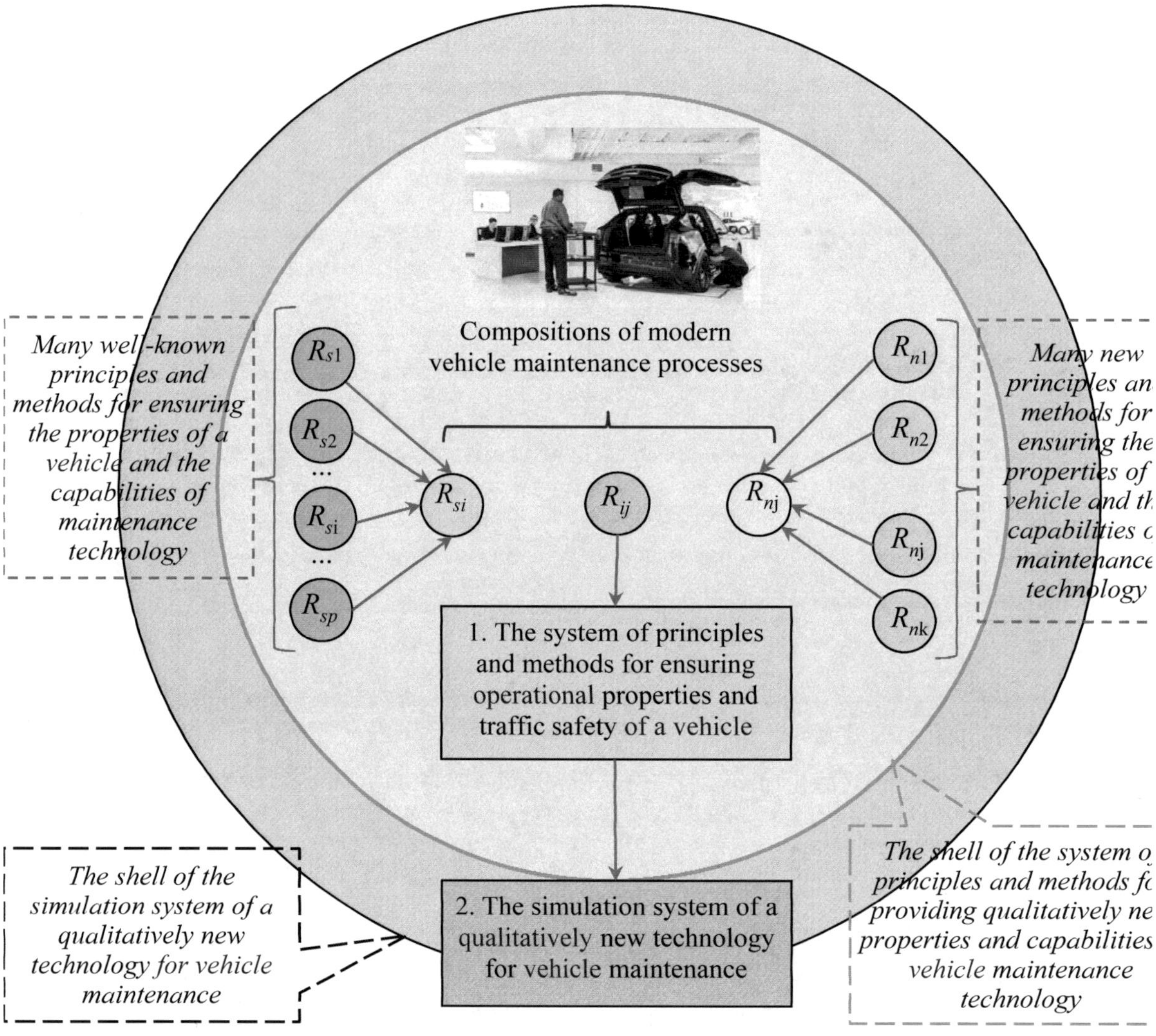

Figure 5. Hypothetical diagram of a general methodology for simulation a qualitatively new technology for maintenance and repair of vehicles (next-generation technology)

Simulation the influence of the technical condition of the chassis of a vehicle on its operational properties by the method of functional resonance analysis showed that it is necessary to pay special attention to modern element-by-element diagnostics, since this significantly affects traffic safety. The fulfilment of these functions is based more on modern knowledge about the design of new generation vehicles and their maintenance technology.

The method of analysis of functional resonance complements the approach to solving the issues of efficiency of transportation process from the point of view of ensuring traffic safety of vehicles during it execution, improving the operational properties of vehicles. It is an innovative method that finds its application in medicine, nuclear energy, and the transport sector. This method is used to analyze the daily activities of transport enterprises in order to create models for specific tasks. The model of the FRAM can be used for specific types of analysis in order to verify the feasibility of the proposed solutions or changes to increase the efficiency of transport enterprises activity, to identify the causes of possible errors or threats, to understand the stages of the implementation and functioning of transport-technological processes of cargo delivery and transportation of passengers. It can be used as a basis for analyzing the risks of traffic accidents and investigating them, or forecasting possible future events.

References

[1] Hollnagel E 2008 *École des Mines de Paris – Centre for Research on Risk and Crises (CRC)*
[2] Braithwaite J, Wears R L and Hollnagel E 2015 *International Journal for Quality in Health Care* **27 (5)** 418–420.
[3] Anderson J E and Kodate N 2015 *Saf. Sci* **809** 105–114.
[4] Clay-Williams R, Hounsgaard J and Hollnagel E 2015 *Implement. Sci.* **10 (1)** p 125.
[5] Pickup L, Atkinson S, Hollnagel E and et al. 2017 *Appl. Ergon.* **59** 234–242.
[6] Raben D C, Bogh S B, Viskum B and et al. 2017 *Cognit. Technol. Work* **19 (4)** 809–817.
[7] Herrera I A and Woltjer R 2010 *Reliability Engineering and System Safety* **95** 1269–1275.
[8] Rutkowska P and Krzyżanowski M 2018 *Scientific Journal of Silesian University of Technology. Series Transport* **101** 159-166.
[9] Davis J P and Hall J W 2003 *Decision Support Systems* **35 (3)** 415–433.
[10] Sakno O P 2013 Managing resources tires of transport by improving monitoring tread wear [Text]: Dis. ... candidate. Sc. Sciences: 05.22.20 Lugansk, p. 216 (in Ukrainian).
[11] Mihaylov A N 2009 Osnovyi sinteza funktsionalno-orientirovannyih tehnologiy mashinostroeniya [Basics of the synthesis of function-oriented engineering technologies]. Moscow: Mashinostroenie, p. 346 (in Russian).
[12] Kostenko A, Mikhaylov A and Poletaykin A 2019 *Mechanics XXI Century* **18** 179–186 (in Russian).

Acknowledgments

The authors would like to express their gratitude for helping to leading scientists Professor Volodymyr Sakhno (National Transport University), Professor Alexander Mihaylov (Donetsk National Technical University) and Erik Hollnagel, Ph.D., Professor, Professor Emeritus.

IOP Conf. Series: Materials Science and Engineering **977** (2020) 012006 doi:10.1088/1757-899X/977/1/012006

Aspect of repeatability in Real Driving Emissions tests regarding to velocity profile

J. Matla[1*], L. J. Sitnik[1], Z. J. Sroka[1], R. M. Włostowski[1]

[1]Wrocław University of Science and Technology
Faculty of Mechanical Engineering
Department of Vehicles Engineering
ul. Wybrzeże Wyspiańskiego 27
50-370 Wrocław, POLAND

[*]jedrzej.matla@pwr.edu.pl

Abstract. Relatively large results discrepancies observed in vehicles emission tests became a reason for change of test procedure. The New European Driving Cycle (NEDC) test was replaced by more dynamic Worldwide Harmonized Light Vehicles Test Procedure (WLTP) on September 2017, however it occurred that emissions obtained in WLTP are still far from real ones. Due to that fact a new Real Driving Emissions (RDE) test was introduced in order to complement WLTP. RDE testing involves emissions measurement while vehicle being driven in real traffic conditions which is the biggest difference from previous tests conducted on chassis dynamometer. Because of that it is suspected that repeatability of RDE tests is significantly lower than in case of WLTP or even does not have to appear. Subject of authors investigation was evaluation of RDE tests repeatability. It included portable emission measuring system (PEMS) installation on vehicle and six test runs along test route designed in accordance with European Commission requirements. Results were then compared regarding to velocity profile of vehicle. Performed analysis involved statistical distribution of velocity profiles and their kurtosis and skewness comparison. It was found that tests repeatability occurs for some velocity ranges, however it is not satisfactory. Conclusions present the concept of further research on this problem.

1. Introduction

Studies on vehicles emissions began over 60 years ago, however normalized test procedures rapid development started along with establishment of first agencies and law regulations in early seventies of twentieth century. Since then many approaches were taken in order to create universal normalized emission test. One of the first tests implemented at large scale were EPA Federal Test Procedure in United States and New European Driving Cycle in Europe. Such state of matter for decades was problematic for manufacturers who had to test their vehicles according to various regulations depending on world region. As a result the Worldwide Harmonized Light Vehicle Test Procedure was created and approved by most of countries with highly developed automotive industry [1].

Over the years NEDC was criticized mainly for its long idling time, very stable velocity changes as well as constant cruising speeds. In comparison, the WLTP incorporated higher maximal and average speeds, longer duration and travel distance, however the most important difference was change of velocity profile (fig.1) and more accurate measurement conditions. All of those factors resulted in more realistic emissions values nonetheless still being obtained in precisely controlled laboratory conditions.

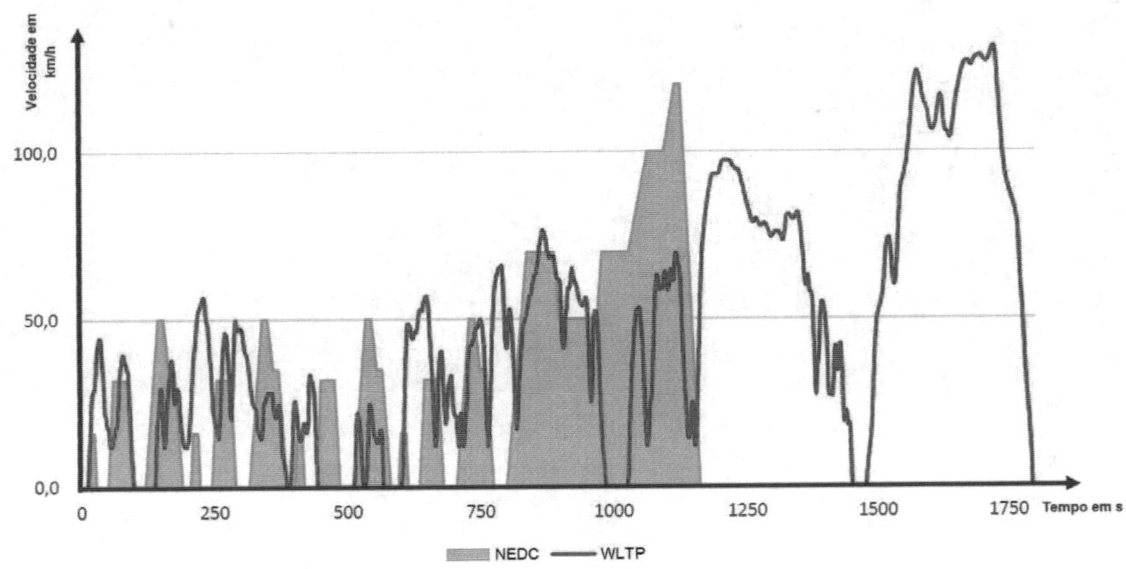

Figure 1. NEDC and WLTP velocity profiles [2]

In order to compromise WLTP results a Real Driving Emissions test was introduced. During RDE test vehicle is driven on public roads and over a wide range of constantly changing conditions such as weather, traffic or road inclination. Pollutant emissions are measured in real time by a set of analyzers fitted by a test car called PEMS. Moreover during test it is permitted to use auxiliary devices being a part of car equipment which in combination with individual driving style makes RDE test results significantly different from data obtained in WLTP [3]. Taking into account many variables as well as analyzers margin of error so called conformity factors were introduced by European Commission's regulatory committee. Conformity factors are defined as not-to-exceed limits and their task is to correlate WLTP and RDE test results.

2. Test setup

Presented experiment being a part of wider research concentrates its interest on a velocity profiles of RDE tests repeated along pre-defined route. Due to that fact test setup consisted of a full set of portable emission measuring system mounted on a vehicle as it is presented in the figure 2. Such configuration allowed for precise record of a car velocity independently on its own instruments. Data used in presented paper came from GARMIN GPS16HVS system mounted on the vehicle's roof and being capable to communicate with up to 12 satellites at the same time.

Test vehicle was classified as a M1 category according to European Parliament regulation [4] which makes RDE testing of it fall under LDV test procedure described in details by European Commission JRC [5]. Further requirements analysis enabled to classify it as having certain mileage thus being suitable to apply in-service conformity testing procedures. Vehicle technical condition assessment covering computer diagnostic together with its complete service history inquiry allowed to ensure minimal risk of potential failures and their effect on collected data.

Figure 2. Test vehicle with installed PEMS

3. Test route

Test course developed for the purposes of testing was created with regard to LDV on-road testing requirements of EC JRC presented in table 1. The route consisted of two loops encircling city of Wrocław and its neighbouring districts, having starting point coinciding with its finish. Total length of test course confirmed with GPS data was around 78,4km while elevation gain being approximately 45m.

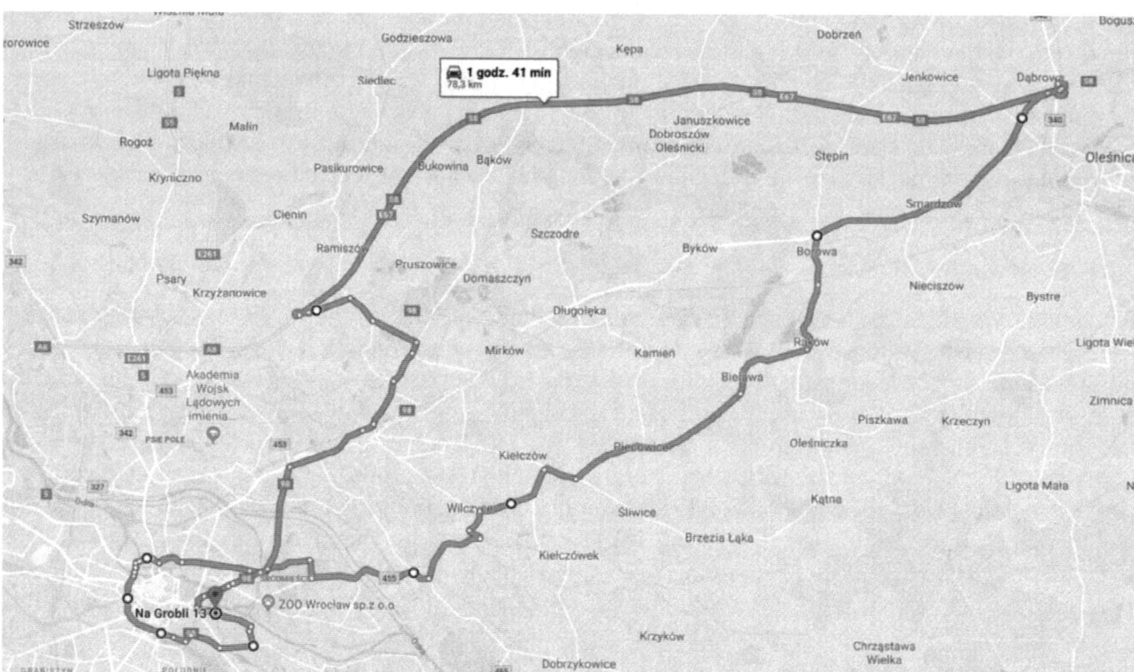

Figure 3. Test route [6]

Table 1. General trip requirements for RDE test [7]

Distance	Urban	> 16 km
	Rural	>16 km
	Motorway	>16 km
Trip composition	Urban	29% to 44% of total distance
	Rural	23% to 43% of total distance
	Motorway	23% to 43% of total distance
Average speeds	Urban	15 to 40 km/h
	Rural	60 to 90 km/h
	Motorway	>90 km/h (including >100 km/h for 5 min. at least)
Total duration	Between 90 and 120 min	

EC JRC regulations defines also additional criteria connected with driving dynamics not mentioned in table 1. According to moving average windows method called EMROAD, parameters such as acceleration points or relative positive acceleration are described by adequate formulas [8]. Due to velocity profile being the subject of presented research, mentioned parameters were not considered in further steps.

4. Software tools

Purposes of presented research required usage of various software in order to acquire and process obtained data. In order to control measuring equipment and record time course of velocity an AVL M.O.V.E. System Control application was used. It allowed for simultaneous acquisition of data from various measuring devices such as GPS and vehicle ECU during whole test, giving the operator live preview of parameters such as engine speed, ambient conditions or actual velocity.

Undoubtedly big advantage of mentioned system control application was its ability of storage data from all devices and sensors in one file. Moreover it featured also user-guided test sequences, which allowed for continuous monitoring of test conformity with RDE requirements. Thanks to that property it was possible to minimize number of invalid test runs. Software was constantly recalculating parameters such as trip duration, composition, average velocities, travelled distance and driving dynamics. Results were presented to the driver in form of table with values changing its color to green every time one of them became compliant with RDE test requirements, as it is presented in the figures 4a and 4b.

Once all test runs were finished, results were further subjected to post-processing with AVL Concerto 5 software provided together with PEMS equipment. Data channels related to time and GPS ground speed were selected and exported to ASCII format files in order to make them suitable for subsequent MS Excel calculations.

a)

b)

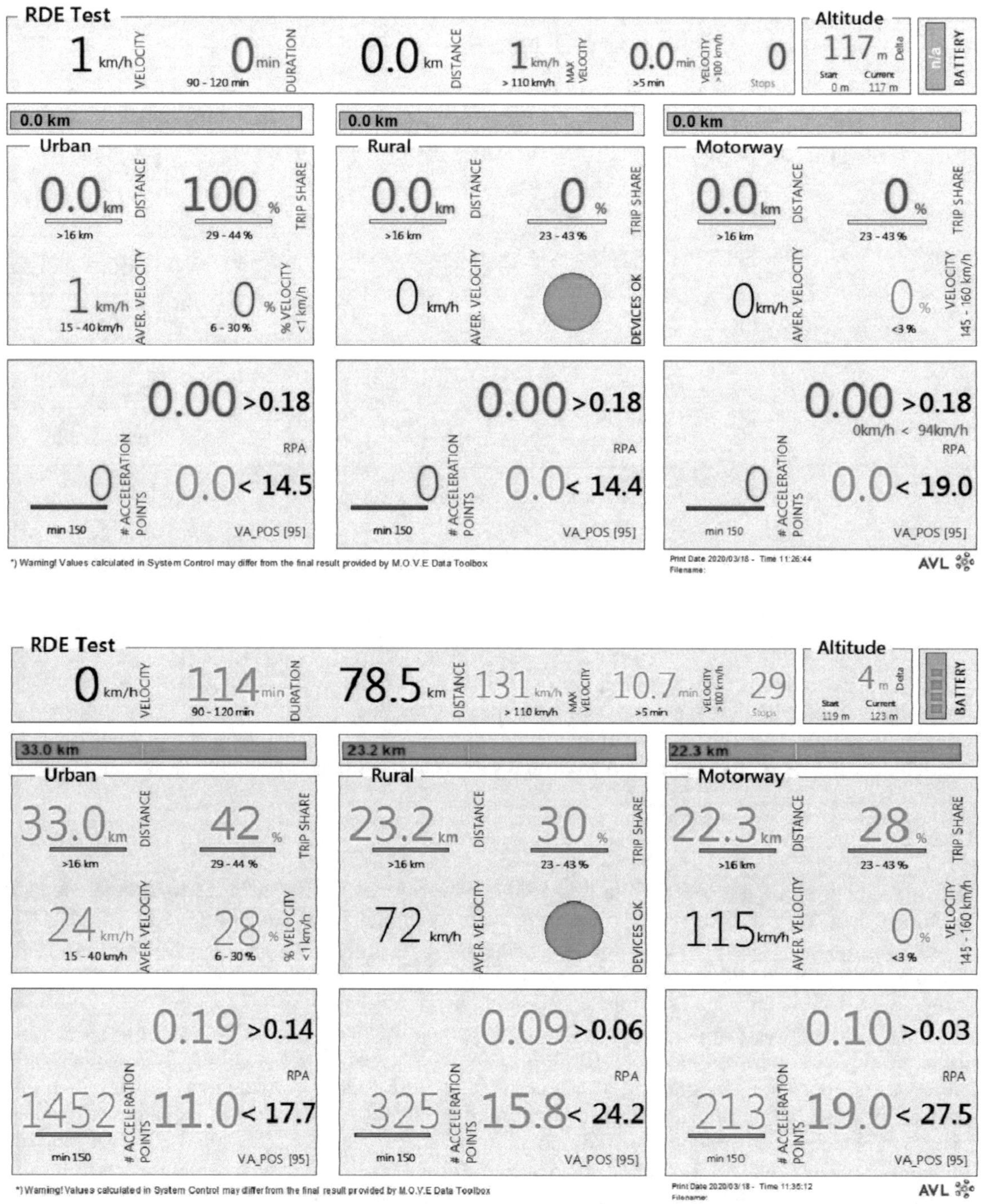

Figure 4. RDE test parameters: a) at the beginning and b) at the end of test run

5. Results and analysis

Obtained results consisted of vehicle GPS ground speed sampled every 500ms. Using simple formulas it was possible to calculate instantaneous and cumulative distance so velocity profiles of test runs could been created (see figure 5).

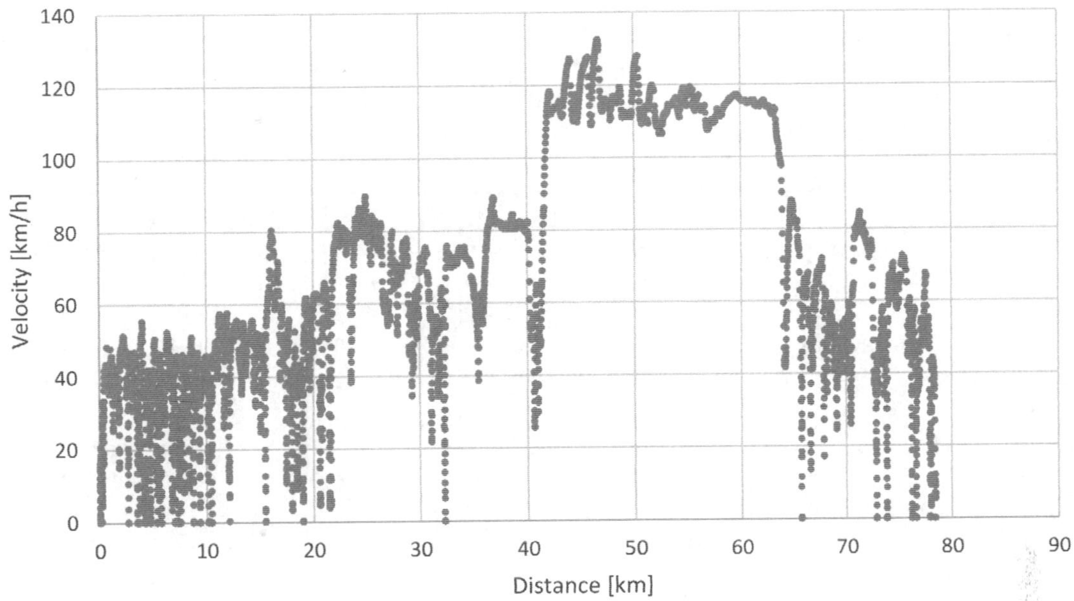

Figure 5. Velocity profile of RDE test no.1

Having all velocity profiles created next step was data preparation to statistical analysis. In order to do so sampling points were grouped into intervals regarding to instantaneous velocity. Each interval covered range of 10km/h excluding the highest value. For example an interval denoted as "10" was mathematically described as <0,10) etc. As a result graph of velocity distribution for each test was created as it is presented in the figure 6.

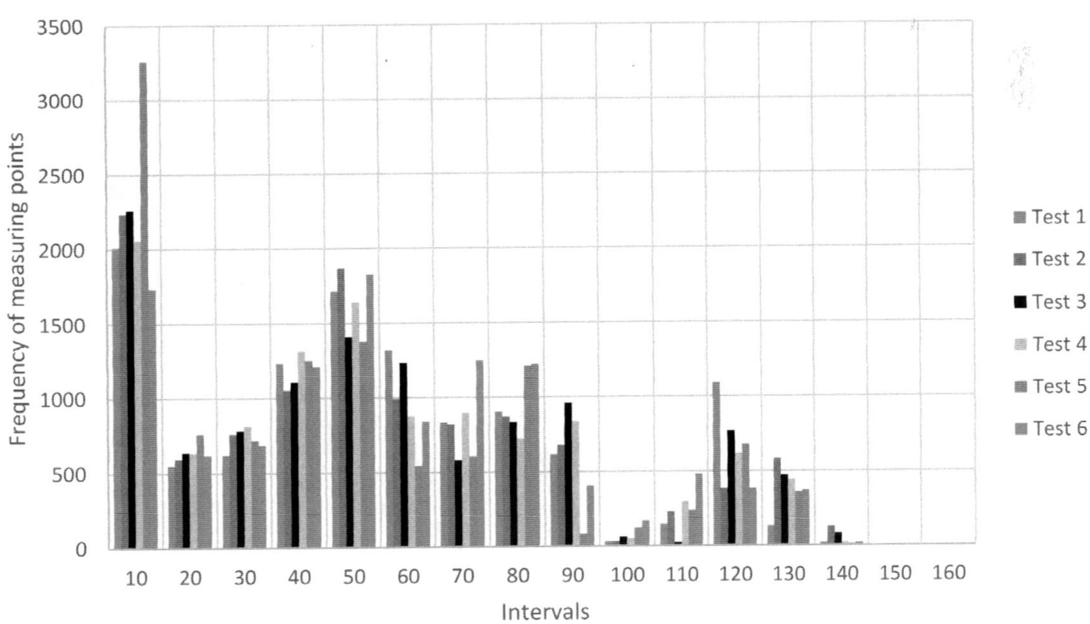

Figure 6. Velocity distribution graph

At this point it was observed that intervals "100" and "140" exhibited relatively low frequencies for all test runs, while in case of others ex. interval "10" or "50" it was hard to recognize results repeatability. In order to investigate results correlation all test runs were subjected to more detailed statistical analysis, which results were demonstrated in table 2.

Table 2. Statistics of RDE test runs

Parameter:	Test 1	Test 2	Test 3	Test 4	Test 5	Test 6
Average velocity [km/h]	48.830	48.116	48.305	47.635	41.396	50.371
Standard error of the mean	0.320	0.330	0.341	0.326	0.300	0.319
Median	45.900	45.400	45.200	43.750	39.000	47.100
Population standard deviation	34.392	35.788	36.844	35.545	35.048	33.760
Population variance	1182.821	1280.810	1357.503	1263.423	1228.382	1139.760
Kurtosis	-0.494	-0.353	-0.571	-0.533	-0.296	-0.497
Skewness	0.405	0.520	0.484	0.481	0.671	0.333
Range	132.600	139.900	136.800	133.000	131.100	131.800
Minimum	0	0	0	0	0	0
Maximum	132.600	139.900	136.800	133.000	131.100	131.800
Count	11565	11739	11687	11852	13658	11206
Confidence (95.0%)	0.627	0.647	0.668	0.640	0.588	0.625

Further statistical parameters assessment enabled to indicate rather minor differences between tests except for test no.5. At first glance difference in average velocity being lower than other tests was recognized. Later on it was found that other parameters departing significantly from the rest were kurtosis, skewness and confidence level. Worth mentioning is fact that at the same time standard deviation remained at comparable level (see figure7).

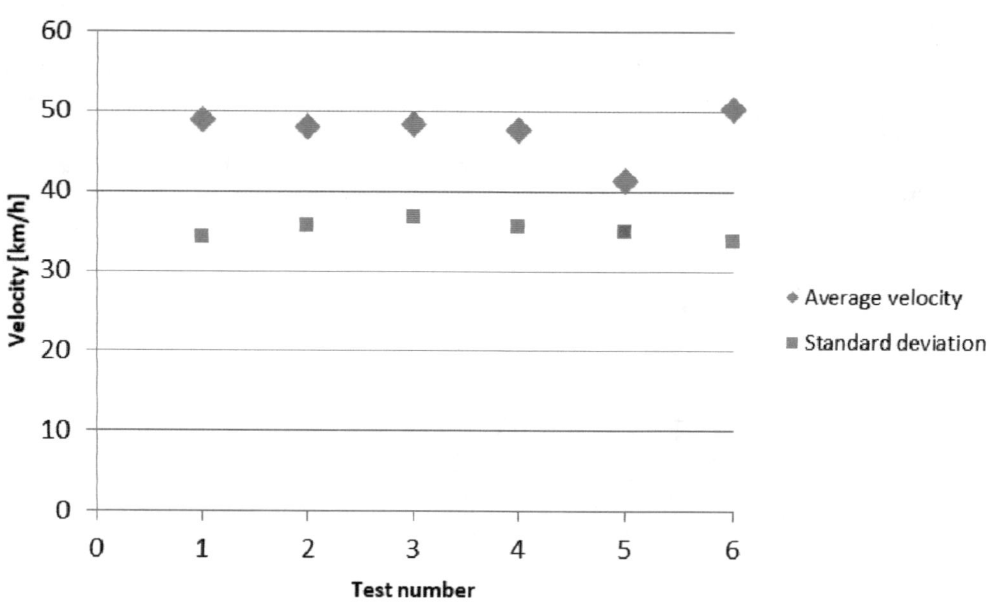

Figure 7. Selected test runs parameters

In the last step it was checked how neglection of test no. 5 will affect overall data repeatability. For this purpose distribution of skewness and kurtosis for all tests was investigated as parameters characterizing statistical distribution. Results were presented in table 3, where dataset 1 stands for all 6 tests and dataset 2 excludes test no. 5.

Table 3. Influence of test no.5 neglection on data variation

	Kurtosis		Skewness	
Parameter:	Dataset 1	Dataset 2	Dataset 1	Dataset 2
Average	-0.457	-0.490	0.482	0.445
Standard error of the mean	0.044	0.037	0.047	0.034
Median	-0.495	-0.497	0.483	0.481
Population standard deviation	0.108	0.083	0.114	0.075
Population variance	0.012	0.007	0.013	0.006
Kurtosis	-1.136	2.651	1.165	-0.385
Skewness	0.787	1.428	0.607	-0.901
Range	0.275	0.218	0.338	0.186
Minimum	-0.571	-0.571	0.333	0.333
Maximum	-0.296	-0.353	0.671	0.520
Count	6	5	6	5
Confidence (95.0%)	0.087	0.072	0.091	0.066
Coefficient of variation	-0.237	-0.169	0.237	0.169
Relative standard deviation	100.000	71.296	100.000	71.139

6. Conclusions

Presented research focusing its interest on velocity profile investigation of RDE tests allowed for improvement in test repeatability by means of simple statistical analysis. While interpreting presented results it is important to remember that test distance and shape were assumed equal at the beginning.

Despite fact that all of six test runs were qualified by AVL System Control application as valid it occurred that it is still possible to observe differences between them. Such state of matter is caused by factors randomly varying during RDE tests however limited to some extent by regulations. Mentioned factors are connected with ambient conditions, traffic and individual driving style.

In this place authors formulated two questions in order to maximize tests repeatability: what limits of statistic parameters variations should be assumed and should the tests with significantly different velocity profiles be excluded from consideration?

Results showed that in both cases velocity profiles should be described with statistical distribution regardless of its form, due to the fact that each distribution may be described by its skewness and kurtosis. Among six conducted tests the one with extreme values of skewness and kurtosis was chosen to be neglected and then two datasets were created. First dataset consisted of all data while second one excluded results of selected test. Statistical analysis of both datasets indicated almost 30% decrease of data variation in case of dataset 2. All of that lead to conclusion that test no.5 is characterized by statistical distribution different from others.

As a result authors presented a method basing on simple preliminary calculations, which may be helpful to assess RDE tests repeatability and decide about test consideration or rejection. In order to verify and develop presented method it is required to collect more data, that is why further research covering wide range of RDE parameters are in progress.

References

[1] Regulation (EC) No 715/2007 of the European Parliament and of the Council of 20 June 2007 on type approval of motor vehicles with respect to emissions from light passenger and commercial vehicles (Euro 5 and Euro 6) and on access to vehicle repair and maintenance information

[2] https://www.fuchs.com/pt/en/company/news/news-entry/2203-differences-between-wltp-and-nedc/ accessed on 12.03.2020

[3] Marotta A, Pavlovic J, Serra S, Anagnostopoulos K, Ciuffo B, Fontaras G, Tsiakmakis S, Zacharof N 2015 *A Comparison of Gaseous Emissions from Light-Duty Vehicles under the NEDC and the WLTP test procedures* Transportation Research Board 94th Annual Meeting, At Washington DC, United States

[4] Regulation (EU) No 168/2013 of the European Parliament and of the Council of 15 January 2013 on the approval and market surveillance of two- or three-wheel vehicles and quadricycles

[5] Valverde Morales V, Bonnel P 2018 *On-road testing with Portable Emissions Measurement Systems (PEMS) - Guidance note for light-duty vehicles*, EUR 29029 EN, Publications Office of the European Union, Luxembourg, ISBN 978-92-79-77345-7

[6] https://www.google.com/maps, accessed on 6.01.2020

[7] Commission Regulation (EU) 2016/646 of 20 April 2016 amending Regulation (EC) No 692/2008 as regards emissions from light passenger and commercial vehicles (Euro 6)

[8] Commission Regulation (EU) 2018/1832 of 5 November 2018 amending Directive 2007/46/EC of the European Parliament and of the Council, Commission Regulation (EC) No 692/2008 and Commission Regulation (EU) 2017/1151 for the purpose of improving the emission type approval tests and procedures for light passenger and commercial vehicles, including those for in-service conformity and real-driving emissions and introducing devices for monitoring the consumption of fuel and electric energy

Abbreviations

ASCII – American Standard Code for Information Interchange
AVL – Anstalt für Verbrennungskraftmaschinen List
ECU – Engine Control Unit
EPA – Environmental Protection Agency
EU – European Union
GPS – Global Positioning System
ISC – In-Service Conformity
JRC – Joint Research Centre
LDV – Light Duty Vehicle
M.O.V.E. – Mobile Onboard Vehicle Evaluation
NEDC – New European Driving Cycle
PEMS – Portable Emissions Measuring System
RDE – Road Driving Emissions
WLTP – Worldwide harmonised Light-duty vehicles Test Procedure

Acknowledgements

Testing equipment used in presented research was purchased by Wrocław University of Science and Technology as a part of "Complex GEO-3EM ENERGY ECOLOGY EDUCATION" financed by European Regional Development Fund within the Regional Operational Program of the Lower Silesian Voivodship for the years 2014-2020.
Authors are also deeply indebted to express their gratitude to Mrs Anna and Mr Leszek Leśniak for provision of test vehicle.

IOP Conf. Series: Materials Science and Engineering **977** (2020) 012007 doi:10.1088/1757-899X/977/1/012007

Study of urban fuel consumption of a car with a petrol internal combustion engine

M Magdziak-Tokłowicz[1,2]

[1] Mechanical Department, Wrocław University of Science and Technology, Poland, Wrocław, Braci Gierymskich 164
[2] GEO-3EM Research Center, Wrocław University of Science and Technology, Poland, Wrocław, Na Grobli 13

E-mail: monika.magdziak-toklowicz@pwr.edu.pl

Abstract. This article presents analysis and an attempt to forecast the actual fuel consumption of the vehicle in the urban cycle. The most important factors affecting fuel consumption were discussed, in particular the influence of the driver's driving style. Each driver drove a designated, the same route in the urban area outside the rush hour. Based on the readings from the OBD on-board computer and the mobile application, the driver's occurrence points on the universal engine characteristics were read. On the axes of torque and rotational speed, the most frequently occurring engine work fields for each driver were determined. The author divided the driving styles of the driver according to the most common occurrence on the universal characteristics of the engine and depending on the specific fuel consumption. The author's method has shown that it is possible to point out drivers who will certainly exceed the average fuel consumption after driving a few kilometers of the road. This method shows the possibility of using the on-board computer and OBD readings to optimize fuel consumption and match each driving style to the appropriate external and universal characteristics of the vehicle engine to use as little fuel as possible, and thus contribute to reducing CO2 emissions to the environment and consumption of fossil fuels. This method can be successfully used on all vehicles with an internal combustion engine.

1. Introduction

The development of global industry and the progressing globalization of life, as well as the increase in the mobility of people dynamically affect all aspects of automotive in the world. In 2019, the world production of passenger cars amounted to almost 92 million units. China is the undisputed leader, in which over 28% of the total was produced, followed by the USA, followed by Japan and Germany [1]. Road transport is still the leading means of transporting people and goods, but any land, sea or air transport is associated with the problem of environmental pollution caused by the use of internal combustion engines in these means of transport. In most countries on Earth, various types of ordinances have already been introduced that define the conditions for the admission of motor transport means to use, the need to test their technical condition, as well as approval methods [2-6]. In order to further reduce the emission of harmful components of exhaust gases, it is necessary to make society and the broadly understood industry aware of the catastrophic effects of environmental pollution. Possible effects of the greenhouse effect were discussed at the end of the 20th century, and then the European

Union at the beginning of the 21st century began to consider reducing CO2 emissions by implementing stringent emission standards for harmful exhaust emissions from motor vehicles. All restrictions have a common goal to clean the air and reduce fuel consumption. To meet the limits of permissible emission standards for individual compounds, new constructions and technologies for the construction of internal combustion engines and vehicles are emerging. It is also important to check the functioning of new components and engine components so that it meets stringent standards throughout its entire life cycle. In addition to retrofit in vehicle designs, work continues in the automotive industry that aims to introduce and disseminate the idea of energy saving. First of all, the use of new fuels, hybrid and electric drives as well as hydrogen power supply or fuel cells should be mentioned [7-9].

However, despite all restrictions, combustion-powered vehicles will continue to be produced in the coming years. The total conversion of vehicle fleets into all-electric vehicles is not possible due to the insufficient energy resources of all countries and the lack of vehicle charging infrastructure. For this reason, the author focused on possible methods of reducing fuel consumption by internal combustion vehicles.

2. Fuel consumption

Fuel consumption in vehicles can be defined in two terms:

- hourly fuel consumption (seconds) in the unit: kg/s determines the mass of fuel consumed by the engine per unit of time. This value is determined by weight or volume measurement;
- specific fuel consumption in the unit: hp/ h or W/s or kW/h - is a measure of the use of the fuel supplied. It determines the mass of fuel consumed by the engine per unit of power and time.

The concept of average fuel consumption in real conditions is a general and estimated parameter. At each stage of the calculation, the personal and environmental conditions of the tests should be described, the type of route and pavement should be shown, due to the fact that too many parameters affect the change in fuel consumption. The parameters affecting the increase or reduction of fuel consumption are presented in table No. 1. An important factor is the type of engine under consideration, whether it is naturally aspirated, turbocharged, downsizing or rightsizing. It is possible to improve the usability of the combustion engine efficiency by using e.g. thermal activation of the combustion chamber of a reciprocating internal combustion engine, which returns portions of heat in one cycle, which significantly improves performance engine work and many other solutions [10-12].

Table 1. Parameters affecting fuel consumption.

Constructional factors	Environmental factors	Human factors
vehicle mass and inertia of rotating parts	atmospheric conditions (air temperature, density, viscosity)	driving style of the driver
vehicle shape (and shape changing systems)	wind direction and speed	health
construction of tires	terrain shape	frame of mind
powertrain efficiency	road surface state and art.	
resistance of rotational parts (bearings, brakes, etc.)		

It is recognized that the driver and driving style have the greatest impact on fuel consumption. Three types of drivers are defined:

- aggressive,
- moderate or optimal,
- conservative.

Driving styles can be combined depending on the parameters that are used for testing.

3. Description of research

The test object was a 2000cc gasoline engine with a 4-cylinder in-line 16-valve engine in a D-segment vehicle shown in the figures below.

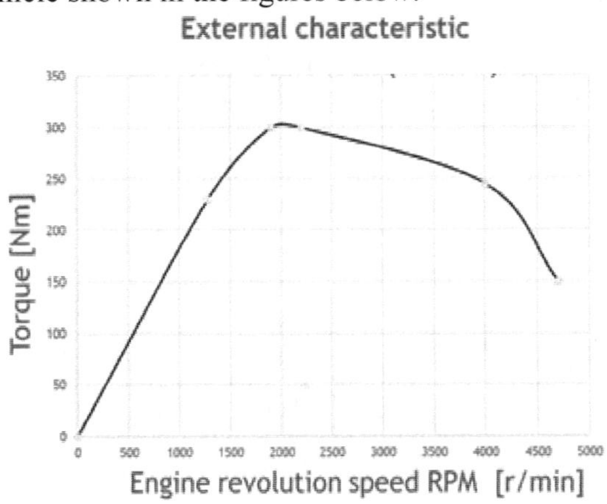

Figure 1. External characteristic of selected car, 2.0, 203 hp, average sfc= 190 g/kWh

Figure 2. Selected engine, 4-cylinder in-line engine 16v [13]

The study covered five drivers aged 30-40 covering almost 13 km route from point A-B-A presented on the map below. Tests were carried out outside peak hours in city. Data was read using a mobile application, while measurements were made using a device connected to OBD (On-Board Diagnostic). The following parameters were read: time [s], engine revolution speed [r/min], torque [%], fuel consumption [cc/min], vehicle speed [km/h], the throttle opening [%]. The data sampling frequency was 1 second. Data has been converted to get power [kW] and specific fuel consumption sfc [g/kWh].

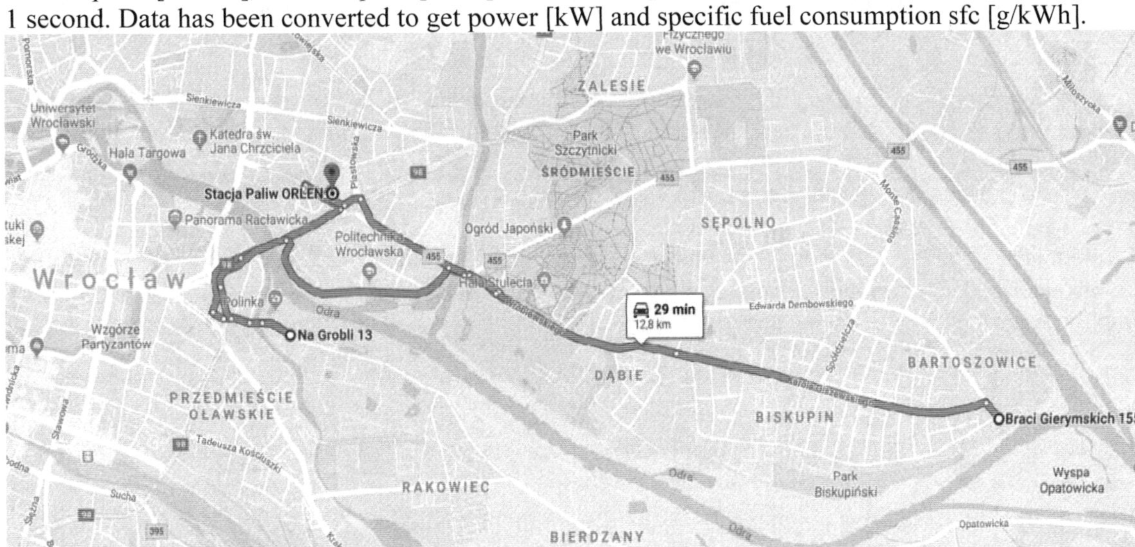

Figure 3. Selected route of the car in the city of Wroclaw in Poland.

4. Data analysis

The data analysis consisted of the implementation of the universal engine characteristics for each driver's passes. Points corresponding to the momentary engine load have been plotted on the charts. The number of points depended on the driving time of the driver, due to the fact that the sampling frequency was 1

IOP Conf. Series: Materials Science and Engineering **977** (2020) 012007 doi:10.1088/1757-899X/977/1/012007

second. The abscissa axis of each point was the engine speed, and the ordinate- the engine torque. The points marked on the charts allowed to identify the character and driving style of each driver. The first driver's chart was determined to be focused, the dispersion of points on the characteristics was insignificant. The driving style of the driver was defined as conservative. The first driver in 98% of driving time did not exceed 2300 [r/min], engine load also in the lower torque range. This location of most points on the characteristics resulted in a high specific fuel consumption of sfc=203 [g/kWh]. The average specific fuel consumption for a given engine is around 186 [g/kWh]. The center of the points was determined by calculating the means of the coordinates of the torque and engine speed.
The average point had coordinates T = 48.76 [Nm] and RPM = 1612.46 [r/min]- figure 4.

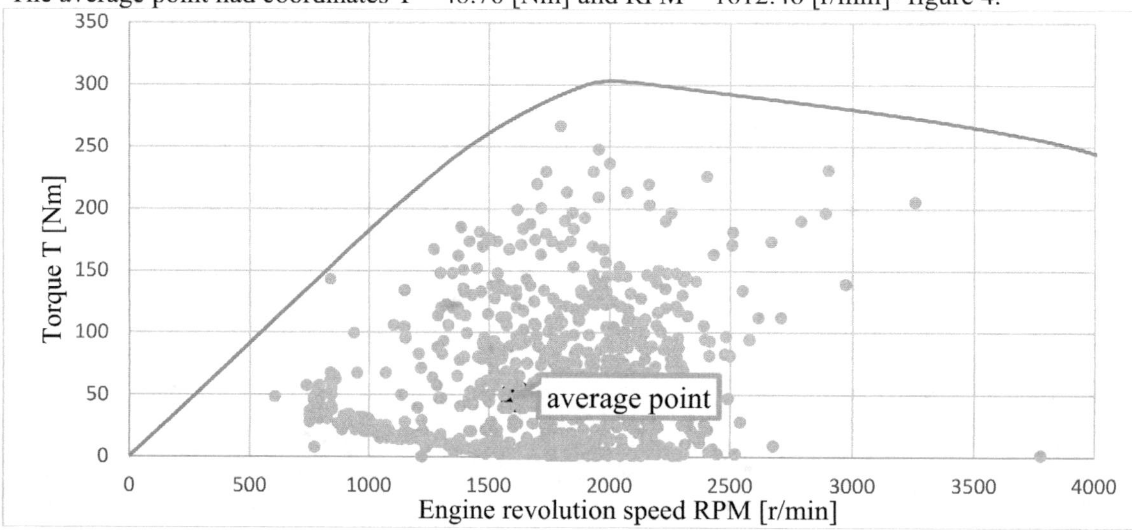

Figure 4. 1-st driver, engine load work map for the first driver. The coordinates of the points of torque and engine revolution speed are focused on the general characteristics. The average specific fuel consumption is sfc = 203 [g/kWh].

Analysis of the data of all drivers showed the dependence of the location of the average point determining the center of points of all registered on the fuel consumption. Analyzing the fifth driver, significant points were dispersed in Figure 5. The points had greater distances from the center of the average point, which resulted in a shift to a larger range of engine revolutions, average RPM = 2054 [r/min] and significantly lower specific fuel consumption of sfc = 162 [g/kWh].

Statistical analysis showed that the correlation coefficient of the average point coordinate dependence on specific fuel consumption exists. The confidence interval estimates at 95% that the average point will be within the specified fuel consumption range. The average point has significantly lower T and RPM coordinates in the specific fuel consumption group below sfc = 190 [g/kWh].

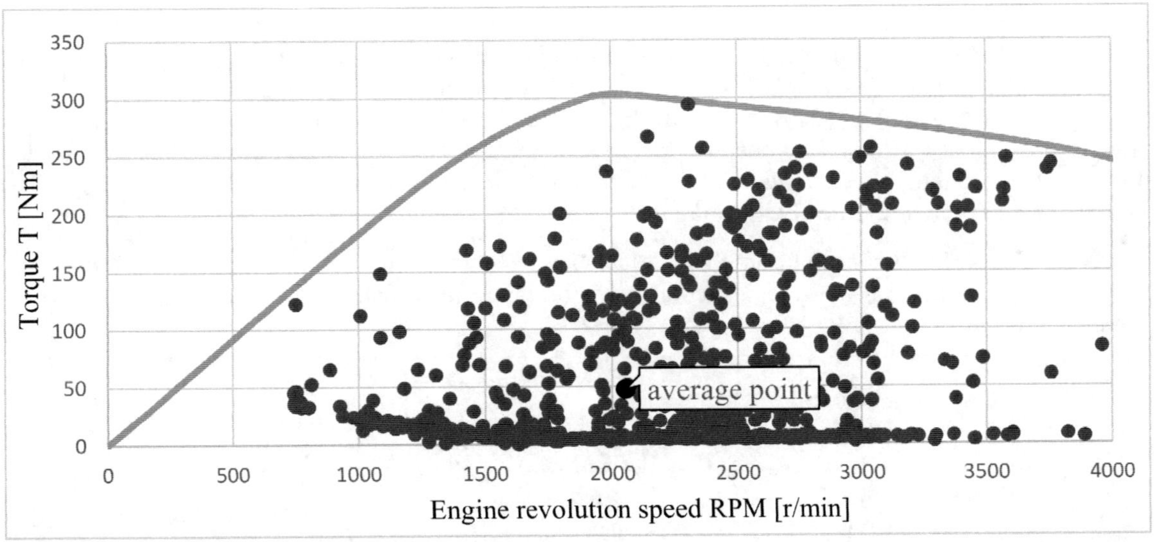

Figure 5. 5-th driver, engine load work map for the fifth driver. The coordinates of the points of torque and engine speed are clearly dispersed on the general characteristics. The average specific fuel consumption is sfc =162 [g/kWh].

The figures below present aggregate data of average points for each driver along with their fuel consumption. The relationship between the mean point location on the engine map and fuel consumption is visible.

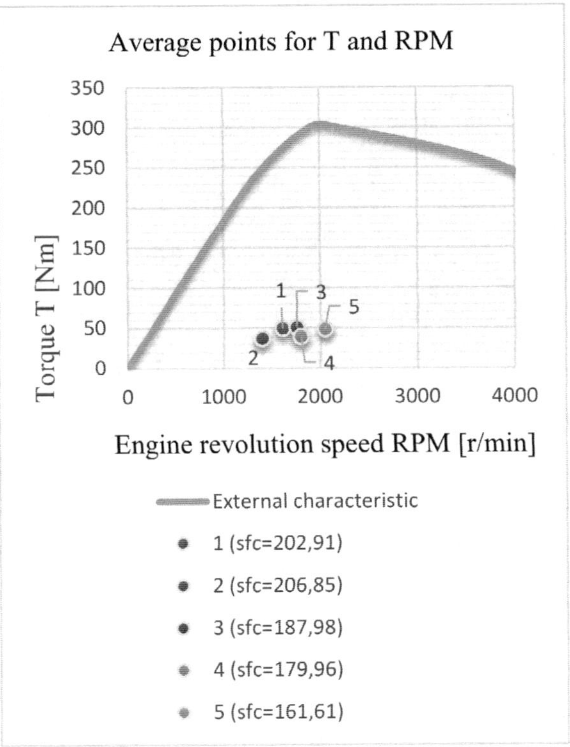

Figure 6. Average points of the coordinates of torque and engine revolution speed for five drivers.

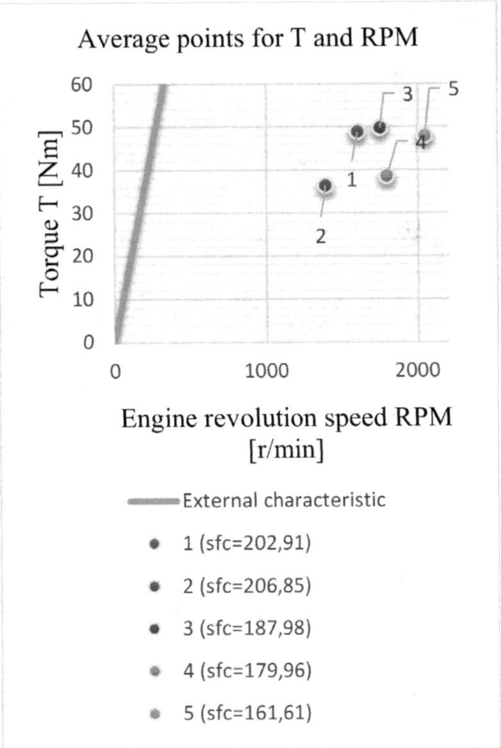

Figure 7. Distribution of average points on the universal engine characteristics for five drivers.

In connection with the above, an attempt was made to forecast fuel consumption based on the location of the coordinates of the universal characteristic points and the average point. Average points and specific fuel consumption were determined for each driver after 10 minutes and 15 minutes by a car.

This forecasting method turned out to be a simple tool for quickly determining whether a given driver after a few minutes of driving tends to exceed the average fuel consumption or not. The relationship between the location of the average fuel consumption point has been demonstrated for each driver. The presented results are presented below in table 2.

Table 2. Specific fuel consumption for all drivers after 10 minutes by car, after 15 minutes and the average of the entire trip. The average coordinates of the torque points and engine speed are shown in the table.

Driver	Average T after 10 min driving [Nm]	Average RPM after 10 min [r/min]	Average T [Nm]	Average RPM [r/min]	Specyfic fuel consumption sfc after 10 min [g/kWh]	Specyfic fuel consumption sfc after 15 min [g/kWh]	Average specyfic fuel consumption sfc [g/kWh]
1	55.88	1618.45	48.76	1612.46	217.43	208.83	202.91
2	37.44	1357.42	36.26	1402.14	204.82	209.84	206.85
3	56.65	1970.32	49.66	1759.82	197.62	189.47	187.98
4	38.17	1899.09	38.47	1803.36	190.86	176.61	176.96
5	50.94	2197.23	47.81	2054.18	171.42	160.31	161.61

5. Conclusion

- Based on the universal characteristics, it is possible to determine the driving styles of each driver after placing the engine load points on the charts.
- These graphs show the dispersion of points - dynamic driving or their focus - conservative driving style.
- Statistical analysis showed that there is a correlation coefficient between the mean of the point coordinate and the specific fuel consumption. The confidence interval estimates at 95% that the average point will be within the specified fuel consumption range. The average point has significantly lower T and RPM coordinates in a given fuel consumption group below sfc = 190 [g/kWh].
- The method of forecasting average specific fuel consumption showed which driver will almost certainly exceed fuel consumption after a few minutes of driving the vehicle.
- This method also shows that every driver drives in different speed ranges, which shows the need to adapt a person to a given vehicle and engine in the future.
- One engine is installed in millions of cars and used by different drivers, which reduces the chances of matching the driver's driving style to a given vehicle. The method can be used to convert the engine map for a specific driver.
- Autonomous vehicles, thanks to neural networks, could use good driver points distribution maps to learn the network of correct and economical driver driving styles.

Acknowledgments
Thanks to the Authorities of Wroclaw University of Science and Technology for construction of the GEO-3EM Research Center, in which it was possible to carry out all kinds of experiment.

References

[1] International Organization of Motor Vehicle Manufacturers *Production statistic* Avaible: http://www.oica.net/category/production-statistics/2019-statistics/ Accessed on 10.07.2020

[2] Commission of the European Communities. Proposal for a Regulation of the European Parliament and of the Council: Setting emission performance standards for new light commercial vehicles as part of the Community's integrated approach to reduce CO_2 emissions from light-duty vehicles COM (2009) 593 final 2009/0173(COD)

[3] Merkisz J, Pielecha J, Radzimirski S 2012 *Emisja zanieczyszczeń motoryzacyjnych w świetle nowych przepisów Unii Europejskiej* (WKŁ, Warszava) p 220 (In Polish)

[4] Merkisz J and Radzimirski S 2009 Stan obecny i przewidywane zmiany w europejskich przepisach o emisji zanieczyszczeń z samochodów ciężarowych i autobusów *Transport Samochodowy* (Wydawnictwo ITS) (In Polish)

[5] *Worldwide Emissions Standards. Passenger Cars and Light Duty Vehicles* Delphi 2012/2013 (www.delphi.com)

[6] Katrašnik T 2007 Hybridization of powertrain and downsizing of IC engine – A way to reduce fuel consumption and pollutant emissions – Part 1 *Energy Conversion and Management* (Elsevier) **48**(5) pp 1411-23

[7] He Y, Chowdhury M, Pisu P, Ma Y 2012 An energy optimization strategy for power-split drivetrain plug-in hybrid electric vehicles *Transportation Research Part C: Emerging Technologies* (Elsevier) **22** pp 29-41

[8] Chen H et al. 2015 Modeling temporal variations in global residential energy consumption and pollutant emissions *Applied Energy* (Elsevier) **148** pp 820-29

[9] Chen PY, Chen ST, Hsu CS, Chen CC 2016 Modeling the global relationships among economic growth, energy consumption and CO_2 emissions *Renewable and Sustainable Energy Reviews* (Elsevier) **65** pp 420-431

[10] Sroka Z and Dziedzioch D Mechanical load of piston applied in downsized engine 2015 *Archives of Cyvil and Mechanical Engineering* (Elsevier) **15**(3) pp 663-7

[11] Sroka Z 2013 Wybrane zagadnienia teorii tłokowych silników spalinowych w aspekcie zmian objętości skokowej (Oficyna Wydawnicza Politechniki Wrocławskiej) (In Polish)

[12] Sroka Z and Sadlak Z 2018 Thermal activation of the combustion chamber of a reciprocating internal combustion engine *Journal of Thermal Science* **27** pp 449-55

[13] https://www.netcarshow.com/volvo/2011-s60/ Accessed on 10.07.2020

IOP Conf. Series: Materials Science and Engineering **977** (2020) 012008 doi:10.1088/1757-899X/977/1/012008

Alignment Optimization of 3D Model of Propulsion Shaft System

Stefan Tenev[1]

[1] Department of Mechanics and Machine Elements, Technical University of Varna, 1 Studentska str., Varna 9010, Bulgaria

stefan.i.tenev@gmail.com

Abstract. Three dimensions models of propulsion shafts are presented in this work. A shaft alignment optimization is computed for a crankshaft, intermediate shaft and propeller shaft. "Design Study" sensors are used to report the best solutions of bearing reactions according to the vertical displacement. Shear forces and bending moments are listed for the flanges. The finite element analysis investigated a slope and deflection of the propulsion shaft system. Pictures of stress and displacement of the shafts, bolts and flanges are pointed in the paper. Stress concentrations and a factor of safety is calculated in the 3D dimension.

1. Introduction

Nowadays, the shipbuilding tendency is followed by increasing the vessel cargo capacity, which is associated to the rise of main drive power. The drive shafts then become increasingly sensitive to the disturbances affecting to the bearing's vertical displacement. However, the shaft alignment procedures remain an important problem requiring accurate analyzes [1, 2]. The positive values of supporting reactions in the sliding bearings is the main criteria of shaft alignment. A software method is used to optimize the elastic shafts line [3] by monitoring and compliance of the shaft alignment criteria.

This paper examines the loading of propulsion shaft system of a bulk carrier 21100 DWT. An assemble model is created of the shafts and a methodology of alignment optimization is developed. In figure 1 is presented:

- position 1 – AFT Stern Tube Bearing;
- position 2 – FWD Stern Tube Bearing;
- position 3 – Flanges of propeller and intermediate shaft;
- position 4 – Intermediate bearing;
- position 5 – Flanges of intermediate shaft and crankshaft;
- position 6 to 11 – crankshaft bearings;
- Ø440 – basic diameter of propeller shaft;
- Ø370 – basic diameter of intermediate shaft.

The correct static alignment of shafts is developed from practice to get the smooth dynamic operations of system [4, 5]. This leads to the reduction of excessive vibration of line, as well as the transmission of gearbox noise.

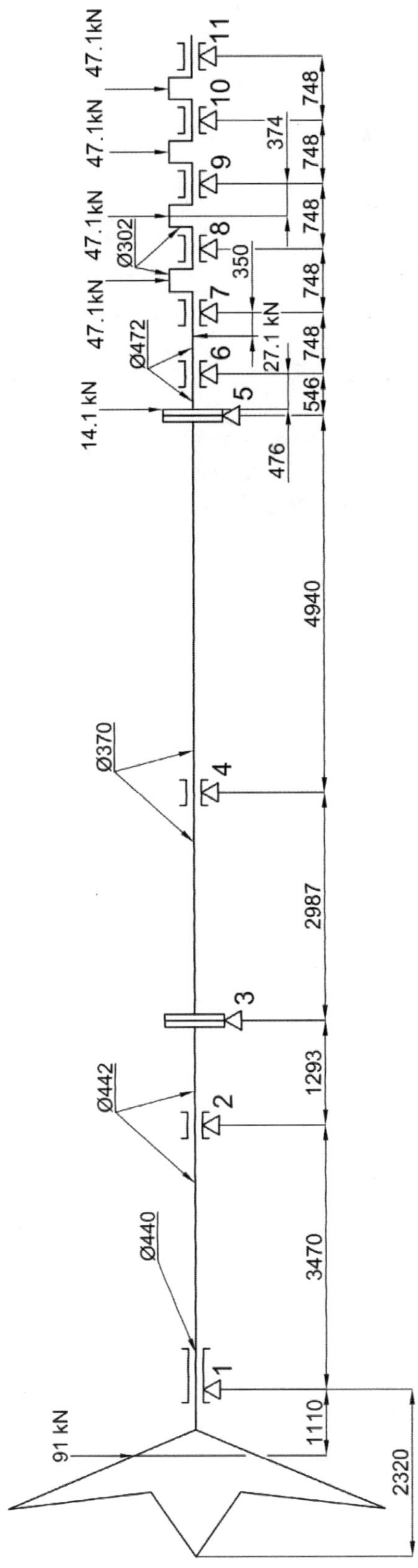

Figure 1. Shaft alignment scheme

The basic criteria, requirements and restrictions for shaft alignment are usually determined by classification societies, shipbuilders, designers and other control authorities [6]. Depending on which parameter is applicable, are considered:

- location of the permanent bearings, temporary bearings and location of the reaction measurement lifting points;
- propeller and intermediate shaft bearings reaction forces;
- crankshaft and transmission bearings reaction forces;
- shafts, bearings, clutches and flange mounting bolts load;
- slope and deflection of the propeller shaft, intermediate shaft and crankshaft of engine;
- propeller shaft slope accorded to the reference line of shaft;
- radial, axial and angular shaft flanges positions;
- engine base deflection and its bending;
- vessel hull deformation.

The shaft alignment is satisfactory if the above mentioned parameters can be controlled within certain limits for all operating conditions of the vessel (fully loaded or unloaded vessel, lowered or into open water, temperature changes affecting shafts, propulsion on maximum power).
The shaft alignment procedure begins once the conditions are met:

- vessel's construction temperature is stabled (usually the alignment is done in the early morning);
- massive parts such as superstructure, main engine, etc. are installed;
- all elements of the hull construction and equipment are presented;
- stern blocks are fully welded;
- leak tests are completed.

It is preferable for the shaft alignment procedures to start in a dry dock (bearing positioning, slope, etc.) and to provide a sufficient information for the production personnel.
The alignment procedures are extended and require many software operations, mathematical calculations and theoretical analyzes. In this regard, the present paper is focused on the study and optimization of some parameters described above. The extreme cases that can lead to damage and failure of the propulsion system of vessel are analyzed.

Figure 1. Shaft alignment scheme

IOP Conf. Series: Materials Science and Engineering **977** (2020) 012008 doi:10.1088/1757-899X/977/1/012008

The goal is to monitor and/or optimize by a software the following criteria [7]:

- vertical displacement of the FWD stern tube bearing and the intermediate bearing;
- crankshaft, intermediate shaft and propeller shaft criteria:
  - reaction forces and bending moments of bearings,
  - slope and deflection of shaft,
  - stress and displacement of shaft;
- radial, axial and angular shaft flanges positions;
- factor of safty.

2. 3D models of crankshaft, intermediate shaft and propeller shaft based on the technical documentation of a bulk carrier 21100 DWT

A crankshaft 3D model (figure 2) is presented as a shaft with variable cross section [8].

Figure 2. Crankshaft 3D model

A 3D model of intermediate shaft is pictured in figure 3 [9].

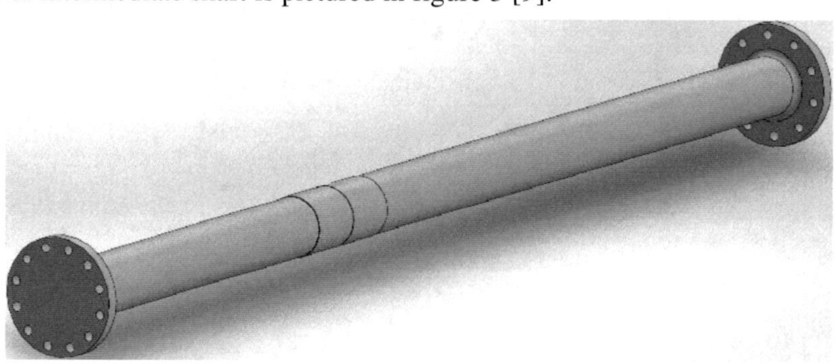

Figure 3. 3D model of intermediate shaft

A 3D model of propeller shaft is shown in figure 4 [9].

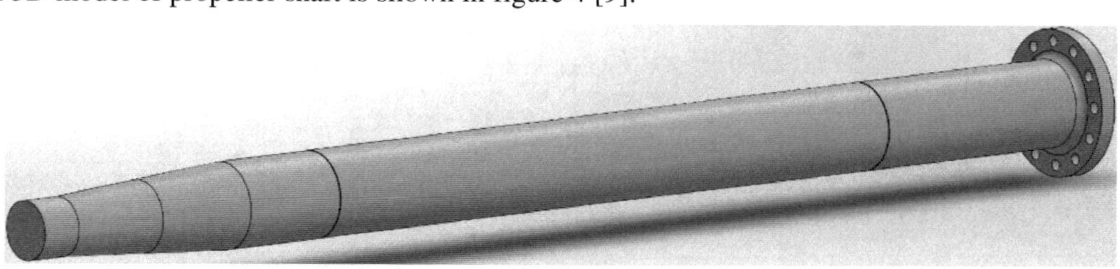

Figure 4. 3D model of propeller shaft

The 3D models of shafts are assembled into one model and links between them are created to meet the real conditions for joining the shafts figure 5.

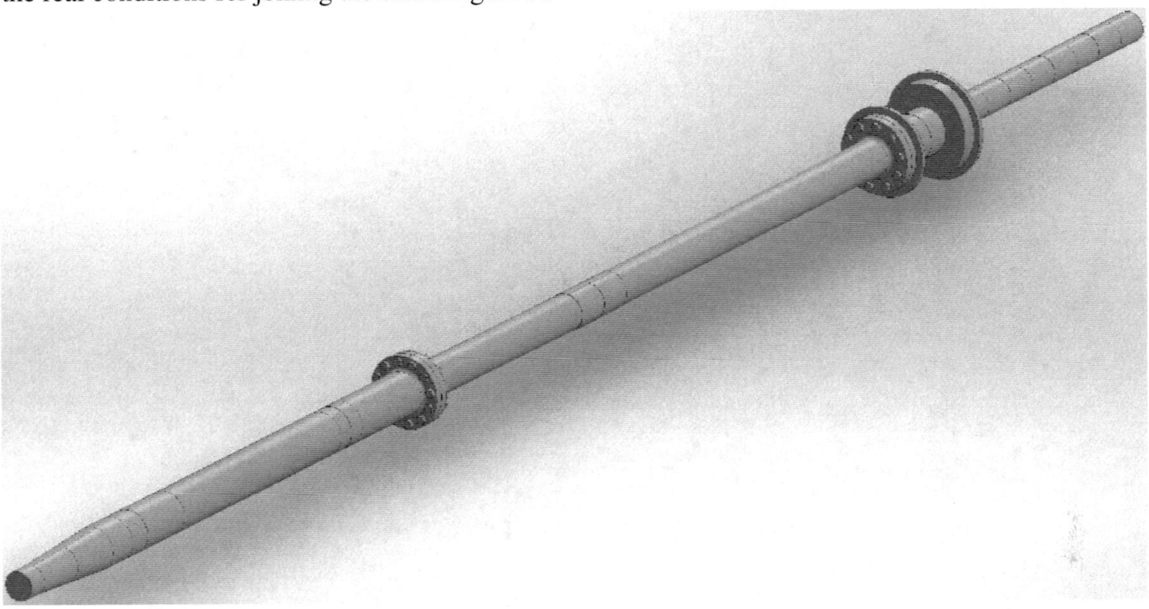

Figure 5. Assemble 3D model of shafts

Following changes are made before the simulation of shafts alignment and load:

- bolts are located on the flanges of shafts;
- auxiliary circles and lines are added for visualizing the working surfaces of intermediate bearing (at the alignment stage of shaft, points limiting surfaces are used as an indicator of angular rotation of the elastic line of shaft);
- auxiliary circles and lines are added for visualizing the working surfaces of AFT stern tube bearing and FWD stern tube bearing;
- auxiliary circles and lines are added for visualizing the acting points of propeller mass ***(91 kN)***, turning wheel mass ***(14,1 kN)***, chain wheel mass ***(27.1 kN)*** and moving masses ***(51.25 kN per point)***, see figure 1;
- auxiliary lines are added for visualizing the points at which movements of flanges are monitored.

3. Configure simulation mode

In the static simulation of shafts, the following settings must be made:

- material properties:
  - Young's modulus – $E = 2*10^{11}\ Pa$,
  - Poisson's ratio – $\upsilon = 0.3$,
  - shear modulus – $G = 8*10^{10}\ Pa$,
  - density – $\rho = 7850\ kg/m^3$,
  - yield strength – $R_{eH} = 3.4*10^8\ Pa$;
- restraints (figure 6):
  - fixing displacement by axis ***Z*** (illustrated in red color),
  - fixing displacement by axis ***Y*** (illustrated in orange color and characterizing the positions of crankshaft bearings, intermediate bearing, AFT and FWD stern tube bearings),

IOP Conf. Series: Materials Science and Engineering **977** (2020) 012008 doi:10.1088/1757-899X/977/1/012008

- fixing displacement by axis ***X*** (illustrated in green color);

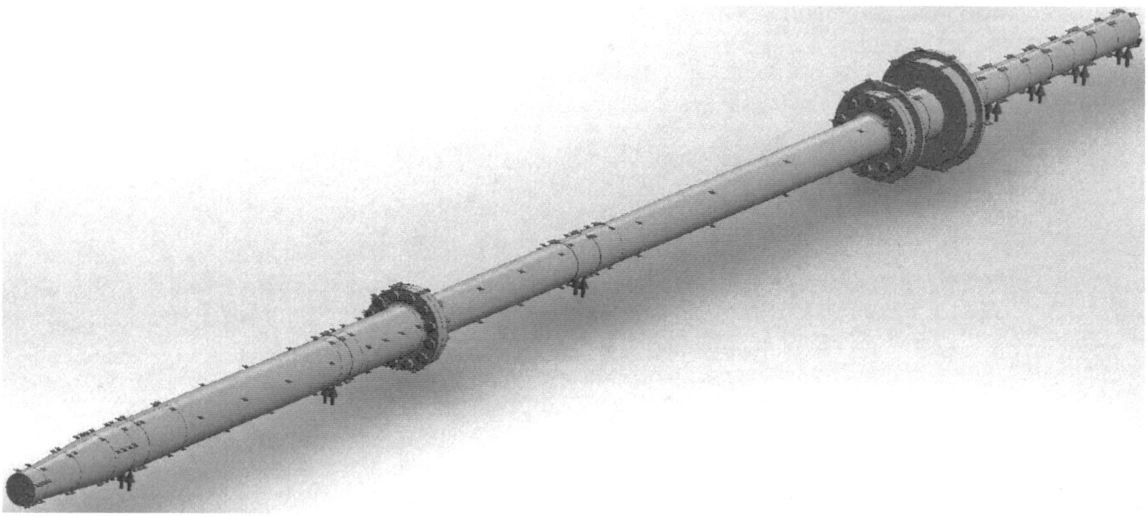

Figure 6. Boundary conditions of shafts

- external loads (figure 7):

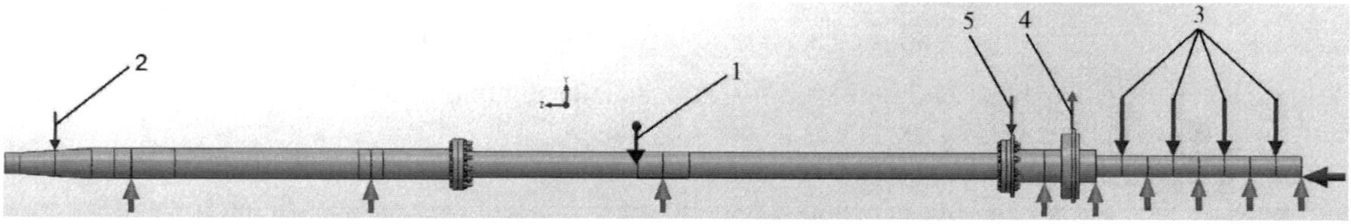

Figure 7. External loads of shafts

- gravity – position 1,
- propeller mass $F_{v1} = 91036.8\ N$ – position 2,
- moving masses $F_m = 47100\ N$ – position 3,
- chain wheel mass $F_{v2} = 27100\ N$ – position 4,
- turning wheel mass $F_g = 14100\ N$ – position 5;
- mesh of finite elements:
- for crankshaft element size are $S_{min} = 20\ mm$ and $S_{max} = 58\ mm$ (figure 8),

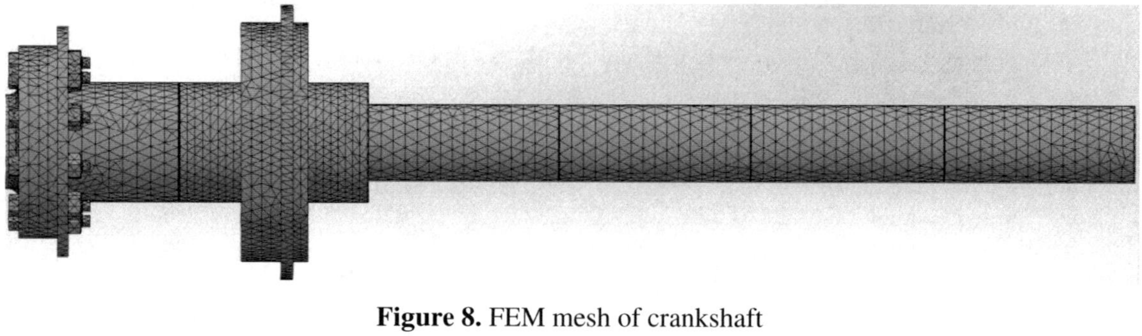

Figure 8. FEM mesh of crankshaft

- for intermediate shaft element size are S_{min} = *20 mm* and S_{max} = *45 mm* (figure 9),

Figure 9. FEM mesh of intermediate shaft

- for propeller shaft element size are S_{min} = *20 mm* and S_{max} = *62 mm* (figure 10).

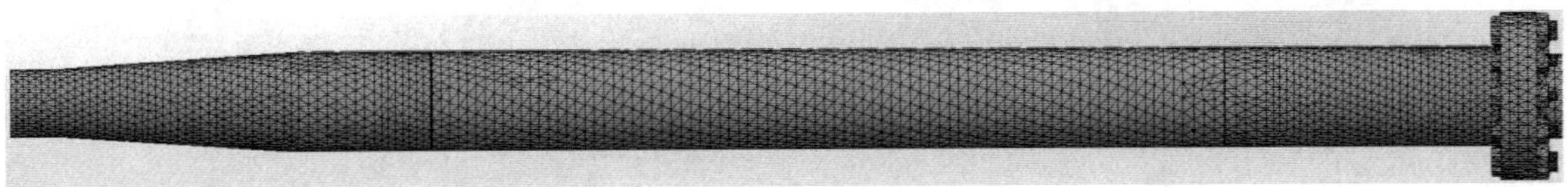

Figure 10. FEM mesh of propeller shaft

4. Software shaft alignment of bulk carrier propulsion system

For most vessels, a vertical displacement of shaft's bearings by 10% of a millimeter can lead to significant changes in the support reactions. Therefore, the process of shaft alignment takes a long time in order to determine satisfactory set of parameters that meet all criteria for the shaft alignment. The main parameter to be defined is the vertical location of FWD stern tube bearing and intermediate bearing. In addition, it must be verified that the alignment is feasible for the specific shaft geometry, material properties, installation restrictions and other requirements related to the drive shaft and the impact of surrounding systems [10].

The shaft alignment problem is different and has an infinite number of bearing vertical displacements satisfying the alignment requirements. For a optimization purpose is provided a set of acceptable solutions meeting the constraints imposed, the alignment parameters and the relevant criteria [6]. Numerous solutions are needed to engineers choose of desired alignment.

Three parameters sets are introduced for the shaft alignment optimization application: variables, constraints, and goals. Report date sensors are created for limitations and purposes of the optimization.

- variables – vertical displacement of bearing *№2* and *№4*;
- constraints:
  - all bearing reactions have to be positive,
  - bearing reaction *№1* have to be less than ***232320 N*** [11],
  - every bearing reaction from №6 to №11 have to be greater than ***9300 N*** and less than ***186000 N*** [11],
  - max bending stress of the crankshaft, intermediate shaft and propeller shaft have to be less than allowable,
  - max stress of the crankshaft, intermediate shaft and propeller shaft have to be less than yield strength,
  - reporting shear stress of the crankshaft, intermediate shaft and propeller shaft,
  - reporting shear forces and bending moments of the crankshaft and propeller flange,
  - reporting displacement of the crankshaft and propeller flange,
  - reporting axis displacement of bearing *№1, №2, №4, №6* and *№7,*
  - reporting radial displacement of bearing *№1, №2* and *№4,*
  - reporting deflection of the intermediate and propeller shaft;
- goals – getting min factor of safety;
- sensors:
  - report sensor of the reaction forces,
  - report sensor of the max and min bending stresses of crankshaft, intermediate shaft and propeller shaft,

- report sensor of the max and min shear stresses of crankshaft, intermediate shaft and propeller shaft,
- report sensor of the maximal stresses of crankshaft, intermediate shaft and propeller shaft,
- report sensor of the max and min bending stresses of crankshaft and propeller flange,
- report sensor of the shear forces of crankshaft and propeller flange,
- report sensor of the max and min displacements by axis ***Y*** and ***Z*** of crankshaft and propeller flange,
- report sensor of the max and min cross – section displacement of bearing *№1, №2, №4, №6* and *№7* by axis ***Y*** and ***Z***,
- report sensor of the max and min deflection of intermediate and propeller shaft.

72 sensors have been created to meet the variables, constraints, goals and sensors of the optimizing shaft alignment. A part of them are shown in figure 11.

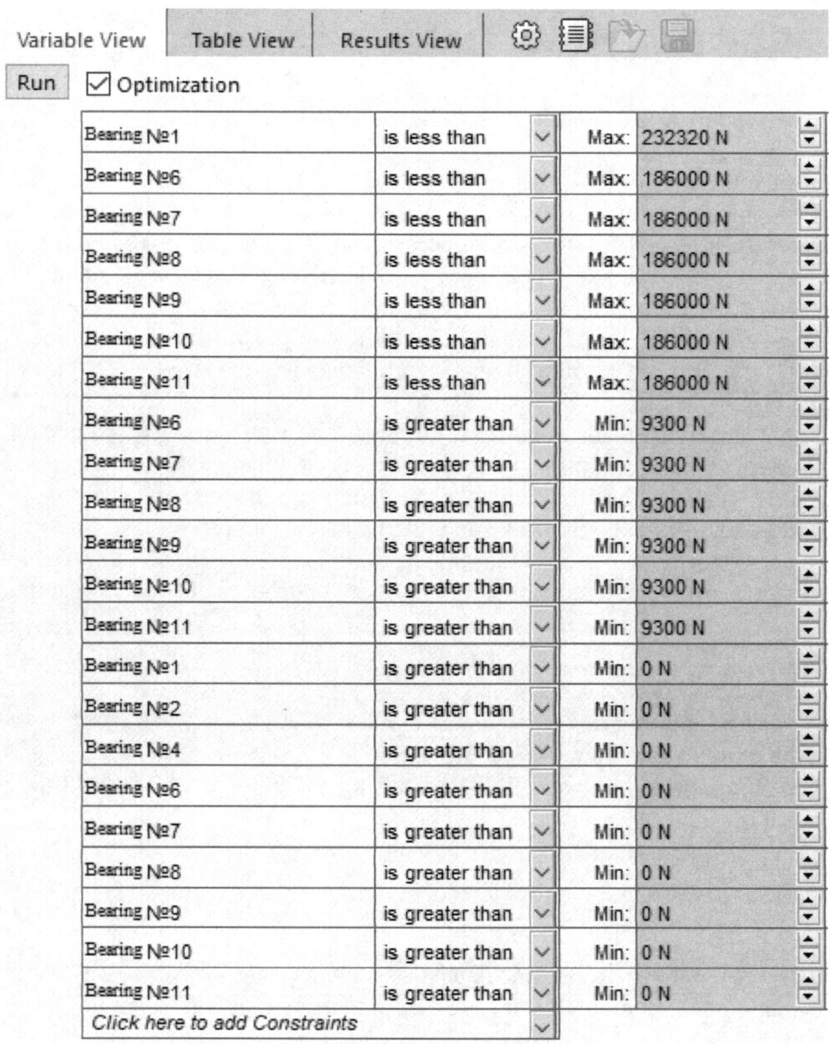

Bearing №1	is less than	Max: 232320 N
Bearing №6	is less than	Max: 186000 N
Bearing №7	is less than	Max: 186000 N
Bearing №8	is less than	Max: 186000 N
Bearing №9	is less than	Max: 186000 N
Bearing №10	is less than	Max: 186000 N
Bearing №11	is less than	Max: 186000 N
Bearing №6	is greater than	Min: 9300 N
Bearing №7	is greater than	Min: 9300 N
Bearing №8	is greater than	Min: 9300 N
Bearing №9	is greater than	Min: 9300 N
Bearing №10	is greater than	Min: 9300 N
Bearing №11	is greater than	Min: 9300 N
Bearing №1	is greater than	Min: 0 N
Bearing №2	is greater than	Min: 0 N
Bearing №4	is greater than	Min: 0 N
Bearing №6	is greater than	Min: 0 N
Bearing №7	is greater than	Min: 0 N
Bearing №8	is greater than	Min: 0 N
Bearing №9	is greater than	Min: 0 N
Bearing №10	is greater than	Min: 0 N
Bearing №11	is greater than	Min: 0 N
Click here to add Constraints		

Figure 11. Constraints

5. Shaft alignment optimization results

The simulation is made of ***38 displacement scenarios***. The vertical displacement is set to the bearing ***№2 (2.1 mm ÷ 2.6 mm)*** and the bearing ***№4 (1.3 mm ÷ 1.8)*** for duration ***37 minutes*** and data ***10.27 GB***. Data is extracted of the ***scenario №25*** with vertical displacement ***2.1 mm*** of ***bearing №2*** and vertical displacement ***1.7 mm*** of ***bearing №4***. The slope of ***bearing №1*** – ***0.9516*10⁻³ rad***.

The following figures are presented to show deformation and stress in the individual shaft sections as follows:

- crankshaft displacements (figure 12);

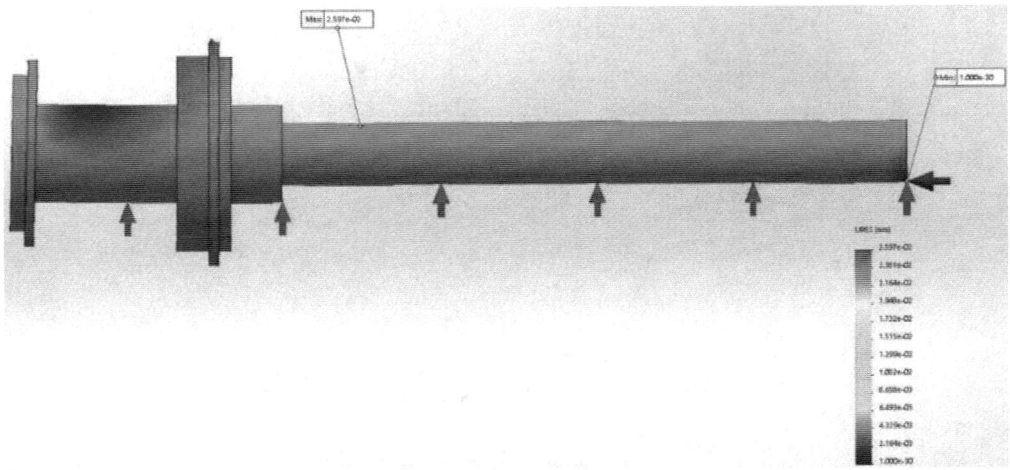

Figure 12. Crankshaft displacements

- crankshaft stress concentrations (figure 13);

Figure 13. Stress concentrations of crankshaft

- bolt displacements and stresses (figure 14);

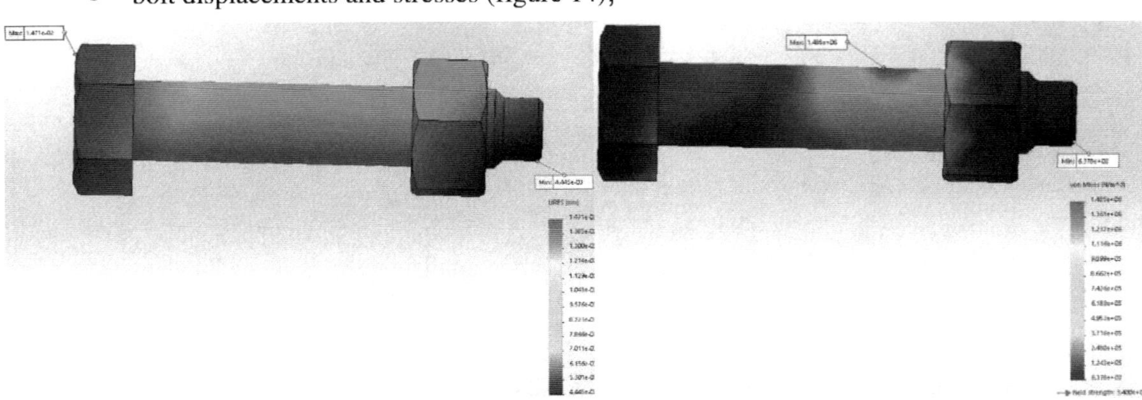

Figure 14. Bolt displacement and stresses

- intermediate shaft displacement (figure 15);

Figure 15. Intermediate shaft displacement

- propeller shaft displacement (figure 16);

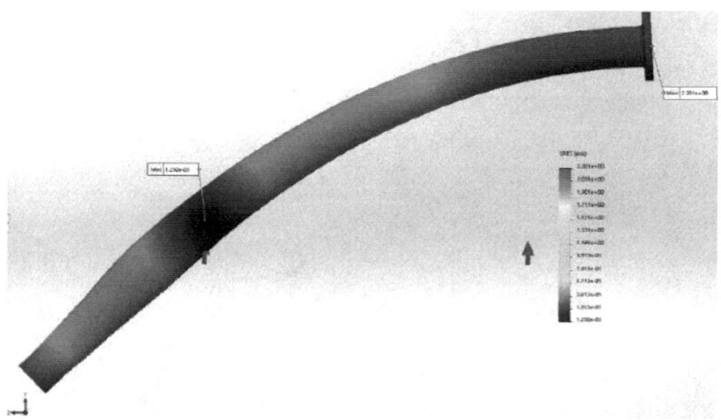

Figure 16. Propeller shaft displacement

- propulsion shaft system displacement (figure 17)

Figure 17. Propulsion shaft system displacement

6. Conclusion

A methodology has been developed for a software alignment optimization of a propulsion shaft system. The method advantage is the deformations and stress calculation of loaded 3D shaft models, associated to bearing supports displacement. Load data is presented of the additional elements, such as bolt connections. Stress concentrators are pictured of the crankshaft.

The method can be used to simulate the loading of additional elements, such as stern tube elements, seals, caps, technological holes, thread etc., in parallel with the shaft alignment optimization. It is possible to optimize the model geometry, to monitor the working capacity criterion, to set temperature differences and many others.

Some of the ***scenario №25 values*** are written in table 1.

Table 1.Values of scenario №25

Sensors	Values	Unit
Bearing reaction №1	136543	N
Bearing reaction №2	48631.4	N
Bearing reaction №4	40406.6	N
Bearing reaction №6	13896.3	N
Bearing reaction №7	35330.4	N
Bearing reaction №8	57771.7	N
Bearing reaction №9	49658.4	N
Bearing reaction №10	58555.7	N
Bearing reaction №11	18896.1	N
Crankshaft flange shear forces	10.64	kN
Crankshaft flange bending moment	28.223	kNm
Factor of safety	2.56	-

7. Reference

[1] Sestan A, Cvitanic V and Vulic N 2006 Shafting alignment calculation and validation criteria *5thICCSM* (Trogir/Split, Croatia) p 8

[2] Sverko D 2006 Shaft alignment optimization *17th Symposium on Theory and Practice of Shipbuilding* (Opatija, Croatia) pp 257-283

[3] Batrak R, Berind D and Batrak Y 2013 Computer application for shaft alignment design *International Conference on Computer Applications in Shipbuilding* (Busan, Korea) The Royal Institution of Naval Architects p 8

[4] Sverko D 2003 *Design concerns in propulsion shafting alignment* American Bureau of Shipping p 14

[5] Yuan Q, Chen Q, Lei M and Wang M 2016 Shafting alignment computing method of new multibearing rotor system under specific installation requirement *Mathematical Problems in Engineering* **2016** p 12 http://dx.doi.org/10.1155/2016/1647575

[6] Murawski L 2016 The influence of shaft line alignment accuracy on the operational reliability of marine propulsion systems *Journal of KONES Powertrain and Transport* **23 (1)** pp 247-254 DOI: 10.5604/12314005.1213580 ISSN: 1231-4005

[7] Rao S S, Boddapati V, Sumanth K and Manideep J 2015 Design of marine propulsion shafting system for 53000 dwt bulk carrier *IJMERR 2015* **4 (2)** pp 171-180 ISSN: 2278 – 0149

[8] Tenev St 2020 Load value verification of crankshaft 3D model according to stiffness matrix boundary conditions *Transport, ecology - sustainable development* (Varna) vol 27 Technical university - Varna p 9

[9] Tenev St 2020 *3D Shaft model alignment optimization of bulk carrier 21100 DWT* (Varna) Technical university of Varna p 117 (in print)

[10] Jarysz-Kaminska E 2010 Sighting through as part of shaft alignment procedure *DIAGNOSTYKA* **4 (56)** pp 65-68

[11] https://marine.man-es.com/

IOP Conf. Series: Materials Science and Engineering **977** (2020) 012009 doi:10.1088/1757-899X/977/1/012009

Impact of Shear Forces on Variable Cross Section Shaft Stiffness

Stefan Tenev[1]

[1] Department of Mechanics and Machine Elements, Technical University of Varna, 1 Studentska str., Varna 9010, Bulgaria

stefan.i.tenev@gmail.com

Abstract. Influence of shear forces is presented in this paper for static analysis of a shaft with variable cross-section. A stiffness matrix method is used to evaluate displacement and reaction forces on the shaft in regard to the slope deflection equations. The Timoshenko beam theory is addressed according to the shear coefficient and the shear correction factor. Bending moments are calculated.

1. Introduction

The most common numerical analysis tool and widely used in software products to simulate loaded structures, spatial elements and systems is the finite element method ***(FEM)***. Combining many advantages, the method is preferred by scientists, engineers and mathematicians. The main advantage of ***FEM*** is used to study various model shapes with a different dimension number and to calculate stress and displacement of an element or a system of elements. Different material models with different characteristics can be studied. The ***FEM*** provides opportunity to be converted basic energy principles and calculate differential equations in a matrix form. Additional procedures for subsequent calculations allow a determination of integrated result at different points in the finite element shape. After processing the results, a detailed picture and a timeline study decision is obtained. An advantage is data evaluation and the ability to refined survey parameters to reach the user-defined level of accuracy. The ***FEM*** is basic tool to determined stress and displacement.

The classical beam bending theory ***(Euler – Bernoulli theory)*** [1] provided approximate calculations of stress and strain depending of a load type. The theory can suggest that the cross-section normal to the beam centroid axis remained normal (orthogonal) during and after the bending deformation (figure 1). In this case, the shear force deformations are ignored and may not participate in the mathematical model calculation of examined beams. Good results can be obtained for considerable beams length in comparison with cross section geometric parameters [2]. Calculations of short shafts deformations are a different case. When a shaft diameter value is approximated to its length, the shear forces are significant and play an important role in the determining shaft elasticity.

The impact of shear forces and the bending moment of short beams are taken from the ***Timoshenko*** model [3, 4]. He is proposing a shear coefficient k as a criterion for compatibility of the actual shear stress and the beam deformation. k is introduced by ***Timoshenko*** to account the curvature cross section with respect to the central axis (figure 1) [5-13]**.** Later the ***Cowper*** mathematical expressions for the shear coefficient is presented, regard to the ***Poisson's ratio*** [7].

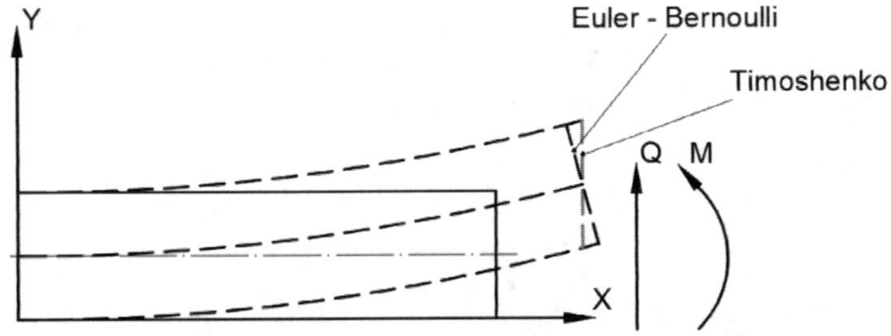

Figure 1. Comparison of Euler – Bernoulli theory and Timoshenko theory

2. Stiffness matrix expression according to slope deflection method

Relationships of individual elements stiffness [14] are created by the stiffness matrix method to calculate a load and deformation in a system of elements. In applying the method, the system must be modeled as simple sections connected to each other by nodes [15]. The material hardness properties of the elements are used in the equations and characterizes behavior of the entire idealized structure. Unknown displacements and system loads can be determined by solving the equations (1, 2). The method is a base of most commercial software sources using the finite elements.

The basic form of the equations is [16]:

$$F_i = k_i d_i + Q_i, \tag{1}$$

where

i – section number; F_i – vector of the section's characteristic forces, which are unknown internal forces; k_i – section stiffness matrix which characterizes the section's resistance against deformations; d_i – vector of the section's displacements or deformations characteristic; Q_i – vector of the section's forces characteristic caused by the external effects (such as the known forces and the temperature changes) applied to a member while $d_i = 0$.

For a system of elements connected with nodes (continuous beams), the equation (1) after integration has following observations:

$$R = k_C d + R_0, \tag{2}$$

where R – vector of the nodal forces, representing the external forces applied to the system's nodes; k_C – system stiffness matrix, which is established by assembling the section's stiffness matrices k_i; d – vector of the system's nodal displacements that can define all possible deformed configurations of system subject to arbitrary the nodal forces R; R_0 – vector of the equivalent nodal forces, representing all external effects other than the nodal forces which are already included in the preceding nodal force vector R.

A simple section is shown in (figure 2)

Figure 2. Simple idealized section

Where Q_1 – internal force acting on the left side of section; ϑ_1 – deflection on the left side of section; M_2 – internal bending moment acting on the left side of section; θ_2 – slope on the left side of section; Q_3 – internal force acting on the right side of section; ϑ_3 – deflection on the right side of section; M_4 – internal bending moment acting on the right side of section; θ_2 – slope on the right side of section.

A stiffness matrix of simple section is made on a base of possible deformations. Four options are considered, if the shear force and the bending moment are acting on the section (figure 3). Stiffness values are determined by the slope deflection equations.

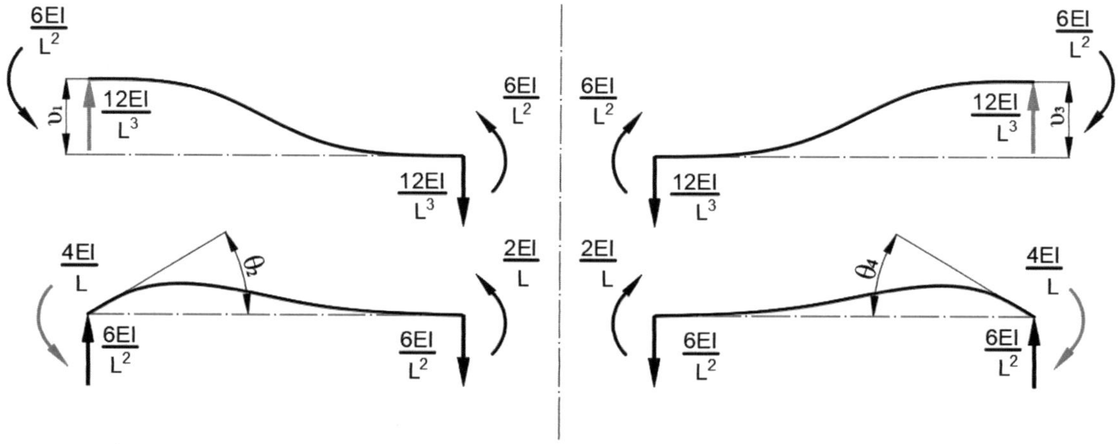

Figure 3. Possible deformations

Due to the deformation of bending moment and the shear forces of simple section, the stiffness matrix becomes:

$$k_i = \begin{bmatrix} \frac{12EI}{(1+\alpha)L^3} & \frac{6EI}{(1+\alpha)L^2} & \frac{-12EI}{(1+\alpha)L^3} & \frac{6EI}{(1+\alpha)L^2} \\ \frac{6EI}{(1+\alpha)L^2} & \frac{(4+\alpha)EI}{(1+\alpha)L} & \frac{-6EI}{(1+\alpha)L^2} & \frac{(2-\alpha)EI}{(1+\alpha)L} \\ \frac{-12EI}{(1+\alpha)L^3} & \frac{-6EI}{(1+\alpha)L^2} & \frac{12EI}{(1+\alpha)L^3} & \frac{-6EI}{(1+\alpha)L^2} \\ \frac{6EI}{(1+\alpha)L^2} & \frac{(2-\alpha)EI}{(1+\alpha)L} & \frac{-6EI}{(1+\alpha)L^2} & \frac{(4+\alpha)EI}{(1+\alpha)L} \end{bmatrix} \quad (3),$$

where

$\alpha = \frac{12EI}{GA_S L^2} = 24(1+\vartheta)\frac{A}{A_S}\left(\frac{r}{L}\right)^2$ – shear correction factor; E – Young's modulus; G – Shear modulus; I – moment of inertia; L – length of section; ϑ –Poisson's ratio; r – radius of gyration; A – area of cross – section; $A_s = \frac{A}{k}$ – area of deformed cross – section of shear forces; $k = \frac{3}{4}$ – Timoshenko shear coefficient for circular cross – section [11, 17, 18].

More than two supports beam has a total stiffness matrix k_C and must be structured as follows:

$$k_C = \begin{bmatrix} [k_{uu}] & [k_{ur}] \\ [k_{ru}] & [k_{rr}] \end{bmatrix} \quad (4)$$

where k_{uu} – symmetric matrix of unknown system deformations;
k_{ru} – matrix of load effect over support reactions;
k_{ur} – transpose matrix of k_{ru};
k_{rr} – symmetric matrix of support displacement effect over reactions forces.
k_{ur} and k_{rr} are used in structures with additional displacement of supports.

3. Reaction force and bending moment calculations of shaft with variable cross – section according to shear forces impact

Variable moment of inertia of a shaft is studied in the paper (figure 4).

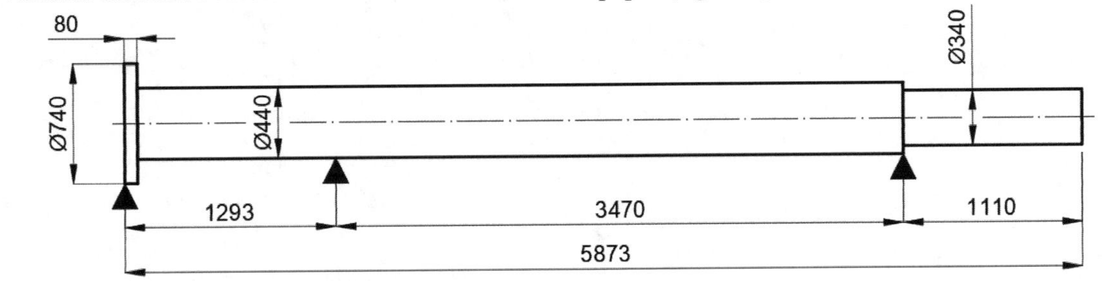

Figure 4. Shaft with variable cross - section

Scheme of simple sections is introduced in (figure 5).

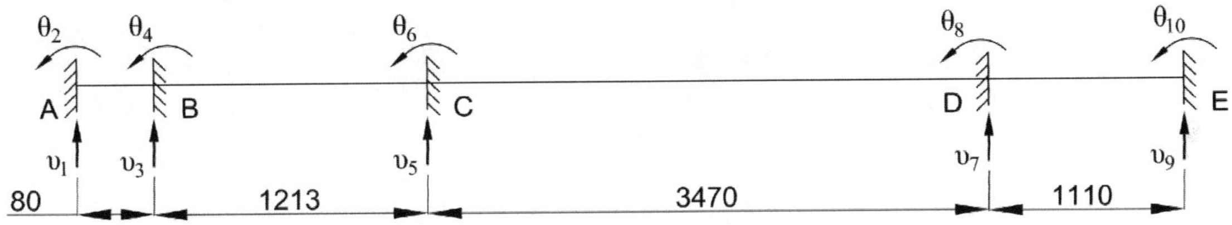

Figure 5. Simple sections of shaft

Number of joints in figure 5 are $n_j = 5$ and number of restraints are $n_r = 3$ $(\vartheta_1, \vartheta_5, \vartheta_7)$.
Then degrees of freedom of the shaft is $n = 2n_j - n_r = 2 * 5 - 3 = 7$ $(\theta_2, \theta_4, \theta_6, \theta_8, \theta_{10}, \vartheta_3, \vartheta_9)$.
Basic parameters are written in table 1 to start calculation of the shaft.

Table 1. Basic parameters of shaft

Section	Joints		Possible deformations				Diameter (m)	Length (m)	Area of cross - section (m^2)	Moment of inertia (m^4)	Shear correction factor
			υ	θ	υ	θ	D	L	A	I	α
AB	A	B	ϑ_1	θ_2	ϑ_3	θ_4	0.74	0.08	0.430084	0.0147196	500.540625
BC	B	C	ϑ_3	θ_4	ϑ_5	θ_6	0.44	1.213	0.1520531	0.0018398	0.7697321
CD	C	D	ϑ_5	θ_6	ϑ_7	θ_8	0.44	3.47	0.1520531	0.0018398	0.0940594
DE	D	E	ϑ_7	θ_8	ϑ_9	θ_{10}	0.34	1.11	0.090792	0.000656	0.5488678

Total shaft stiffness matrix based on (3) is redistributed by (4) of possible displacements and has a form shown in figure 6. Where k_{uu} values are in orange box and k_{ru} – yellow box.

IOP Conf. Series: Materials Science and Engineering **977** (2020) 012009 doi:10.1088/1757-899X/977/1/012009

$$\theta_2 \quad \theta_4 \quad \theta_6 \quad \theta_8 \quad \theta_{10} \quad \vartheta_3 \quad \vartheta_9 \quad \vartheta_1 \quad \vartheta_5 \quad \vartheta_7$$

$$k_C := \begin{bmatrix}
\frac{(4+\alpha_{AB})E I_{AB}}{(1+\alpha_{AB})L_{AB}} & \frac{(2-\alpha_{AB})E I_{AB}}{(1+\alpha_{AB})L_{AB}} & 0 & 0 & 0 & \frac{-6 E I_{AB}}{(1+\alpha_{AB})L_{AB}^2} & 0 & \frac{6 E I_{AB}}{(1+\alpha_{AB})L_{AB}^2} & 0 & 0 \\
\frac{(2-\alpha_{AB})E I_{AB}}{(1+\alpha_{AB})L_{AB}} & \left[\frac{(4+\alpha_{AB})E I_{AB}}{(1+\alpha_{AB})L_{AB}}\right]+\left[\frac{(4+\alpha_{BC})E I_{BC}}{(1+\alpha_{BC})L_{BC}}\right] & \frac{(2-\alpha_{BC})E I_{BC}}{(1+\alpha_{BC})L_{BC}} & 0 & 0 & \left[\frac{-6 E I_{AB}}{(1+\alpha_{AB})L_{AB}^2}\right]+\left[\frac{6 E I_{BC}}{(1+\alpha_{BC})L_{BC}^2}\right] & 0 & \frac{6 E I_{AB}}{(1+\alpha_{AB})L_{AB}^2} & \frac{-6 E I_{BC}}{(1+\alpha_{BC})L_{BC}^2} & 0 \\
0 & \frac{(2-\alpha_{BC})E I_{BC}}{(1+\alpha_{BC})L_{BC}} & \left[\frac{(4+\alpha_{BC})E I_{BC}}{(1+\alpha_{BC})L_{BC}}\right]+\left[\frac{(4+\alpha_{CD})E I_{CD}}{(1+\alpha_{CD})L_{CD}}\right] & \frac{(2-\alpha_{CD})E I_{CD}}{(1+\alpha_{CD})L_{CD}} & 0 & \frac{6 E I_{BC}}{(1+\alpha_{BC})L_{BC}^2} & 0 & 0 & \left[\frac{-6 E I_{BC}}{(1+\alpha_{BC})L_{BC}^2}\right]+\left[\frac{6 E I_{CD}}{(1+\alpha_{CD})L_{CD}^2}\right] & \frac{-6 E I_{CD}}{(1+\alpha_{CD})L_{CD}^2} \\
0 & 0 & \frac{(2-\alpha_{CD})E I_{CD}}{(1+\alpha_{CD})L_{CD}} & \left[\frac{(4+\alpha_{CD})E I_{CD}}{(1+\alpha_{CD})L_{CD}}\right]+\left[\frac{(4+\alpha_{DE})E I_{DE}}{(1+\alpha_{DE})L_{DE}}\right] & \frac{(2-\alpha_{DE})E I_{DE}}{(1+\alpha_{DE})L_{DE}} & 0 & \frac{-6 E I_{DE}}{(1+\alpha_{DE})L_{DE}^2} & 0 & \frac{6 E I_{CD}}{(1+\alpha_{CD})L_{CD}^2} & \left[\frac{-6 E I_{CD}}{(1+\alpha_{CD})L_{CD}^2}\right]+\left[\frac{6 E I_{DE}}{(1+\alpha_{DE})L_{DE}^2}\right] \\
0 & 0 & 0 & \frac{(2-\alpha_{DE})E I_{DE}}{(1+\alpha_{DE})L_{DE}} & \frac{(4+\alpha_{DE})E I_{DE}}{(1+\alpha_{DE})L_{DE}} & 0 & \frac{-6 E I_{DE}}{(1+\alpha_{DE})L_{DE}^2} & 0 & 0 & \frac{6 E I_{DE}}{(1+\alpha_{DE})L_{DE}^2} \\
\frac{-6 E I_{AB}}{(1+\alpha_{AB})L_{AB}^2} & \left[\frac{-6 E I_{AB}}{(1+\alpha_{AB})L_{AB}^2}\right]+\left[\frac{6 E I_{BC}}{(1+\alpha_{BC})L_{BC}^2}\right] & \frac{6 E I_{BC}}{(1+\alpha_{BC})L_{BC}^2} & 0 & 0 & \left[\frac{12 E I_{AB}}{(1+\alpha_{AB})L_{AB}^3}\right]+\left[\frac{12 E I_{BC}}{(1+\alpha_{BC})L_{BC}^3}\right] & 0 & \frac{-12 E I_{AB}}{(1+\alpha_{AB})L_{AB}^3} & \frac{-12 E I_{BC}}{(1+\alpha_{BC})L_{BC}^3} & 0 \\
0 & 0 & 0 & \frac{-6 E I_{DE}}{(1+\alpha_{DE})L_{DE}^2} & \frac{-6 E I_{DE}}{(1+\alpha_{DE})L_{DE}^2} & 0 & \frac{12 E I_{DE}}{(1+\alpha_{DE})L_{DE}^3} & 0 & 0 & \frac{-12 E I_{DE}}{(1+\alpha_{DE})L_{DE}^3} \\
\frac{6 E I_{AB}}{(1+\alpha_{AB})L_{AB}^2} & \frac{6 E I_{AB}}{(1+\alpha_{AB})L_{AB}^2} & 0 & 0 & 0 & \frac{-12 E I_{AB}}{(1+\alpha_{AB})L_{AB}^3} & 0 & \frac{12 E I_{AB}}{(1+\alpha_{AB})L_{AB}^3} & 0 & 0 \\
0 & \frac{-6 E I_{BC}}{(1+\alpha_{BC})L_{BC}^2} & \left[\frac{-6 E I_{BC}}{(1+\alpha_{BC})L_{BC}^2}\right]+\left[\frac{6 E I_{CD}}{(1+\alpha_{CD})L_{CD}^2}\right] & \frac{6 E I_{CD}}{(1+\alpha_{CD})L_{CD}^2} & 0 & \frac{-12 E I_{BC}}{(1+\alpha_{BC})L_{BC}^3} & 0 & 0 & \left[\frac{12 E I_{BC}}{(1+\alpha_{BC})L_{BC}^3}\right]+\left[\frac{12 E I_{CD}}{(1+\alpha_{CD})L_{CD}^3}\right] & \frac{-12 E I_{CD}}{(1+\alpha_{CD})L_{CD}^3} \\
0 & 0 & \frac{-6 E I_{CD}}{(1+\alpha_{CD})L_{CD}^2} & \left[\frac{-6 E I_{CD}}{(1+\alpha_{CD})L_{CD}^2}\right]+\left[\frac{6 E I_{DE}}{(1+\alpha_{DE})L_{DE}^2}\right] & \frac{6 E I_{DE}}{(1+\alpha_{DE})L_{DE}^2} & 0 & \frac{-12 E I_{DE}}{(1+\alpha_{DE})L_{DE}^3} & 0 & \frac{-12 E I_{CD}}{(1+\alpha_{CD})L_{CD}^3} & \left[\frac{12 E I_{CD}}{(1+\alpha_{CD})L_{CD}^3}\right]+\left[\frac{12 E I_{DE}}{(1+\alpha_{DE})L_{DE}^3}\right]
\end{bmatrix}
\begin{matrix} \theta_2 \\ \theta_4 \\ \theta_6 \\ \theta_8 \\ \theta_{10} \\ \vartheta_3 \\ \vartheta_9 \\ \vartheta_1 \\ \vartheta_5 \\ \vartheta_7 \end{matrix}$$

Figure 6. Total stiffness matrix of shaft

Internal reactions of the shaft are:

$$IR_{AB} = \begin{bmatrix} \frac{q_{AB}L_{AB}}{2} \\ \frac{q_{AB}L_{AB}^2}{12} \\ \frac{q_{AB}L_{AB}}{2} \\ -\frac{q_{AB}L_{AB}^2}{12} \end{bmatrix} = \begin{bmatrix} 1324.81 \\ 17.66 \\ 1324.81 \\ -17.66 \end{bmatrix} \begin{matrix} N \\ Nm \\ N \\ Nm \end{matrix},$$

$$IR_{BC} = \begin{bmatrix} \frac{q_{BC}L_{BC}}{2} \\ \frac{q_{BC}L_{BC}^2}{12} \\ \frac{q_{BC}L_{BC}}{2} \\ -\frac{q_{BC}L_{BC}^2}{12} \end{bmatrix} = \begin{bmatrix} 7101.74 \\ 1435.74 \\ 7101.74 \\ -1435.74 \end{bmatrix} \begin{matrix} N \\ Nm \\ N \\ Nm \end{matrix},$$

$$IR_{CD} = \begin{bmatrix} \frac{q_{CD}L_{CD}}{2} \\ \frac{q_{CD}L_{CD}^2}{12} \\ \frac{q_{CD}L_{CD}}{2} \\ -\frac{q_{CD}L_{CD}^2}{12} \end{bmatrix} = \begin{bmatrix} 20315.77 \\ 11749.29 \\ 20315.77 \\ -11749.29 \end{bmatrix} \begin{matrix} N \\ Nm \\ N \\ Nm \end{matrix},$$

$$IR_{DE} = \begin{bmatrix} \frac{q_{DE}L_{DE}}{2} \\ \frac{q_{DE}L_{DE}^2}{12} \\ \frac{q_{DE}L_{DE}}{2} \\ -\frac{q_{DE}L_{DE}^2}{12} \end{bmatrix} = \begin{bmatrix} 3880.43 \\ 717.88 \\ 3880.43 \\ -717.88 \end{bmatrix} \begin{matrix} N \\ Nm \\ N \\ Nm \end{matrix},$$

where q – a distributed load equal to the gravity force regard to the mass of shaft sections.

Slope and deflection of each joint on the shaft are pictured on figure 7.

IOP Conf. Series: Materials Science and Engineering **977** (2020) 012009 doi:10.1088/1757-899X/977/1/012009

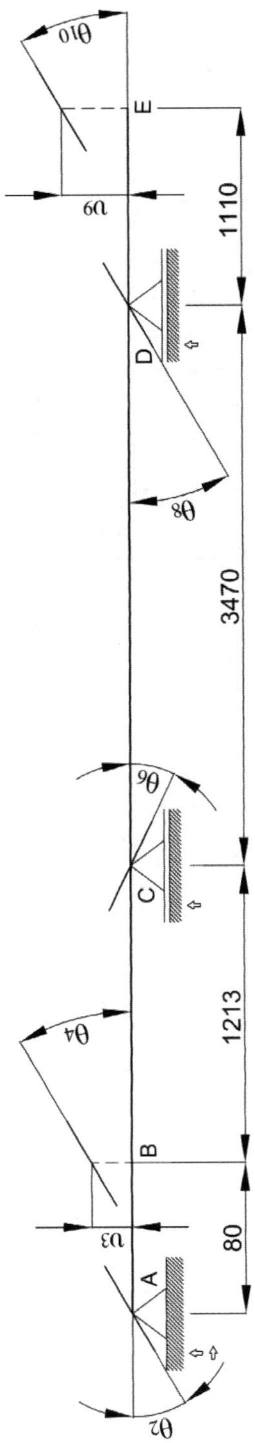

Bending moments and shear forces are calculated for every possible joint displacement.

$$JL_m = \begin{bmatrix} -17.66 \\ -1418.07 \\ -10313.56 \\ 11031.41 \\ 717.88 \\ -8426.54 \\ -3880.43 \end{bmatrix} \begin{matrix} Nm \\ Nm \\ Nm \\ Nm \\ Nm \\ N \\ N \end{matrix}.$$

Deformation vector ($\theta = k_{uu}^{-1} * JL_m$) is:

$$\theta = \begin{bmatrix} 1.4429579 * 10^{-6} \\ 1.4425257 * 10^{-6} \\ -1.2571074 * 10^{-5} \\ 2.4606417 * 10^{-5} \\ 1.2458872 * 10^{-5} \\ 1.2151947 * 10^{-7} \\ 1.5350089 * 10^{-5} \end{bmatrix} \begin{matrix} rad \\ rad \\ rad \\ rad \\ rad \\ m \\ m \end{matrix}.$$

Deformation values of each section are presented in:

$$d_{AB} = \begin{bmatrix} 0 \\ 1.4429579 * 10^{-6} \\ 1.2151947 * 10^{-7} \\ 1.4425257 * 10^{-6} \end{bmatrix} \begin{matrix} m \\ rad \\ m \\ rad \end{matrix};$$

$$d_{BC} = \begin{bmatrix} 1.2151947 * 10^{-7} \\ 1.4425257 * 10^{-6} \\ 0 \\ -1.2571074 * 10^{-5} \end{bmatrix} \begin{matrix} m \\ rad \\ m \\ rad \end{matrix};$$

$$d_{CD} = \begin{bmatrix} 0 \\ -1.2571074 * 10^{-5} \\ 0 \\ 2.4606417 * 10^{-5} \end{bmatrix} \begin{matrix} m \\ rad \\ m \\ rad \end{matrix};$$

$$d_{DE} = \begin{bmatrix} 0 \\ 2.4606417 * 10^{-5} \\ 1.5350089 * 10^{-5} \\ 1.2458872 * 10^{-5} \end{bmatrix} \begin{matrix} m \\ rad \\ m \\ rad \end{matrix}.$$

Figure 7. Slope and deflection on shaft

Joint reactions ($F_i = k_i d_i + Q_i$) are calculated and the values are:

$$F_{AB} = \begin{bmatrix} 485.59 \\ 0 \\ 2164.02 \\ -67.14 \end{bmatrix} \begin{matrix} N \\ Nm \\ N \\ Nm \end{matrix}; \quad F_{BC} = \begin{bmatrix} -2164.02 \\ 67.14 \\ 16367.49 \\ -11306.5 \end{bmatrix} \begin{matrix} N \\ Nm \\ N \\ Nm \end{matrix};$$

$$F_{CD} = \begin{bmatrix} 22332.84 \\ 11306.5 \\ 18298.71 \\ -4307.27 \end{bmatrix} \begin{matrix} N \\ Nm \\ N \\ Nm \end{matrix}; \; F_{DE} = \begin{bmatrix} 7760.85 \\ 4307.27 \\ 0 \\ 0 \end{bmatrix} \begin{matrix} N \\ Nm \\ N \\ Nm \end{matrix}.$$

Reaction forces and bending moments are:

$$\boldsymbol{R_A = 485.59\ N,\ R_C = 38700.34\ N,\ R_D = 26059.56\ N,}$$

$$\boldsymbol{M_A = M_C = M_D = 0\ Nm.}$$

4. Conclusions

A matrix approach has been applied to investigate the impact of shear forces over the reaction force and the bending moment of shaft joints. A shear correction factor is used to calculate the exact values of the slope and deflection according to the total stiffness matrix of system. Significant shear effect is studied. The presented method can be used for different boundary conditions and for a verification of 3D shaft models loads.

5. Reference

[1] Deshpande S S and Wankhade R L 2013 Analysis of thick beams using first order shear deformation theory *ICSEM 2013* (Rourkela, India) p 18

[2] Ivanova E and Vasilev T 2018 Influence of Elastic Element Design Parameters on Buckling Effect for Different „Periflex“ Shaft Couplings *Scientific Bulletin of Naval Academy* **21** pp 335-341 ISSN: 2392-8956

[3] Costa F P 2008 *On timoshenko's beams coefficient of sensibility to shear effect* (DCET, UESC, 45662-000 Ilheus, BA, Brasil) Computational and Applied Mathematics Group, Exact Sciences and Technology Department vol 9 no 3 pp 447-457

[4] Dimitrov D M, Kostov K P and Chirikov V A 2015 Universal experimental relation for natural friquencies of transverse vibration of stubby free-free beams *Diagnostics, Resource and Mechanics of materials and structures* **4** pp 42-54

[5] Gruttmann F and Wagner W 2001 *Shear correction factors in Timoshenko's beam theory for arbitrary shaped cross–sections* Computational Mechanics vol 27 pp 199-207

[6] Sohani F and Eipakchi H R 2018 *Analytical solution for modal analysis of Euler-Bernoulli and Timoshenko beam with an arbitrary varying cross-section* (Shahrood, I. R. Iran) Mathematical models in engineering vol 4 pp 164-174 DOI https://doi.org/10.21595/mme.2018.20116

[7] Marinetti A and Oliveto G *On the evaluation of the shear correction factors: a boundary element approach* (University of Catania, Italy) Department of Civil and Environmental Engineering p 10

[8] Franco-Villafane J A and Mendez-Sanchez R A 2016 On the accuracy of the Timoshenko beam theory above the critical frequency: best shear coefficient *Journal of Mechanics* **32** pp 515-518 DOI: https://doi.org/10.1017/jmech.2015.104 ISSN: 1727-7191

[9] Severn R T 1970 Inclusion of shear deflection in the stiffness matrix for a beam element *Journal of Strain Analysis* **5 (4)** p 3

[10] Stephen N G 1980 Timoshenko's shear coefficient from a beam subjected to gravity loading *Journal of Applied Mechanics* **47** pp 121-127 DOI: 10.1115/1.3153589

[11] Ike C C 2019 Timoshenko beam theory for the flexural analysis of moderately thick beams – variational formulation, and closed form solution *TECNICA ITALIANA-Italian Journal of Engineering Science* **63 (1)** pp 34-35 https://doi.org/10.18280/ti-ijes.630105

[12] Freund J and Karakoc A 2015 Shear and torsion correction factors of Timoshenko beam model for generic cross sections *Research on Engineering Structures & Materials* **1** pp 19-27 DOI: http://dx.doi.org/10.17515/resm2015.19me0827 ISSN: 1243-3716

[13] Kugi A, Mang H A and Steinboeck A 2013 Energy-consistent shear coefficients for beams with circular cross sections and radially inhomogeneous materials *International Journal of Solids and Structures* **50** pp 1859-1868 DOI: 10.1016/j.ijsolstr.2013.01.030

[14] Slavov S D, Yordanov K K and Dimitrov D M 2019 Stiffness design of machine tools structures by topology management optimization approach *In IOP Conference Series: Materials Science and Engineering* vol 564 IOP Publishing p 012071
[15] Gattas J and Albermani F *Introduction to Direct Stiffness Method* (Australia) The University of Queensland p 61
[16] Nikishkov G P 2004 *Introduction to the finite element method* (Aizu-Wakamatsu 965-8580, Japan) University of Aizu p 45
[17] Gavin H 2012 *Geometric stiffness effects in 2D and 3D frames* Duke University CEE 421L. Matrix Structural Analysis p 14
[18] Chugh A K 1977 Stiffness matrix for a beam element including transverse shear and axial force effects *International journal for numerical methods in engineering* **11** pp 1681-1697

IOP Conf. Series: Materials Science and Engineering **977** (2020) 012010 doi:10.1088/1757-899X/977/1/012010

Load Value Verification of Crankshaft 3D Model according to Stiffness Matrix Boundary Conditions

Tihomir Vasilev[1]

[1] Department of Mechanics, Nikola Vaptsarov Naval Academy, 73 Vasil Drumev str., Varna 9002, Bulgaria

t.vasilev@naval-acad.bg

Abstract. A crankshaft 3D model of bulk carrier 21100 DWT is studied. Moving masses, a chain wheel mass and a turning wheel mass are applied. A matrix displacement method is used to validate simulation loads of the crankshaft model. Boundary conditions and mesh control are entered for the model study in order to minimize the difference between values of analytic calculations and the 3D simulation loads. A percentage difference is presented between an analytical model and a simulation model. Pictures of stress and displacements of the shaft and the flange are pointed in the paper.

1. Introduction

Defining and observation of crankshaft performance is a complex task required an extensive research. The engine power that is transformed into a rotary movement of the crankshaft generate a torque with considerable loads. The crankshaft has to withstand of bending, shear and torsional stresses. Further the stresses and deformations [1] arisen from the torsional vibrations, regard to the constantly abruptly acceleration of the rotary movement.

A stiffness matrix method is used to calculate reaction forces and bending moments of the moving masses section (figure 1). The method is applied for the simply and symmetrically loaded section of crankshaft. The calculation results are enough condition to satisfy the load values of crankshaft 3D model.

3D models are made [2], regarded to a section with diameter $\emptyset = 302\ mm$ and the entire crankshaft model. The first one is studied to validate the stiffness matrix load values of crankshaft section. The second one is studied to get values of the reaction forces and the bending moment, according to the boundary conditions and the mesh control used in the validation study.

In this paper the crankshaft of main engine of bulk carrier 21100 DWT is represented as a straight shaft with variable cross section. The forces of moving masses including crank throw are placed in the middle of rod bearing journals. The chain force minus chain wheel mass and the turning wheel mass are pointed on the diameters $\emptyset = 1080\ mm$ and $\emptyset = 903\ mm$ of crankshaft (figure 1).

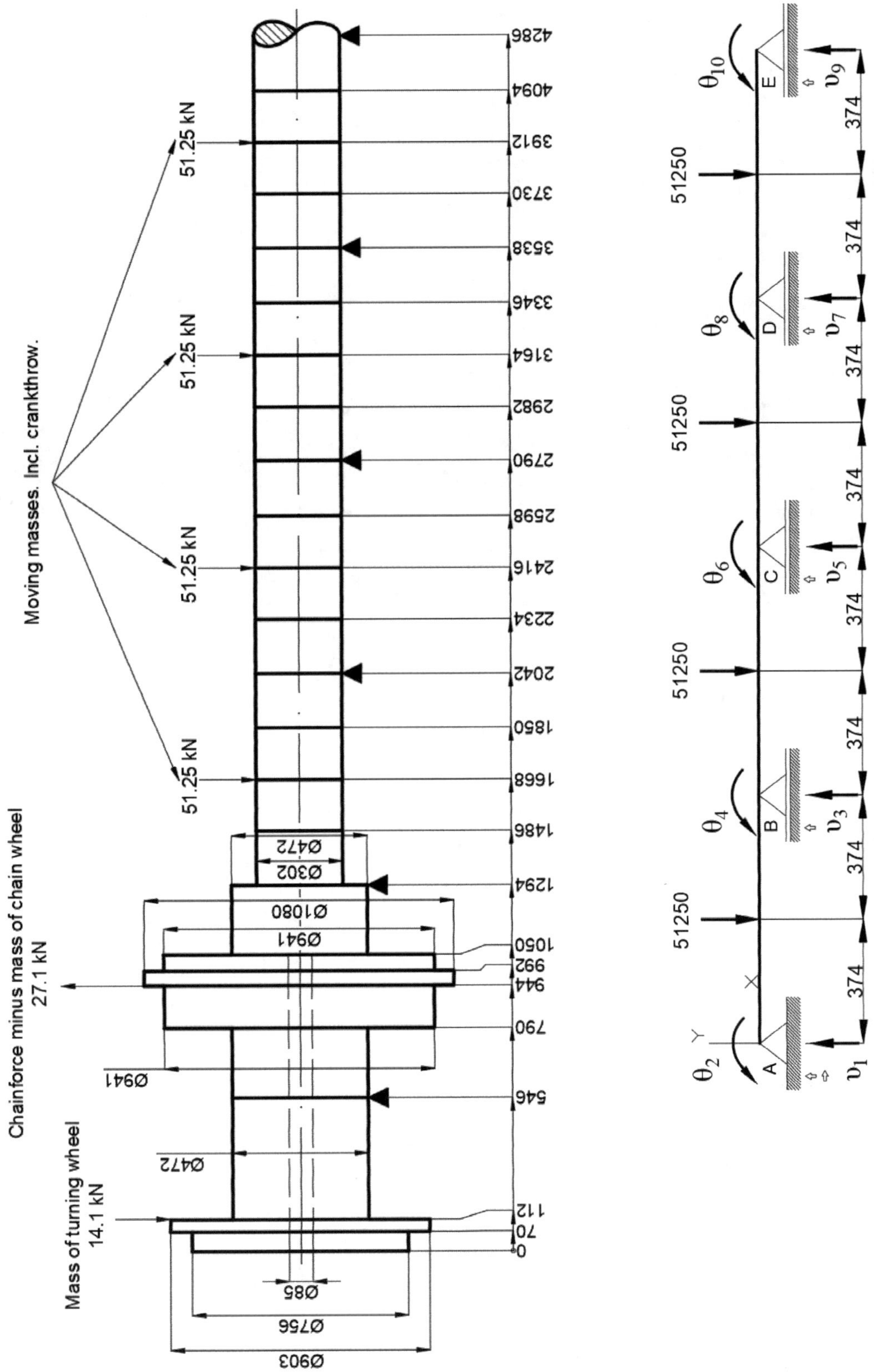

Figure 1. Crankshaft scheme (left side) and displacement of moving masses section (right side)

2. Load values verification with stiffness matrix method

Resources are used basis on [3] and the moving masses section accorded to the possible joints displacement is presented in figure 1.

The total shaft stiffness matrix is illustrated on figure 2 [4, 5].

$$
k_C := \begin{array}{cccccccccc|c}
\theta_2 & \theta_4 & \theta_6 & \theta_8 & \theta_{10} & \vartheta_1 & \vartheta_3 & \vartheta_5 & \vartheta_7 & \vartheta_9 & \\
\frac{(4+\alpha)EI}{(1+\alpha)\cdot L} & \frac{(2-\alpha)EI}{(1+\alpha)\cdot L} & 0 & 0 & 0 & \frac{6EI}{(1+\alpha)\cdot L^2} & \frac{-6EI}{(1+\alpha)\cdot L^2} & 0 & 0 & 0 & \theta_2 \\
\frac{(2-\alpha)EI}{(1+\alpha)\cdot L} & 2\frac{(4+\alpha)EI}{(1+\alpha)\cdot L} & \frac{(2-\alpha)EI}{(1+\alpha)\cdot L} & 0 & 0 & \frac{6EI}{(1+\alpha)\cdot L^2} & 0 & \frac{-6EI}{(1+\alpha)\cdot L^2} & 0 & 0 & \theta_4 \\
0 & \frac{(2-\alpha)EI}{(1+\alpha)\cdot L} & 2\frac{(4+\alpha)EI}{(1+\alpha)\cdot L} & \frac{(2-\alpha)EI}{(1+\alpha)\cdot L} & 0 & 0 & \frac{6EI}{(1+\alpha)\cdot L^2} & 0 & \frac{-6EI}{(1+\alpha)\cdot L^2} & 0 & \theta_6 \\
0 & 0 & \frac{(2-\alpha)EI}{(1+\alpha)\cdot L} & 2\frac{(4+\alpha)EI}{(1+\alpha)\cdot L} & \frac{(2-\alpha)EI}{(1+\alpha)\cdot L} & 0 & 0 & \frac{6EI}{(1+\alpha)\cdot L^2} & 0 & \frac{-6EI}{(1+\alpha)\cdot L^2} & \theta_8 \\
0 & 0 & 0 & \frac{(2-\alpha)EI}{(1+\alpha)\cdot L} & \frac{(4+\alpha)EI}{(1+\alpha)\cdot L} & 0 & 0 & 0 & \frac{6EI}{(1+\alpha)\cdot L^2} & \frac{-6EI}{(1+\alpha)\cdot L^2} & \theta_{10} \\
\frac{6EI}{(1+\alpha)\cdot L^2} & \frac{6EI}{(1+\alpha)\cdot L^2} & 0 & 0 & 0 & \frac{12EI}{(1+\alpha)\cdot L^3} & \frac{-12EI}{(1+\alpha)\cdot L^3} & 0 & 0 & 0 & \vartheta_1 \\
\frac{-6EI}{(1+\alpha)\cdot L^2} & 0 & \frac{6EI}{(1+\alpha)\cdot L^2} & 0 & 0 & \frac{-12EI}{(1+\alpha)\cdot L^3} & 2\frac{12EI}{(1+\alpha)\cdot L^3} & \frac{-12EI}{(1+\alpha)\cdot L^3} & 0 & 0 & \vartheta_3 \\
0 & \frac{-6EI}{(1+\alpha)\cdot L^2} & 0 & \frac{6EI}{(1+\alpha)\cdot L^2} & 0 & 0 & \frac{-12EI}{(1+\alpha)\cdot L^3} & 2\frac{12EI}{(1+\alpha)\cdot L^3} & \frac{-12EI}{(1+\alpha)\cdot L^3} & 0 & \vartheta_5 \\
0 & 0 & \frac{-6EI}{(1+\alpha)\cdot L^2} & 0 & \frac{6EI}{(1+\alpha)\cdot L^2} & 0 & 0 & \frac{-12EI}{(1+\alpha)\cdot L^3} & 2\frac{12EI}{(1+\alpha)\cdot L^3} & \frac{-12EI}{(1+\alpha)\cdot L^3} & \vartheta_7 \\
0 & 0 & 0 & \frac{-6EI}{(1+\alpha)\cdot L^2} & \frac{-6EI}{(1+\alpha)\cdot L^2} & 0 & 0 & 0 & \frac{-12EI}{(1+\alpha)\cdot L^3} & \frac{12EI}{(1+\alpha)\cdot L^3} & \vartheta_9
\end{array}
$$

Figure 2. Total stiffness matrix of moving masses section

The internal reactions are

$$
IR_{AB} = IR_{BC} = IR_{CD} = IR_{DE} = \begin{bmatrix} \frac{F}{2} \\ \frac{FL}{8} \\ \frac{F}{2} \\ -\frac{FL}{8} \end{bmatrix} = \begin{bmatrix} \frac{51250}{2} \\ \frac{51250 * 0.748}{8} \\ \frac{51250}{2} \\ -\frac{51250 * 0.748}{8} \end{bmatrix} = \begin{bmatrix} 25625 \\ 4791.88 \\ 25625 \\ -4791.88 \end{bmatrix} \begin{matrix} N \\ Nm \\ N \\ Nm \end{matrix}.
$$

The deformation vector accorded to the bending moments are calculated for every possible joint displacement.

$$
JL_m = \begin{bmatrix} -4791.88 \\ 0 \\ 0 \\ 0 \\ 4791.88 \end{bmatrix} Nm, \theta = \begin{bmatrix} -1.7705 * 10^{-5} \\ 1.87 * 10^{-6} \\ 0 \\ -1.87 * 10^{-6} \\ 1.7705 * 10^{-5} \end{bmatrix} rad.
$$

Slope and deflections are pointed in figure 3.

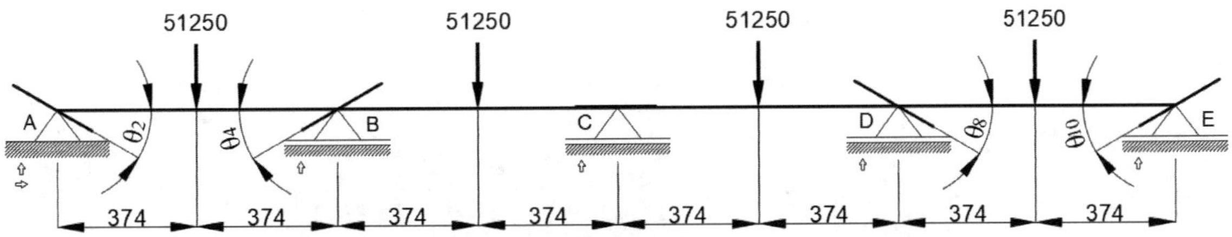

Figure 3. Slope and deflection of moving masses section

After calculation processes based on [3], the deformations and the reactions forces of each joint are:

$$d_{AB} = \begin{bmatrix} 0 \\ -1.7705 * 10^{-5} \\ 0 \\ 1.87 * 10^{-6} \end{bmatrix} \begin{matrix} m \\ rad \\ m \\ rad \end{matrix}, \ d_{BC} = \begin{bmatrix} 0 \\ 1.87 * 10^{-6} \\ 0 \\ 0 \end{bmatrix} \begin{matrix} m \\ rad \\ m \\ rad \end{matrix},$$

$$d_{CD} = \begin{bmatrix} 0 \\ 0 \\ 0 \\ -1.87 * 10^{-6} \end{bmatrix} \begin{matrix} m \\ rad \\ m \\ rad \end{matrix}, \ d_{DE} = \begin{bmatrix} 0 \\ -1.87 * 10^{-6} \\ 0 \\ 1.7705 * 10^{-5} \end{bmatrix} \begin{matrix} m \\ rad \\ m \\ rad \end{matrix}.$$

$$F_{AB} = \begin{bmatrix} 18526.68 \\ 0 \\ 32723.32 \\ -5309.54 \end{bmatrix} \begin{matrix} N \\ Nm \\ N \\ Nm \end{matrix}, \ F_{BC} = \begin{bmatrix} 26463.26 \\ 5309.54 \\ 24786.74 \\ -4682.52 \end{bmatrix} \begin{matrix} N \\ Nm \\ N \\ Nm \end{matrix},$$

$$F_{CD} = \begin{bmatrix} 24786.74 \\ 4682.52 \\ 26463.26 \\ -5309.54 \end{bmatrix} \begin{matrix} N \\ Nm \\ N \\ Nm \end{matrix}, \ F_{DE} = \begin{bmatrix} 32723.32 \\ 5309.54 \\ 18526.68 \\ 0 \end{bmatrix} \begin{matrix} N \\ Nm \\ N \\ Nm \end{matrix}.$$

Finally, the reaction forces and the bending moments are:

$$\boldsymbol{R_A = 18526.68\ N, R_B = 59186.58\ N, R_C = 49573.48\ N, R_D = 59186.58\ N\ R_E = 18526.68\ N, M_A = M_B = M_C = M_D = M_E = 0\ Nm.}$$

Achieving the results of stiffness matrix calculations is related to correct boundary conditions and a finite element mesh [6] when simulating the loaded 3D crankshaft model.

An important condition is the places of fix (supports places). This should be carried out on a sufficiently small surface. The small surfaces restraints can avoid the concentration of stress and the significant deformations in the supports (if the supports are theoretically fixed in a point or line).

The restraints of entire model are placed over the auxiliary surfaces (figure 4).

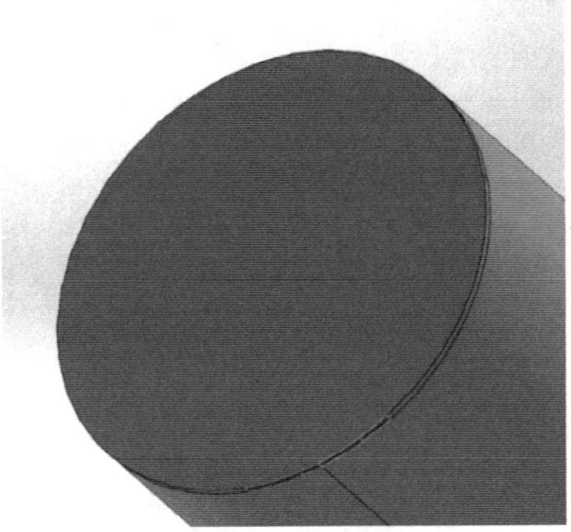

Figure 4. Auxiliary surfaces

IOP Conf. Series: Materials Science and Engineering **977** (2020) 012010 doi:10.1088/1757-899X/977/1/012010

A symmetric 3D model of moving masses section is made (figure 5) to reduce the time of simulation study.

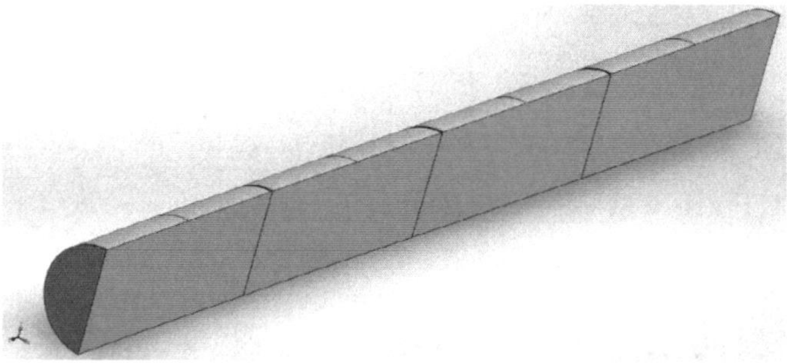

Figure 5. Symmetric 3D model of moving masses section

Restraints has been made to (figure 6) [7]:

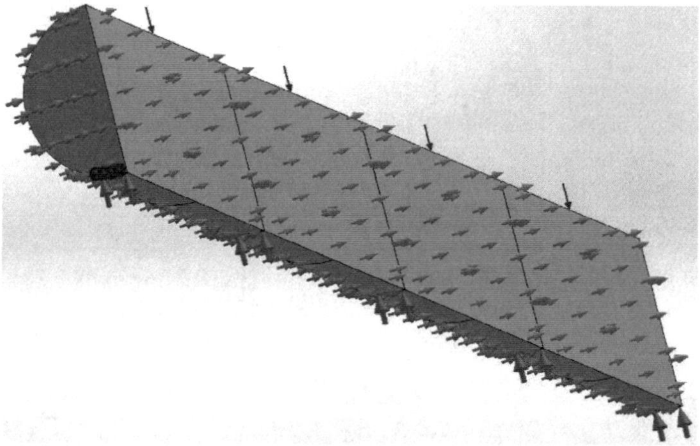

Figure 6. Restraints of symmetric 3D model of moving masses section

- fixing displacement by axis ***Z*** (illustrated in red color);
- fixing displacement by axis ***Y*** (illustrated in orange color);
- fixing displacement of the semi-cylindrical surfaces by axis ***X*** (illustrated in green color);
- fixing displacement of the diametrical surfaces by coordinate system ***YZ*** (illustrated in yellow color).

The fixed model thus approximates sufficiently to the boundary conditions [8] characterizing the mathematical calculation explained above and determines the proper arrangement of finite elements [9]. The mesh of finite elements is represented on figure 7.

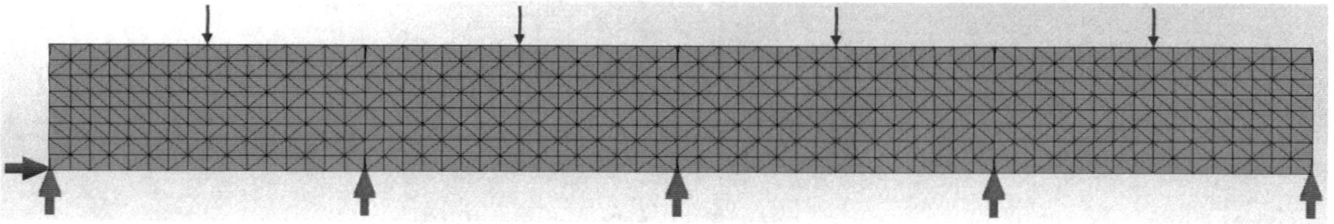

Figure 7. FEM of symmetric 3D model of moving masses section

The displacements are:

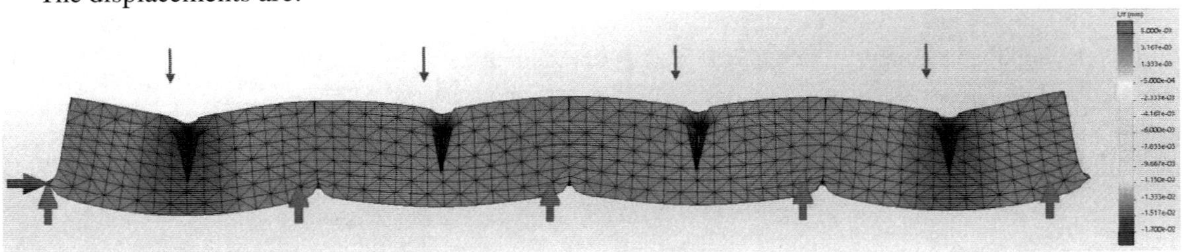

Figure 8. Displacements of symmetric 3D model of moving masses section

The results of support reactions and bending moments by the stiffness matrix method are presented, as well as the simulation data of loaded symmetric 3D model of moving masses section (table 1).

Table 1. Results of the support reactions and bending moments

IR	Results		
	Stiffness matrix method	Symmetric 3D model of moving masses section	Percentage difference
R_A *(N)*	18526.68	18577	0.27
R_B *(N)*	59186.58	59118	0.12
R_C *(N)*	49573.48	49596	0.05
R_D *(N)*	59186.58	59145	0.07
R_E *(N)*	18526.68	18548	0.12
M_A *(Nm)*	0	0	0
M_B^L*(Nm)*	-5309.54	-5301.1	0.16
M_B^R *(Nm)*	5309.54	5301.1	0.16
M_C^L *(Nm)*	-4682.52	-4694.6	0.26
M_C^R *(Nm)*	4682.52	4694.6	0.26
M_D^L *(Nm)*	-5309.54	-5316.8	0.14
M_D^R *(Nm)*	5309.54	5316.8	0.14
M_E *(Nm)*	0	0	0

The table values of percentage difference between the stiffness matrix and the 3D model is small enough and it can be assumed that the model boundary conditions with FE mesh is sufficient to verify the loaded 3D model of moving masses section. About those considerations, the crankshaft model is created (figure 9) and its load simulation is performed with relatively uniform boundary conditions and FE mesh values [10].

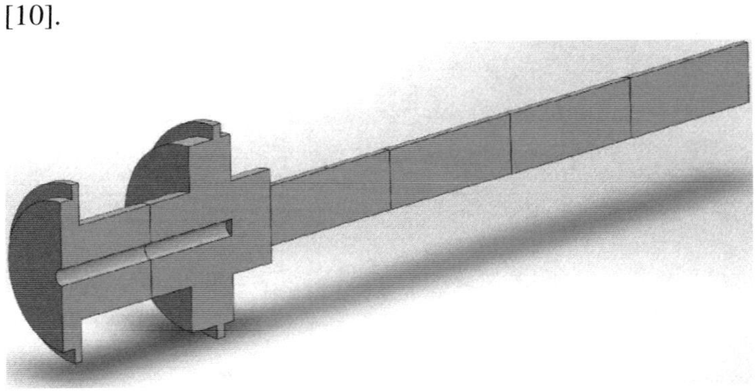

Figure 9. Crankshaft 3D model

IOP Conf. Series: Materials Science and Engineering **977** (2020) 012010 doi:10.1088/1757-899X/977/1/012010

The boundary and load conditions of crankshaft are shown in figure 10 and they are:

- fixing displacement by axis ***Z*** (marked with red arrows);
- fixing displacement by axis ***Y*** (marked with orange arrows);
- fixing displacement of the cylindric faces by axis ***X*** (marked with green arrows);
- load of the chain wheel mass and the turning wheel mass (marked with pink arrows);
- load of the moving masses (marked with pink arrows);
- gravity (marked with blue arrow).

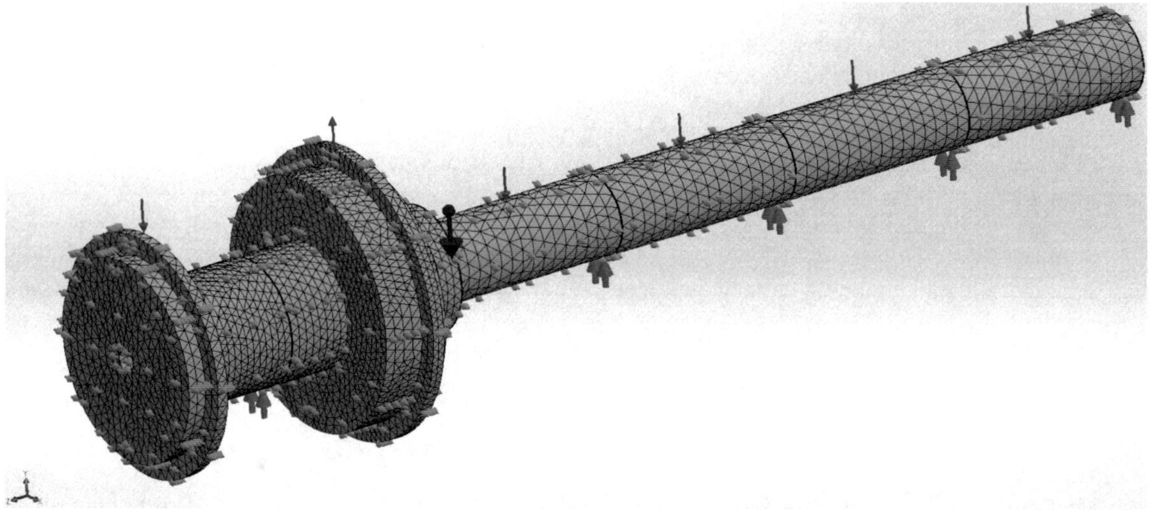

Figure 10. Boundary conditions and FE mesh of crankshaft 3D model

Displacements of crankshaft 3D model is pictured in figure 11.

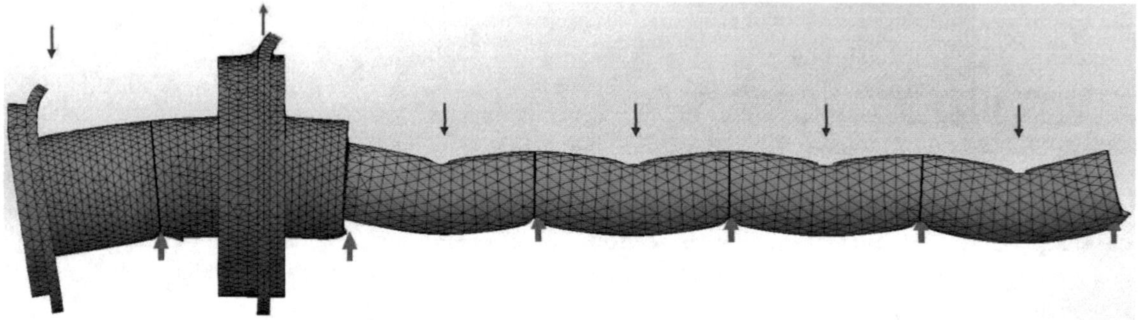

Figure 11. Displacement of crankshaft 3D model

The values of crankshaft load simulation are given in table 2 and table 3.

Table 2. Reaction forces of crankshaft 3D model

Reaction forces	R_A	R_B	R_C	R_D	R_E	R_F
Values (*N*)	29656	14582	56721	55050	63391	20300

Table 3. Bending moments of crankshaft 3D model

Bending moment	M_A	M_B	M_C	M_D	M_E	M_F
Values *(Nm)*	-9461	-3687.6	-4858.3	-5089.4	-5554.4	0

3. Conclusion

Stiffness matrix method is used according to shear forces influence to calculate reaction forces and bending moments of crankshaft moving masses section. The section is represented as simple sections with constant moment of inertia.

The created boundary conditions, finite element mesh and loading modes of symmetric 3D model of moving masses section are optimal for obtaining approximately the same values of rection forces and bending moments accorded to stiffness matrix results.

The percentage difference is below 0.30 and satisfy the crankshaft 3D model verification.
The reaction forces and bending moments are shown in table 2 and table 3 regarded to the procedures applicated in the stiffness matrix method.

4. Reference

[1] Tenev St, Hristov Hr and Stoyanov Sv 2016 A variant of generalized model for four-point lifting beam checking *Mechanics of Machines* **24** p 7-10 ISSN: 0861-9727 (Тенев Ст, Христов Хр и Стоянов Св 2016 Вариант на обобщен модел за проверка на четириточкова товароподемна греда Механика на Машините 24 с 7-10 ISSN: 0861-9727)

[2] Hristov Hr, Stoyanov Sv, Tenev St, Spasov N and Mehmedov I 2016 Regarding to selection of optimal construction of four-support lifting beam *Transport, Ecology - Sustainable Development* (Varna) vol 23 Technical university of Varna p 68-75 ISSN: 2367-6299 (Христов Хр, Стоянов Св, Тенев Ст, Спасов Н и Мехмедов И 2016 Относно избора на оптимална конструкция на четириопорна товароподемна греда ТРАНСПОРТ, ЕКОЛОГИЯ - УСТОЙЧИВОРАЗВИТИЕ (Варна) Том 23 Технически университет Варна с 68-75 ISSN: 2367-6299)

[3] Tenev St 2020 Impact of Shear Forces on Variable Cross Section Shaft Stiffness *Transport, ecology - sustainable development* (Varna) vol 27 Technical university - Varna p 8

[4] https://community.ptc.com/t5/PTC-Mathcad/PTC-Mathcad-User-Forum/m-p/230255

[5] Gattas J and Albermani F *Introduction to Direct Stiffness Method* (Australia) The University of Queensland p 61

[6] Tenev St 2014 Comparative analysis of SEGE coupling load characteristic according to the Mooney-Rivlin calculation model for highly elastic materials *Transport, Ecology - Sustainable Development* (Varna) vol 21 Technical university of Varna p 437-445 ISSN: 2367-6299 (Тенев Ст 2014 Сравнителен анализ на характеристиката на натоварване на съединител СЕГЕ по модел на пресмятане Mooney-Rivlin за високоеластични материали *ТРАНСПОРТ, ЕКОЛОГИЯ - УСТОЙЧИВОРАЗВИТИЕ ЕКОВАРНА 2014* (Варна) Том 21 Технически университет Варна с 437-445 ISSN: 2367-6299)

[7] Tenev St, Hristov Hr and Ivanova El 2015 Some features of SEGE-type elastic coupling modeling *Mechanical engineering and Machine science* **10** p 87-90 ISSN: 1312-8612 (Тенев Ст, Христов Хр и Иванова Ел 2015 Някои особености при моделиране на еластичен съединител тип СЕГЕ *Машиностроене и Машинознание* **10** с 87-90 ISSN: 1312-8612)

[8] Yastrebov V A 210 *Introduction to computational contact mechanics* (Paris, France) Centre des Matériaux, MINES ParisTech p 308

[9] Chang M H and Juang S H 2014 Theoretical design study on shafting alignment calculation for high speed craft *Brodogradnja/Shipbuilding* **65 (3)** pp 31-47 ISSN: 0007-215X

[10] Stoyanov Sv, Hristov Hr, Tenev St and Spasov N 2011 Joining Influence of SEGME-type elastic coupling over movement of rubber elastic element under load *Transport, Ecology - Sustainable Development* (Varna) vol 18 Technical university of Varna p 574-580 ISBN: 954-20-00030

(Стоянов Св, Христов Хр, Тенев Ст и Спасов Н 2011 Влияние на свързването на еластичен съединител тип СЕГМЕ върху преместването на гумен еластичен елемент под товар ТРАНСПОРТ, ЕКОЛОГИЯ - УСТОЙЧИВОРАЗВИТИЕ (Варна) Том 18 Технически университет Варна с 574-580 ISBN: 954-20-00030)

IOP Conf. Series: Materials Science and Engineering **977** (2020) 012011 doi:10.1088/1757-899X/977/1/012011

Energy efficiency improvement in farming equipment, for agricultural holdings

H I Beloev[1], A K Terziev[2], I K Iliev[1], M P Ivanov[2]

[1] University of Ruse, Dept. of Thermal Engineering, Hydraulics and Environmental Engineering, 8 Studentska str., 7017 Ruse, Bulgaria
[2] Deptartment of "Power Engineering and Power Machines", 8 Kliment Ohridski blvd., Technical University of Sofia, 1000 Sofia, Bulgaria

E-mail: aterziev@tu-sofia.bg

Abstract. The use of depreciated and outdated agricultural equipment reflects both, on the increased cost of production (due to the higher energy component), and the greater amount of emissions produced during the operation of that equipment. The presented paper deals with energy and emission analysis of the replacement of existing farming equipment in a holding, with a new one. The analysis considers the baseline tractor and harvester and their specific agricultural operations, related to the processing of the soil and the harvesting of the production. The analysis is based on the existing energy consumption of the old machines and accounting information on the operation of the new equipment, one year after its commissioning. The study shows that the estimated energy savings from the measures of 17.2%, with a similar carbon footprint reduction. Based on the technical characteristics, CO, HC, HC + NOx and NOx emissions were also analyzed. The percentage of estimated carbon emission savings is higher than 43.5%.

1. Introduction

Large portion of the existing agricultural equipment in Bulgaria is heavily depreciated and energy consuming, which is responsible for both, the high cost of production of the farm and the significant emissions produced by the machinery. In [1], an analysis of the energy consumption of the agricultural sector in Iraq was carried out, according to which, from 1994 to 2006, energy consumption increased from 29 to about 42%, while the introduction of energy efficiency measures led to savings in the range of 16 to 20%. This shows that the growth in energy efficiency gains is much smaller than in European countries.

When assessing energy consumption in the agricultural sector, the state of the agricultural equipment used must be taken into account, including the attached inventory, the type of soil processing operations, and the conditions under which the activities are carried out. The studies in [2, 12] present an analysis of the fuel consumption of the same soil activities during different seasons. On this basis, with a view to improving energy efficiency, a recommendation is made as to the type and timing of soil cultivation activities. This achieves energy savings of up to 14.24% in winter and up to 14.10% in summer.

In studies [3-5] are shown the influence of different systems (wireless sensors, energy management systems, mobile Sinks collectors etc.) on fuel consumption when using agricultural equipment. These systems not only provide real-time energy consumption monitoring, but also provide optimization recommendations for machine operation. Another way to reduce energy consumption in the agricultural sector is by the use of hybrid electric agricultural tractors [6, 7]. In this way, energy savings of up to 5% are achieved, when performing operations.

In the recent decades, with the increase of the Earth's population, there has been a significant increase in the cultivation of agricultural land. The operation of machines that largely use liquid fuels is associated with the generation of significant carbon emissions that have a negative impact on the environment. In studies [8-10], a critical analysis is made, regarding the percentage of increase in emissions resulting from increased plowing, as well as the type of additional activities carried out.

2. Aim of the presented study

The aim of the presented study is to analyze the energy consumption of different agricultural equipment, depending on the type of activities performed. For this purpose, two farming vehicle machines have been selected - a wheeled tractor and a harvester in an agricultural holding. The main activity of the farm is the cultivation of cereals: wheat, barley, rapeseed, sunflower and corn.

3. Efficiency assessment of the implementation of energy efficient farming equipment

The cultivation of the land is carried out by means of an existing wheeled driven tractor and attached implements, and the harvesting of the finished products is carried out by the harvester available on the holding. In addition, logistic operations within the farm (relocation, loading and unloading activities) are carried out by a front loader.

3.1. Analysis of the energy consumption, when using a wheeled driven tractor

3.1.1. Current status (baseline)

The activities related to the cultivation of the land are carried out by the tractor "UMZ 6L" and the equipment attached to it. The tractor is produced 1990. The power of the machine is 45.6kW (61.9 hp).

The equipment carries out the following agricultural activities: plowing, cultivation, disking, sowing. The average diesel fuel consumption of the tractors provided is approximately as follows: plowing - 3.00 l/da, cultivation - 1.50 l/da, disking - 1.20 l/da and sowing - 1.25 l/da.

The fuel consumption of the individual treatments is high, leading to high fuel consumption compared to modern machines. With current machine use, according to the farm accounting information, the annual fuel consumption is 14 790 l/year diesel, which energy equivalent is 137 355.32 kWh/year (188 123 kWh/year - related to primary energy). Detailed information on the amount of workable land and the type of activities carried out is presented below, in Table. 1.

Table. 1. Consumption by type of activities carried out before the measure

ACTIVITIES	**WORKED AREA**	**CONSUMPTION**		
		SPECIFIC	**TOTAL**	
	da/year	l/da	l/year	kWh/year
Deep plowing	2 900	3.00	8 700.00	80 797.25
Disking	1 450	1.20	1 740.00	16 159.45
Cultivation	2 900	1.50	4 350.00	40 398.62
		TOTAL	**14 790.00**	**137 355.32**

From the table 1 it is seen that, the highest consumption is for deep plowing (8 700 liters), accounting for 58.82% of the total fuel consumed by the tractor. Second in fuel consumption is cultivation (1 740 liters) with a share of 29.41%. The lowest amount of diesel fuel is spent on disking. This expenditure accounts for about 11.7% of total consumption.

3.1.2. Suggested measure

In order to increase the energy efficiency in soil tillage, it is foreseen to purchase a new energy efficient John Deere 8370R tractor. Information on the consumption of the new tractor by type of activity, based on accounting data provided by the farm for the purposes of the assessment, is presented in Table. 2. The tractor is high-performance and energy-efficient. The engine power is 275.9kW (370 hp). Engine emission level is Tier 4.

Table. 2. Consumption by type of activities, carried out after the measure

ACTIVITIES	WORKED AREA	CONSUMPTION		
		SPECIFIC	TOTAL	
	da/year	l/da	l/year	kWh/year
Deep plowing	2 900	2.998	8 693.10	80 733.12
Disking	1 450	0.787	1 141.73	10 603.28
Cultivation	2 900	0.770	2 233.00	20 737.96
		TOTAL	**12 067.82**	**112 074.36**

Table 3 shows a reduced consumption of diesel fuel, compared to that before the measure was introduced, with more significant reduction in disking and cultivation activities. The total consumption is 12 067.82 liters, respectively 112 074.36 kWh, while the plowing activity within its limits retains its maximum share.

3.1.3. Energy saving after the introduction of an energy efficient tractor

The diesel savings, achieved with the introduction of Measure 1, are calculated as the difference between the consumption of the tractor used (from Table 1) and that of the new John Deere 8370R tractor (from Table 2). The total amount of diesel savings achieved from Measure 1 is 2 722 l/year. This amount of saved diesel fuel is equivalent to 25 280 kWh/year saved final energy. Converted to primary energy, these savings are 27 809 kWh/year. Conversions are made by adjusting the value of the saved final energy by coefficient, regarding the losses in extraction, production and transportation, according to Appendix №2 to Regulation №E-RD-04-2/22.01.2016. For the diesel fuel, this ratio is 1.1. Based on the final energy saved in kWh, the CO_2 savings are calculated. For all farm activities, the estimated CO_2 savings for the measure are 6.75 t/year. The ecological equivalent coefficients have been used for the calculations, according to Appendix No. 3 to Regulation No.E-RD-04-2/22.01.2016. For diesel fuel this coefficient is 267 gCO_2/kWh.

The energy saving factor of Measure 1 is estimated at 18.41%. The results of the calculations by types of activities performed are presented in Table. 3.

Table. 3. Savings, achieved by the introduction of the measure

ACTIVITIES	SAVED ENERGY			SAVED CO_2 EMISSIONS
	FINAL		PRIMARY	
	l/year	MWh/year.	MWh/year	t/year
Deep plowing	6.90	0.06	0.07	0.02
Disking	598.27	5.56	6.11	1.48
Cultivation	2 117.00	19.66	21.63	5.25
	2 722.18	**25.28**	**27.81**	**6.75**

3.2. Analysis of the energy consumption, when using a harvester

3.2.1. Current status (baseline)

For harvesting of the agricultural production, mainly wheat and sunflower seeds, the farm uses a grain harvester. The main activities, carried out with this type of agricultural machinery, are classified by previous survey as:

- Harvest of spring cultures (1 400 da);
- Harvest of autumn cultures (1 500 da).

The annual diesel fuel consumption within the scope of the measure (consumption of the farm's own grain harvester used) is 4 630 liters, equivalent to 42 999 kWh. Information on this consumption by type of harvested cultures, based on a previous survey (baseline), is presented in Table 4.

Table. 4. Consumption by types of harvested cultures (before the measure)

ACTIVITIES	WORKED AREA	CONSUMPTION SPECIFIC	CONSUMPTION TOTAL	
	da/year	l/da	l/year	kWh/year
Harvest of spring cultures	1 400	1.70	2 380.00	22 103.16
Harvest of autumn cultures	1 500	1.50	2 250.00	20 895.84
		TOTAL	**4 630.00**	**42 999.00**

Table 4 shows, that both activities are characterized by close annual consumption of diesel fuel. This is explained by the almost identical areas for the two harvested cultures and the small difference in the values of specific consumption for the two activities.

3.2.2. Suggested measure
As an energy efficient measure, it is recommended to replace the existing combine harvester with a new energy efficient John Deere S660. Engine volume is 9l, and the engine power is 239kW (320 hp).

Information on the consumption of the new grain harvester by type of activity, based on accounting data and related to the cultivation area from the baseline, is presented in Table. 5.

Table. 5. Consumption by types of harvested cultures (after the measure)

ACTIVITIES	WORKED AREA	CONSUMPTION SPECIFIC	CONSUMPTION TOTAL	
	da/year	l/da	l/year	kWh/year
Harvest of spring cultures	1 400	1.280	1 792.00	16 642.38
Harvest of autumn cultures	1 500	1.480	2 220.00	20 617.23
		TOTAL	**4 012.00**	**37 259.60**

3.2.3. Energy saving after the introduction of an energy efficient harvester combiner
The savings on diesel fuel, achieved with the commissioning of the new harvester, are calculated as the difference between the consumption of the existing combiner and that of the new energy efficient John Deere S660 combiner.

The total amount of diesel savings achieved from is 618 l/year. This amount of saved diesel fuel is equivalent to 5 739.39 kWh/year saved final energy. Converted to primary energy, these savings are 6 313.33 kWh/year (same conversion methodology presented in 3.1.3).

Based on the final energy saved in kWh, the CO_2 savings are calculated. For all farm activities, the estimated CO_2 savings for the measure are 1.53 t/year.

The energy saving factor of Measure 2 is estimated at 13.35% (here accepted 267 gCO_2/kWh).

The results of the calculations by types of activities performed are presented in Table. 6.

Table. 6. Savings, achieved by the introduction of the measure

ACTIVITIES	SAVED ENERGY FINAL		SAVED ENERGY PRIMARY	SAVED CO_2 EMISSIONS
	l/year	MWh/year	MWh/year	t/year
Harvest of spring cultures	588	5.46	6.01	1.46
Harvest of autumn cultures	30	0.28	0.31	0.07
	618	**5.74**	**6.31**	**1.53**

3.3. Analysis of the energy consumption after implementing the measures
The total energy savings from the implemented measures for one full year operation are estimated at 31 020 kWh. The total energy consumption for all activities of the existing tractor and harvester is 180 354 kWh/year. Thus, the energy saving factor from the introduction of the measures is 17.2%. The annual savings are 8.28 t/year (Table 7).

Table. 7. Savings, achieved by the introduction of the package of measures

ACTIVITIES	SAVED ENERGY FINAL		SAVED ENERGY PRIMARY	SAVED CO_2 EMISSIONS
	l/year	MWh/year	MWh/year	t/year
Wheeled tractor introduction	2 722.18	25.28	27.81	6.75
Harvester combine introduction	618	5.74	6.31	1.53
	3 340.18	**31.02**	**34.12**	**8.28**

4. Analysis of the emissions, after implementing the package of measures

The measures implemented lead in carbon savings of 8.28 t/year.

Stage III/IV emission standards for non-road engines were adopted by the European Parliament in April 2004 (Directive 2004/26/EC), and for agricultural and forestry tractors. Stage III standards, which are further divided into Stages IIIA and IIIB, are phased-in between 2006 and 2013, while Stage IV enters into force in 2014. In addition to the engine categories regulated at Stage I/II, Stage III A standards also cover engines used in inland waterway vessels [11].

The type and amount of emissions released from Stage I, II and IV off-road diesel engines shall not exceed the amount presented in Table 8.

Table. 8. Types of emissions, by power and engine category of agricultural equipment

Category	Net installed capacity kW	CO (g/kWh)	HC (g/kWh)	NO_x (g/kWh)	PM (g/kWh)
	Stage I				
C	$37 \leq P < 75$	6.5	1.3	9.2	0.85
	Stage II				
E	$130 \leq P \leq 560$	3.5	1.0	6.0	0.2
	Stage IV				
Q	$130 \leq P \leq 560$	3.5	0.19	0.4	0.025

The engine of the suggested John Deere 8370R tractor put into service, complies with the Stage IV exhaust emission standard. The engine of the harvester intended for commissioning meets the Stage II exhaust emission standard. In table 8 are presented the emission limit values for the respective diesel engine expressed in grams per kWh of fuel consumed. Based on the amount of fuel consumed before and after implementation of measures are estimated the limit values of emissions released from old and new agricultural equipment, and the estimated savings achieved. The table shows that the percentage of carbon emissions reduced in terms of replacement of equipment is extremely high - over 43%. This is noticeable when the engines are Stage III standard and higher.

Table. 9. Achieved emissions savings, by the replacement of the agricultural equipment

Type of carbon emissions	CO	HC	NOx
Annual emissions released from the UMZ 6L tractor, *t/yr*	0.893	0.179	0.117
Annual emissions released from the new tractor John Deere 8370R, *t/yr*	0.392	0.007	0.015
Annual emissions released from existing harvester, *t/yr*	0.279	0.056	0.037
Annual emissions released from the new tractor John Deere 8370R, *t/yr*	0.130	0.037	0.007
Annual emission savings after the introduction of EEM 1 и 2, ***t/yr***	**0.650**	**0.190**	**0.131**
Percentage of the emission savings, %	**55.4%**	**81.1%**	**85.4%**

5. Conclusion

The use of outdated and depreciated agricultural equipment leads to an increase in the energy component in the cost of agricultural production. In addition, the amount of emissions released is also significant. Depreciation costs are very high and repairs are often required with old farm equipment.

The equipment suggested in the study is high performant and energy efficient. The percentage of energy savings depends on the type of activities performed and the rate of utilization of the equipment.

With the introduction of new wheeled equipment, the percentage of savings for the volume of activities is over 18%, and the replacement of an existing harvester with a new one, results in energy savings of 13.35 %. Thus, with the introduction of the selected set of measures, the annual energy savings are 17.2%.

The reduced energy consumption leads to significant carbon savings. For the measures selected, the estimated annual savings are 8.28 t/year. In addition, the use of engines with higher class, results in savings of CO, HC, HC + NOx and NOx emissions. The estimated savings rate is greater than 50%.

6. References

[1] Avara A and Karami M, Energy and exergy efficiencies in agricultural and utility sectors of Iran compared with other countries, *2010 International Conference on Mechanical and Electrical Technology*, Singapore, 2010, pp. 6-10., doi: 10.1109/ICMET.2010.5598479

[2] Zhang M, Chen F and Zhang H, Effects of tillage treatments on energy-use efficiency of winter wheat and summer corn cropping systems in north China, *2011 International Conference on Remote Sensing, Environment and Transportation Engineering*, Nanjing, 2011, pp. 7549-7552, doi: 10.1109/RSETE.2011.5966118

[3] Shakhov V, On efficiency improvement of energy harvesting wireless sensor networks, *2016 39th International Conference on Telecommunications and Signal Processing (TSP)*, Vienna, 2016, pp. 56-59, doi: 10.1109/TSP.2016.7760828

[4] Rio A, Maurel Y, Barais O and Bugni Y, Benefits of Energy Management Systems on local energy efficiency, an agricultural case study, *2019 IEEE International Conference on Communications, Control, and Computing Technologies for Smart Grids (SmartGridComm)*, Beijing, China, 2019, pp. 1-7, doi: 10.1109/SmartGridComm.2019.8909742

[5] Khan T and Kumar D, Mobile Collector Aided Energy Reduced (MCER) Data Collection in Agricultural Wireless Sensor Networks, *2016 IEEE 6th International Conference on Advanced Computing (IACC)*, Bhimavaram, 2016, pp. 629-633. doi: 10.1109/IACC.2016.122

[6] Jia C, Qiao M and Qu L, Numerical Methods for Optimal Control of Hybrid Electric Agricultural Tractors, *2019 IEEE Transportation Electrification Conference and Expo (ITEC)*, Detroit, MI, USA, 2019, pp. 1-6, doi: 10.1109/ITEC.2019.8790578

[7] Ugale R, Singh G, Baka S and Chaudhari B, Effective energy conservation for the agricultural sector using line start permanent magnet synchronous motors, *TENCON 2009 - 2009 IEEE Region 10 Conference*, Singapore, 2009, pp. 1-5, doi: 10.1109/TENCON.2009.5395904

[8] Zhang X, The function of agricultural carbon sink and its ecological compensation mechanism, *2017 4th International Conference on Industrial Economics System and Industrial Security Engineering (IEIS)*, Kyoto, 2017, pp. 1-4, doi: 10.1109/IEIS.2017.8078586

[9] Zheng L and Yu F, Analysis on the impact of agricultural production of China upon climate change based on carbon budget of arable land, *2010 International Conference on Mechanic Automation and Control Engineering*, Wuhan, 2010, pp. 5091-5094. doi: 10.1109/MACE.2010.5535984

[10] Dong J and Yun K, Land use carbon emissions efficiency analysis in Hubei Province, *2015 Fourth International Conference on Agro-Geoinformatics (Agro-geoinformatics)*, Istanbul, 2015, pp. 168-172, doi: 10.1109/Agro-Geoinformatics.2015.7248110

[11] https://www.transportpolicy.net/standard/eu-nonroad-emissions/

[12] Kamburova V, Iliev I, Velikanov M and Terziev A, Simplified methodology for monitoring and assessment of results of energy saving measures for agricultural equipment. *Sixth International Conference Energy Efficiency and agricultural engineering*, 11-12 November 2015, Ruse, Bulgaria, Printing house, 2015, pp. 309-314.

IOP Conf. Series: Materials Science and Engineering **977** (2020) 012012 doi:10.1088/1757-899X/977/1/012012

Computer program for indicator diagram processing of the internal combustion engine

E Dimitrov[1]

Department of Combustion Engines, Automobile Engineering and Transport, Technical University – Sofia, 8 Kliment Ohridski Blvd., 1756 Sofia, Bulgaria

[1] E-mail: etzd@tu-sofia.bg

Abstract. The article presents computer program for processing an indicator diagram of internal combustion engines developed by the author. The program input data are engine construction and operating parameters and cylinder pressure data array. The input data is saved to user files with *.txt and *.xls extensions. As a result of program execution, the following are calculated: engine indicator parameters; in cylinder working substance temperature; the rate of pressure increase; the current values of the heat-release and net heat-release and other quantities. The program is developed in a *Matlab* environment, compiled into an executable (*.exe) file and works in dialog mode. All calculations results are recorded in a custom file. This file is opened and saved with *Microsoft Excel*. The program is applicable by spark ignition and compression ignition engines working with: gasoline, LPG, CNG, diesel, diesel-LPG dual fuel, diesel-CNG dual fuel and diesel-hydrogen dual fuel.

1. Introduction

Internal combustion engine working cycle is referred with complicated and interconnected physical, chemical, thermo dynamical, gas dynamical and other processes. These processes are performed featuring extreme dynamic. Indicator diagram of internal combustion engine and its next mathematical processing gives the most complete and accurate quantitative assessment of engine process parameters. Internal combustion engine indicator diagram gives dependency of engine cylinder pressure – p_{cyl} upon cylinder volume – V_{cyl} or crankshaft angle – φ. Functions graphics $p_{cyl} = f\,(V_{cyl})$ and $p_{cyl} = f\,(\varphi)$ are called folded and expanded indicator diagram respectively. Engine indicator diagram is recorded with measurement complex which consists of sensors (cylinder pressure sensor, engine speed sensor, crankshaft position sensor); signal amplifiers; data acquisition system; data monitoring system.

Engine test bench laboratory system manufacturers (for example *AVL*, Austria) are offering computer-based systems for recording and processing of engine indicator diagram. Processing of indicator diagram in such systems is fully automated. Such systems have a high price. Furthermore the indicator diagram processing is based on engine operation with the main types of liquid or gaseous fuels. Thus obtained results are not quite correct when operating the engine on gas-diesel cycle. From other hand, research companies with lower financial capacities are developing their own systems for recording engine indicator diagrams [1-5]. These systems are equipped with sensors, amplifiers and measuring tools from different manufacturers. According to specific research purposes, computational programs for engine indicator diagram processing are developed [6-8].

2. Purpose of the article

According to upper text, purpose of current manuscript is to announce developed by the author computational program for processing of experimental obtained internal combustion engine indicator diagram.

3. Mathematical relations

The indicator diagram processing is carried out for the period during which the quantity of the working substance in the engine cylinder does not change, namely from the moment of closing of the inlet valve until the moment of opening of the exhaust valve. During this period the compression, combustion and expansion processes take place. The compression process starts from the moment the inlet valve is closed. The compression end and the start of combustion are determined by the first positive value of the heat-release rate – $dX/d\varphi$. The combustion process ends when the heat-release coefficient – X reaches 0,99 (X = 0,99). If, until the exhaust valve is opened, the heat-release coefficient is less than 0,99 (X < 0,99), the end of combustion is determined by the maximum value of the net heat-release coefficient – X_a^{max}. Further in text the working substance parameters at characteristic points of the indicator diagram are denoted by following indices: start of compression (inlet valve closes) – „***a***“; end of compression and start of burning – „***c***“; end of burning – „***z***“; end of extension (exhaust valve opens) – „***b***“.

During compression are calculated fresh medium temperature – T_{cyl}, K and middle polytrophic index – n_1. Formulas known from thermodynamics are used:

$$T_{cyl} = \frac{p_{cyl} V_{cyl}}{m_{fm}.R_{fm}}; \tag{1}$$

$$n_1 = \frac{\lg p_c - \lg p_a}{\lg V_a - \lg V_c}, \tag{2}$$

where: p_{cyl} is current value of cylinder pressure, Pa; V_{cyl} – current value of cylinder volume, m^3; m_{fm} – in cylinder fresh medium mass, kg; R_{fm} – fresh medium gas constant J/kg.K; p_a, p_c and V_a, V_c are pressure and cylinder volume respectively at compression start and compression end. During compression fresh medium gas constant does not change and is equal neither to that of air or to gas constant of air-fuel mixture.

Cylinder volume is calculated by well-known from internal combustion engine theory and design formula:

$$V_{cyl} = \frac{\pi D^2}{4}\left\{\frac{S}{\varepsilon - 1} + \frac{S}{2}\left[(1-\cos\varphi) + \frac{1}{\lambda}\left(1 - \sqrt{1-\lambda^2.\sin^2\varphi}\right)\right]\right\}, \tag{3}$$

where: D is cylinder bore, m; S – piston stroke, m; λ – ratio of connecting rod length to crank radius; ε – compression ratio; φ – crankshaft angle, rad.

According to engine type, fresh medium mass – m_{fm} is equal neither to cycle air mass – m_{air} (by compression ignition engines) or sum of cycle air mass and cycle fuel mass – m_{fuel} (by spark ignition engines and gas powered diesel engines). Cycle mass air and fuel are calculated by equation:

$$m_{air}(m_{fuel}) = \frac{G_{air}(G_{fuel}).\tau}{i.120.n}, \text{ kg/cycle}, \tag{4}$$

where: G_{air} and G_{fuel} are air and fuel consumption, kg/h; τ – engine strokes; i – number of engine cylinders; n – crankshaft speed, min^{-1}.

Indicator diagram processing during combustion and expansion is based on thermodynamics firs law and is realized by method, published in [9-10]. The essence of methodology is to determine the heat-release rate. After that is calculating the in cylinder working substance temperature. During program developing differential equations suggested by *R. M. Petrichenko* are used:

$$\frac{dX}{d\varphi} = \frac{m_{ws}.c_v.T_{cyl}}{m_{fuel}.H_u}.\left(\frac{d\ln p_{cyl}}{d\varphi} + k.\frac{d\ln V_{cyl}}{d\varphi} + \frac{1}{m_{wm}.c_v.T_{cyl}}.\frac{dQ_w}{d\varphi}\right); \tag{5}$$

$$\frac{dT_{cyl}}{T_{cyl}} = \frac{1}{m_{wm}.c_v.T_{cyl}}.(m_{fuel}.H_u.dX - dQ_w) - (k-1).\frac{dV_{cyl}}{V_{cyl}}, \tag{6}$$

where: m_{ws}, kg, c_v, J/kg.K and k are mass, volume specific heat capacity and adiabatic index of working substance respectively; H_u – fuel lower heating value, J/kg; $dQ_w/d\varphi$ – heat lose rate to over piston space walls, J/deg. Differential equation from net heat-release rate – $dX_a/d\varphi$ is the same at equation (5) without calculating heat loses, i.e. $dQ_w/d\varphi = 0$.

The differential – dQ_w, is calculated from the equation:

$$dQ_w = \alpha_w.F_{cyl}.(T_{cyl} - T_w).dt = \alpha_w.F_{cyl}.(T_{cyl} - T_w).\frac{d\varphi}{6n}, \tag{7}$$

where: α_w is convective heat transfer coefficient, w/m^2.K; F_{cyl} – over piston cylinder area, m^2; T_w – middle wall temperature, K. Convective heat transfer coefficient – α_w is given by *Woschni* formula [11]:

$$\alpha_w = 127{,}93.D^{-0{,}2}.p_{cyl}^{\;0{,}8}.W^{0{,}8}.T_{cyl}^{\;-0{,}53}, \tag{8}$$

where: W is in cylinder working substance characteristic speed, calculated by [11].

Working substance during combustion is a gas mixture, consisting of fresh working mixture and burning gases. This mixture is on continuous change of its composition in proportion to burned fuel quantity, i.e. in proportion of heat-release coefficient. Thermodynamic properties of working substance during combustion (mass specific heat capacities by constant volume and constant pressure – c_v and c_p, gas constant R and adiabatic index k) are calculated based on gas mixture well known formulas:

$$c_v = \frac{m_{fm}}{m_{ws}}.c_v^{\;fm} + \frac{m_{eg}}{m_{ws}}.c_v^{\;eg}, \tag{9}$$

where: c_v, c_v^{fm}, c_v^{eg} and m_{ws}, m_{fm}, m_{eg} are volume specific heat capacities and mass of working substance, fresh working substance and burned gases.

When operating the engine with standard liquid fuels (petrol and diesel fuel) burned gases thermodynamic properties are defined from table data – [12], which depend on temperature and air-fuel ratio. When operating the engine on gas-diesel cycle, thermodynamic properties of exhaust gases are calculated depending on mass fraction of basic components [13], for example:

$$c_v = \sum_{i=1}^{n} m_i.c_{vi}, \tag{10}$$

where m_i and c_{vi} are mass fraction and volume specific heat capacity of corresponding component. Mass fractions of exhaust gas components (carbon dioxide - CO_2, carbon monoxide - CO, water vapor - H_2O, hydrocarbons H_nC_m, free oxygen - O_2 and nitrogen - N_2) are calculated depending of fuel elemental composition and air-fuel ratio – [12-13].

Middle polytrophic index during expansion is calculated by formula (2) as pressures and volumes in points a and c are replaced by these in z and b.

4. Description of the program

The computational program is developed in *Matlab* environment, compiled into executable file and works in dialog mode in the *Microsoft Command Prompt* – figure 1. The program consists of the following main modules: input data; calculation procedure; display and record results. Input module includes: engine geometric parameters; working parameters of engine; crankshaft angle and cylinder pressure data area. Engine geometric parameters are entered in user file *data_eng.xls* – figure 2. The user must enter the following data: cylinder bore D, m; piston stroke S, m; connecting rod length L, m; compression ratio ε; engine stroke τ; number of engine cylinders i; valve timing. The following engine work parameters must be known for processing the indicator diagram: engine speed n, min^{-1}; liquid fuel consumption G_{fuel}, kg/h; gas fuel consumption (only by gas powered spark ignition engines and diesel dual fuel engines) – G_{gas}, kg/h; air consumption G_{air}, kg/h; ignition angle θ, degree before TDC; diesel fuel injection duration φ_{inj}, crankshaft degree; middle cylinder wall temperature T_w, K. The engine work parameters are entered in the user file *eng_work.xls* – figure 3. The crankshaft angle and cylinder pressure data are entered in a text file - *data_id.txt* – figure 4. The respective values must start from the top dead centre (TDC) at the inlet beginning (0 degree) and end one degree before the expansion ending (TDC), i.e. the indicator diagram data are in interval from 0 to 719 crankshaft degrees. After entering the input data, the user has to choose the engine duty cycle type. There are the following options: spark ignition engines working with gasoline, LPG, CNG and hydrogen, and compression ignition engines working with diesel or diesel-LPG dual fuel, diesel-CNG dual fuel and diesel-hydrogen dual fuel.

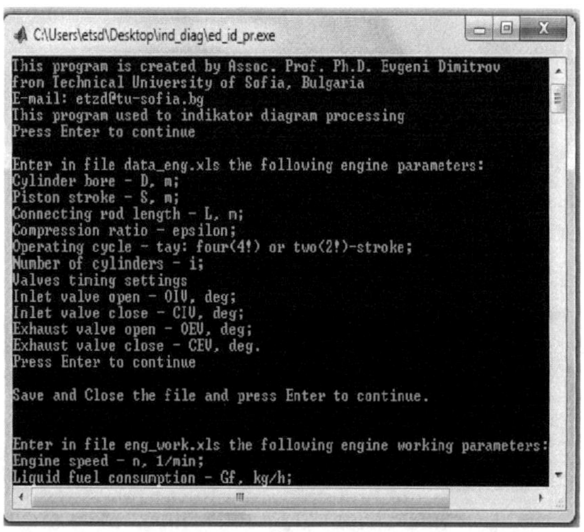

Figure 1. Main window of the program.

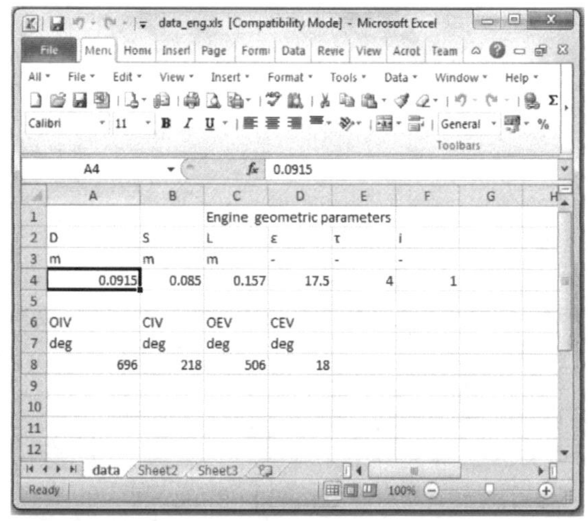

Figure 2. Entering engine geometric parameters.

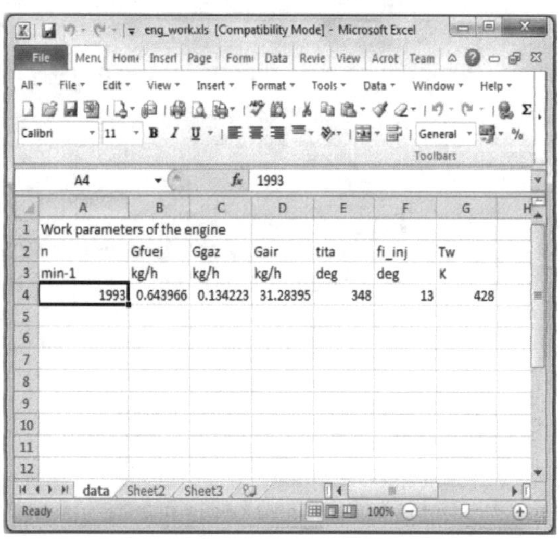

Figure 3. Entering engine work parameters.

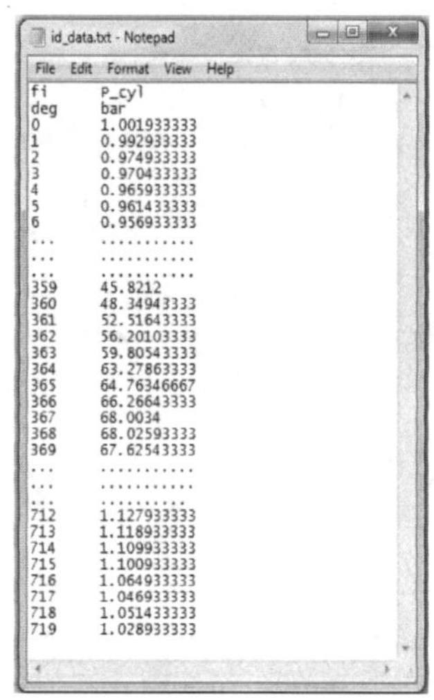

Figure 4. Entering crankshaft angle and cylinder pressure data.

The main part of the calculation procedure is cyclically repetitive mathematical and logical operations. The calculation cycle starts from inlet valve closing and continues until exhaust valve opening. At each step of the calculation, it is checked for the phase of the engine operating cycle. Then, step by step, the process-specific parameters are calculated. For example, during the combustion the following parameters are sequentially calculated: heat-release and net heat-release rate – $dX/d\varphi$ and $dX_a/d\varphi$; heat-release and net heat-release coefficient – X and X_a; work substance temperature –

IOP Conf. Series: Materials Science and Engineering **977** (2020) 012012 doi:10.1088/1757-899X/977/1/012012

T_{cyl}; heat losses – dQ_w and thermodynamic properties of the working substance. The thermodynamic properties of the work substance are calculated by subroutines called from the main program. The values of specific parameters and their corresponding crankshaft angles are determined during performance of the cycle, for example: maximum pressure – p_{max} and φ_{pmax}; maximum temperature – *Tmax* and φ_{Tmax}; maximum pressure increase rate – $(dp/d\varphi)_{max}$ and $\varphi_{(dp/d\varphi)max}$; and others. The crankshaft angles for specific values of heat-release coefficient are also determined. The calculation results are presented (as graph forms) in separate windows – figure 5 to figure 8.

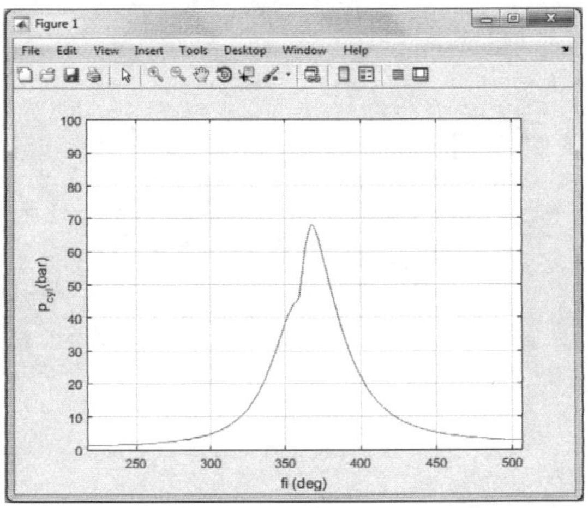

Figure 5. Cylinder pressure change depending on the crankshaft angle – p_{cyl} = f(φ).

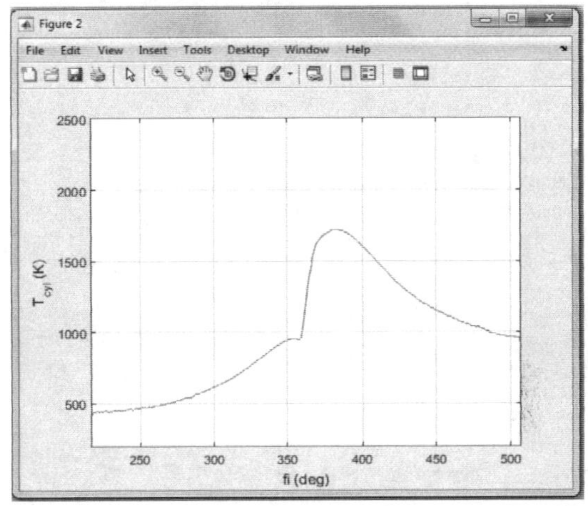

Figure 6. Cylinder temperature change depending on the crankshaft angle – T_{cyl} = f(φ).

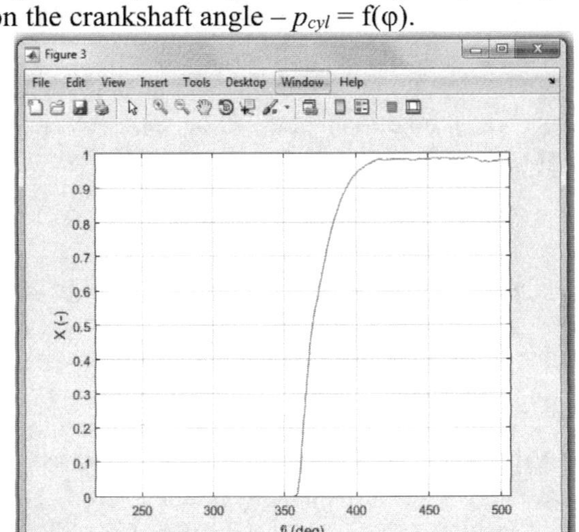

Figure 7. Change of heat-release coefficient depending on the crankshaft angle – X = f(φ).

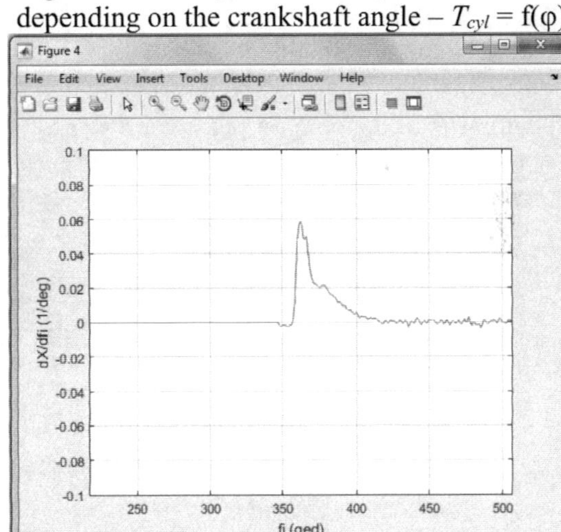

Figure 8. Change of heat-release rate depending on the crankshaft angle – $dX/d\varphi$ = f(φ).

Before starting the main cycle, are calculated the values of characteristic parameters and arrays: mean indicator pressure – p_i, middle polytrophic index – n_1, current cylinder volume – V_{cyl} and its derivative – $dV_{cyl}/d\varphi$, pressure increase rate – $dp/d\varphi$ and others. Specific subroutines have been developed for some of the calculations.

After the calculations, all results (parameters and arrays) are recorded in a custom file *id_full.xls* – figure 9. This file is opened automatically. In order not to lose data by next running the program, the user must save the file *id_full.xls* under another name. After data saving, the program is automatically closes when the user presses a *Enter*.

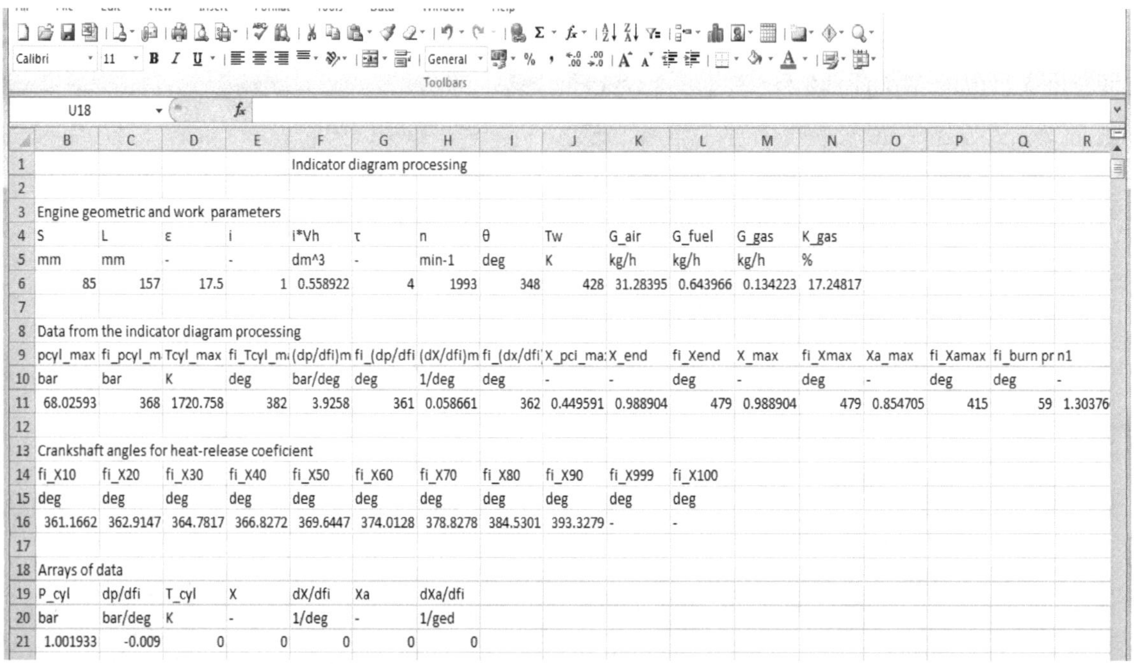

	B	C	D	E	F	G	H	I	J	K	L	M	N	O	P	Q	R
1					Indicator diagram processing												
2																	
3	Engine geometric and work parameters																
4	S	L	ε	i	i*Vh	τ	n	θ	Tw	G_air	G_fuel	G_gas	K_gas				
5	mm	mm	-	-	dm^3	-	min-1	deg	K	kg/h	kg/h	kg/h	%				
6	85	157	17.5	1	0.558922	4	1993	348	428	31.28395	0.643966	0.134223	17.24817				
7																	
8	Data from the indicator diagram processing																
9	pcyl_max	fi_pcyl_m	Tcyl_max	fi_Tcyl_m	(dp/dfi)m	fi_(dp/dfi	(dX/dfi)m	fi_(dx/dfi	X_pci_ma	X_end	fi_Xend	X_max	fi_Xmax	Xa_max	fi_Xamax	fi_burn pr	n1
10	bar	bar	K	deg	bar/deg	deg	1/deg	deg	-	-	deg	-	deg	-	deg	deg	-
11	68.02593	368	1720.758	382	3.9258	361	0.058661	362	0.449591	0.988904	479	0.988904	479	0.854705	415	59	1.30376
12																	
13	Crankshaft angles for heat-release coeficient																
14	fi_X10	fi_X20	fi_X30	fi_X40	fi_X50	fi_X60	fi_X70	fi_X80	fi_X90	fi_X999	fi_X100						
15	deg	deg	deg	deg	deg	deg	deg	deg	deg	deg	deg						
16	361.1662	362.9147	364.7817	366.8272	369.6447	374.0128	378.8278	384.5301	393.3279	-	-						
17																	
18	Arrays of data																
19	P_cyl	dp/dfi	T_cyl	X	dX/dfi	Xa	dXa/dfi										
20	bar	bar/deg	K	-	1/deg	-	1/ged										
21	1.001933	-0.009	0	0	0	0	0										

Figure 9. Recording of data from indicator diagram processing.

5. Conclusion

The author is of the opinion that the presented program is fully functional and can be used both in the educational process and in the processing of experimental research results. The program is developed for non-commercial purpose and it is freely distributed.

References

[1] Ivanov Z, Michajlov V and Kolev A Engine indication system (in Bulgarian), EKO VARNA '2011 *Conference Proceedings*, 590-597

[2] Nikolov V, and Ambarev K 2018 An experimental study of the cylinder pressure at work of diesel engine with generator, *BulTrans-2018 Conference Proceedings*, 77-81

[3] Repka J, Labuda R, Kovalčík A, Bárta D and Skrúcaný T 2014 Indicated pressure of internal combustion engine. Scientific proceedings XXII international scientific-technical conference "trans & MOTAUTO'14" **1** 58-61

[4] Dimitrov R, Bogdanov K, Wrobel R, Serrano L and Mihaylov V 2019 Adjustment parameters of an internal combustion engine working with methane. IOP Conference Series Materials Science and Engineering **664**(2019)012020, doi:10.1088/1757-899X/664/1/012020

[5] Dimitrov A, Bogdanov K and Hristov R 2008 Combustion process parameters and indicator parameters of high-speed diesel engine "ROVER 2.0D" working with added natural gas. Annual Journal of Technical University of Varna **2** 53-58

[6] Ivanov N and Ivanov Z 2004 Investigation the upper position of the piston by means of the compression diagram (in Bulgarian). International scientific-technical conferenc "trans & MOTAUTO'04" **2** 74-78

[7] Dimitrov R, Bogdanov K, Hristov R and Dimitrov A 2013 Calculation parameters of work process of spark ignition engine from indicator diagram (in Bulgarian). RU-Ruse Proceeding **52(4)** 38-42

[8] Bogdanov K and Dimitrov R 2019 Research on the Possibility of Increasing the Power Output

of a High - Speed diesel Engine by Adding Compressed Natural Gas. IOP Conference Series Materials Science and Engineering **614**(2019)012007, doi:10.1088/1757-899X/614/1/012007
[9] Petrichenko R 1983 Physical basis the in cylinder processes of internal combustion engines (in Russian). (Publishing House of St. Petersburg University) p 70-71
[10] Heywood J 1988 Internal Combustion Engine Fundamentals (McGraw-Hil) p 508-511
[11] Merker G, Schwarz C, Stiesch G and Otto F 2006 Simulating Combustion. (Springer) p 149-150
[12] Kolchin A and Demidov B 2008 Calculation of Automobile and Tractor Engines (in Russian). (Moscow Higher School) p 57-61
[13] Dimitrov P 2000 Internal combustion engines – Part I (in Bulgarian). (Publisher of Technical University – Sofia) p 37-39

IOP Conf. Series: Materials Science and Engineering **977** (2020) 012013 doi:10.1088/1757-899X/977/1/012013

Reconditioning of belt conveyor details by vibrating arc overlaying process

I T Todorov[1]

[1]Department of Repair, Reliability, Machines, Mechanisms, Logistic and Chemical Technologies, "Angel Kantchev" University of Ruse, 7017 Ruse

Abstract. The recent paper presents a possibility about reconditioning of rollers and axes from belt conveyors through vibrating arc overlaying process in shielding mixtures of Ar and CO_2 with increased frequency of vibrations of the electrode wire. Work pieces made of steel 45, with dimensions of 50 mm diameter and 250 mm length, were selected for the object of this research. The working regime of overlaying process consists of voltage of 20V, current of 150-180, vibrating amplitude of 1,5 mm, overlaying speed of 1,26 m/min, wire feeding rate of 2,3 m/min, wire electrode outlet of 15 mm and overlaying step at 3 mm/min^{-1}. The shielding mixture consists of 60% Ar + 40% CO_2 fed at rate of 16 l/min. A single factorial experiment was conducted by selecting a wire electrode vibrating frequency as the main variable. The parameters of the arc vibrating process were accepted as criteria by establishing a cybernetic model. It is established that vibrating frequency at rate of 125 Hz responding to the highest duty cycles rate, the shortage of duty cycle and arc burning time length as well as it corresponds to the smaller droplet size and minimal sparkling of molten metal. The durability of the reconditioned rollers in this way and the floating bolts are 5-times higher than new produced ones. The inner roller surface, overlaid by 08Mn2Si wire electrode, possesses higher wear-resistance compared to roller reconditioned by pressing an iron bush.

Key words: belt conveyors, vibrating arc, overlaying process, rollers

1. Introduction

Belt conveyors are widely used in agriculture, industry, energy, mines and other spheres. In the process of using these conveyors, their parts are influenced by too many external factors, leading to their wear. An increase of wear-resistance of working surfaces and shortage of expenses related to spare parts, natural resources, material and energy is possible through reconditioning and it is a key to sustainable economics especially when times of crisis still exist as well as problems about saving natural resources and the protection of environment is very important [1, 3, 4, 6].

The aim of the present research is to increase the durability of the working surfaces of key elements implemented in the structural scheme of belt conveyors by vibrating arc overlaying process with increased wire electrode vibrating frequency.

2. Exposition

A major part of the details, suitable for reconditioning belongs to the group of cylindrical or rotational details. They are made of either steel or iron, possessing a higher surface hardness, accuracy and cleanliness, as well as complex overall shape and high production cost. The wear index rate of these parts ranges from 0.5 mm to 10 mm [2, 3, 5, 7].

The object of the research are the restored parts of the belt conveyors acting as a rotating grill in central pump stations 3 and 4, working at AEC "Kozloduy". The rotating grill is designed as a primary filter for the incoming water from the river Danube in order to cool the reactor. The working conditions are characterized by the presence of a large amount of fine abrasive sand particles into incoming water which causes an extremely fast wear mixed with an active corrosion. This type of wear is best expressed in the steel roller, bronze bushing and steel floating bolt [9].

Before initiating a process of reconditioning it is necessary to determine the structural parameters and wear index rate of the details subject to the process as it refers to the selection of an appropriate method for reconditioning. That affects to surface quality, wear-resistance and durability of the final product. According to recent trends and conceptions, the durability of the reconditioned detail should be at least equal to the same obtained by the newly-produced one [3-6, 8].

The wear index rate, character and distribution of belt conveyor details were determined in advance. The diametrical wear of the rollers on the outer surface is at average rate of 10 mm, as 33% of all rollers show wear rate up to 19 mm. 90% of the rollers lacked the bronze bushings , which led to average wear rate within 10-11 mm, as 18% of all rollers have higher wear rate on their inner diameter. The wear rate of the floating bolt is significantly lower as the average values are in the range from 1,5 to 1,7 mm. [9]

The analysis of the reasons led to such higher wear rates for outer diameter of the roller and bronze bushing showed that the selected materials for these details are not suitable for working in high-abrasive working conditions. To increase wear-resistance and durability of the roller and floating bolt, and thereby increase the durability of entire belt conveyor cleaning system, the possibilities of vibrating arc overlaying process in gas mixtures with an increased wire electrode frequency were used for applying abrasion resistant restorative coatings. For this purpose, it is necessary to carry out experiments to determine the optimal values of the process parameters. The wire electrode vibrating frequency (f), as it is an input factor at the scheme of cybernetic model (fig.1), is selected as the main variable. Increasing the frequency leads to an increase in the number of short circuits per unit time, which in turn is associated with an increase in the number of electric cycles. The number of arc cycles is directly related to the size of the electrode metal droplets passing through the column of the arc - the larger the number of cycles, the smaller the size of the droplets and vice versa. The vibration frequency varies from 25Hz to 150Hz in 25Hz increments and assumes the following values (25, 50, 75, 100, 125, and 150) during single factorial experiments.

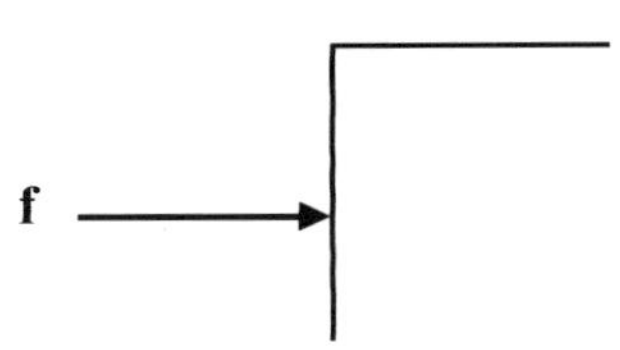

As main criteria for evaluation of vibrating arc overlaying process is selected to be the vector of parameters of the same which includes the following items: duty cycle frequency ($f_ц$), short circuit time length ($t_{кс}$) and arc burning time length ($t_д$). The overlaying of the specimens was carried out on an apparatus for welding in shielding gases with a specially developed vibration arc apparatus with an axial inertial vibrator for a continuous change of wire electrode frequency by means of a potentiometer built into a specially designed direct current source. The welding was performed on cylindrical specimens made of steel 45, having a diameter of 50 mm and a length of 250 mm, corresponding to the average dimensions and mass of the parts to be recovered [3]. Five sectors of 40 mm wide with a 1.6 mm diameter CB08G2C electrode wire were deposited onto each specimen within protective gas mixture of argon and carbon dioxide in 60% Ar and 40% CO2 composition [2]. The working regime parameters were selected in the following mode: vibrations amplitude of 1.5 mm; operating voltage of 20 V; electric current of 150-180 A; loading speed 1.26 m/min; electrode wire feed rate 2.3 m/min; welding step 3 mm/min^{-1}; 15 mm electrode wire outlet; shielded gas consumption 15 l/min.

The study of the process of vibrating arc welding with increased wire electrode vibrating frequency is accompanied by real-time recording of the process parameters - frequency of the electric cycles, the electric voltage and the magnitude of the current. Appropriate shunts were included for measuring and recording the amount of current in the vibration circuit of the vibrating apparatus. The dynamics of changing these parameters were recorded using NATIONAL INSTUMENTS NI USB 6210 analog-to-digital converter. The process waveforms were recorded using the Lab View software. For each change in the frequency of vibration of the electrode wire, 3 entries were made, determining the average values of the output parameters. The recorded data was processed using the Microsoft Office Excel software product. The obtained data for the output parameters of the oscilloscope of the vibrating arc process and micrometry of welded specimens were processed by known statistical methods.

Based on the statistically processed data from the oscillograms of deposited layers, graphical dependences were obtained to evaluate the influence of the wire electrode vibrating frequency onto duty cycles frequency and parameters of the welded coating formation (Fig.2 and Fig.3).

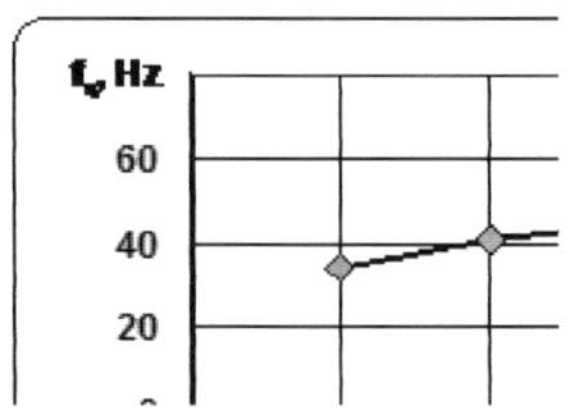

After processing the results of recorded waveforms, it was found that when frequency of vibrations increases, the frequency of the electric cycles also increases as the functional dependence is shown in Fig. 2. The influence caused by wire electrode vibrating frequency has an extreme character reaching a maximum cycle frequency of 76.3 Hz at vibration frequency of 125Hz. As the vibration frequency increases further, a sharp decrease in the cycle frequency is observed. Achieving maximum value is a prerequisite for improving the vibration process in terms of finer drop transfer of the electrode metal, more uniform formation of the individual seams, which in turn is a prerequisite for obtaining a fine-grained structure and higher hardness of the deposited metal [2, 3].

Increasing the frequency of arc cycles $f_{ц}$ influences over their duration, i.e. results in the shortening of time vector parameters - short circuit and arc burning time lengths. Figure 3 shows a graph of the variation of these parameters depending on wire electrode vibration frequency. The graph shows an increase duty cycles number due to gradual decrease in both - short circuit and arc burning time lengths. The short-circuit time affects process stability and the degree of atomization of molten metal. Therefore, the shorter the short-circuit period is, the more stable the electric arc is and the lower the electrode metal losses are. The arc burning time characterizes the enhancement of droplets from the moment of contact of the electrode wire to the metal surface until detachment from wire electrode tip. Therefore, the shorter arc burning time for each cycle is, the smaller the droplet size is, which improves the fusion conditions between the parent and feed metal and the formation of uniformly deposited layers with a fine-grained structure.

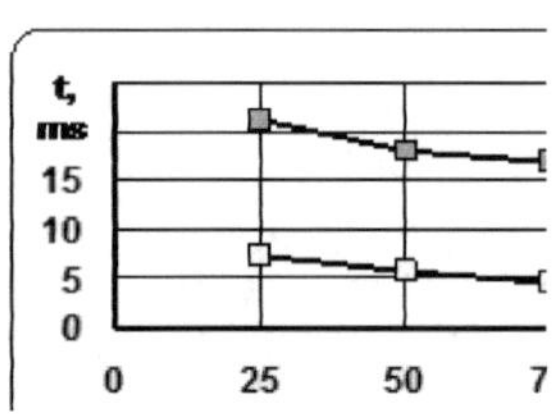

After processing the results, it was found that the alteration of the short circuit and arc burning have an extreme character. Minimum values for both times are obtained at rate of 125 Hz wire vibration frequency. This frequency correlates with Fig.2, which produces the highest frequency of arc cycles. The arc burning time at 125 Hz vibration frequency is reduced almost twice compared to the 25 Hz vibration frequency having values of 9.8 ms and 18 ms respectively. Consequently, the size of metal droplets and level of burning of the alloying elements are reduced in the same order. Spreading the electrode metal during the process of overlaying in a shielding gas environment is essential for forming a high-quality deposited layer with the required purity. There are various methods for reducing the degree of scattering and burning of the electrode metal, one of which is based on the positive influence of wire electrode frequency and its results are presented in Table. 1.

Table.1. Scatter losses of molten metal during vibrating arc overlaying process in gas shielding

№ of specimen	f, Hz	Mass before deposition, g	Mass after deposition, g	Difference ΔG, g	Spatter losses ψ, %
1	25	1700	1729	29	47
2	50	1694	1737	43	21
3	75	1690	1735	45	18
4	100	1712	1761	49	10
5	125	1782	1832	50	8
6	150	1725	1773	48	11,7

The results from Table 1 show that minimal possible scatter losses as well as burning of molten metal appear at wire electrode vibrating frequency of 125 Hz and with the value of 8%. The highest percent losses at rate of 25 Hz can be explained with higher rates of short-circuit and arc burning time lengths mentioned above. With the higher rates, the droplets size increases too and the stability of the process worsens. Another reason is the intensive dissolution of oxygen from shielding mixture which leads to carbon burning and forming of oxides with higher hardness. During separation of gas fumes through the droplet of molten metal, reactive forces are formed which prevent it from separating from the tip and increase its mobility, especially in the case of increased carbon content. This results in an increased velocity of the metal droplets at the moment of their separation from the tip of the electrode wire and their movement in different directions.

Several options were laid down to restore the worn work surfaces of the roller and floating bolt. The outer surface of the roller was reconditioned by vibration arc overlaying with usage of electrode wire 08Mn2Si with a diameter of 1.6 mm, vibration frequency of the electrode wire 125 Hz, while the loaded metal layer was obtained in a protective environment of 60% Ar and 40% CO_2 [2] in order to reduce carbon burning and obtain better formation. The working regime includes voltage of 18-20 V, electric current of 150-180 A, welding speed at rate of 0,4 m/min, electrode wire feed rate 2,3 m/min, welding step 3,5 mm/min^{-1}, electrode wire outlet 15 mm and gas flow 15 l/min.

Two technological variants were used to restore the inner surface of the roller. The first one consists of vibrating arc overlaying, but the second one - by replacing bronze bushing with sleeve made of cast iron 25. The first method was performed within following working conditions: wire electrode 08Mn2Si with a diameter of 1.2 mm in a protective environment of 60% Ar and 40% CO_2; operating voltage 18-20 V, electric current 100-120 A, overlaying speed at rate of 0.6 m/min, wire feed rate 2.3 m/min, welding step 2.5 mm/min^{-1}, electrode wire outlet 15 mm and gas consumption 15 l/min. Two types of electrode wire 08Mn2Si and DUR 350 with a diameter of 1.2 mm and a protective mixture of 60% Ar and 40% CO2 were used for the restoration of the floating bolt with the following parameters: operating voltage 18-20 V, electric current size 100-120 A, welding speed 1 m/min, wire feed rate 2.3 m/min, welding step 2.5 mm/min^{-1}, electrode wire outlet 15 mm and gas flow 15 l/min. The hardened parts were hardened in HDTV to a hardness of 30-35 HRC.

After six months of operation, corresponding to one inter-service period, measurements of reconditioned details were made to determine their wear obtained during exploitation and the results of the measurements are presented in Table. 2. It has been found that the amount of wear of reconditioned rollers and floating bolts is on average five times less than that of new parts over the same service life.

Table 2. Wear index rate of roller and floating bolt, reconditioned by vibrating arc overlaying process within gas shielding mm

Roller			Floating bolt	
Outer surface	Inner surface			
08Mn2Si	08Mn2Si	Iron bush	08Mn2Si	DUR 350
1,2	1,6	2,6	0,32	0,27

The wear of the inner bore of the roller with a crimped cast iron sleeve is of greater value than the welded wire Sv 08G2C due to the higher rigidity of the latter. The recovered floating bolts have less wear on the surface of those bolts overlaid with DUR 350 at the same initial hardness.

3. Conclusions

1. The durability of rollers and floating bolts, reconditioned by vibrating arc overlaying process is five times higher than the newly produced ones.

2. The wear index rate of floating bolt reconditioned with wire electrode DUR 350 is less than one reconditioned with 08Mn2Si within same rate of surface hardness.

3. The reconditioned with 08Mn2Si inner surface of the roller possesses higher wear-resistance compared to roller reconditioned by tapping an iron bushing.

4. References

[1] Dimitrov M Nikolov M Gospodinova N and Pavlov D 2014 Running-in of Repaired Engines Using Friction Modifiers *Industrial Lubrication and Tribology* **No 1 Vol. 66** pp 62-65 ISSN 0036-8792

[2] Nikolov M 2015 Research on the Impact of Amplitude of Vibrations on Electrical Parameters of Vibroarc Weld Overlay in Argon *Acta Technologica Agriculturae* **No 18(2)** pp 46-48 ISSN 1335-2555

[3] Nikolov M 2019 Reconditioning Vibroarc Deposited Layers in Gas Mixture Shielding Atmosphere *Ruse: Publish house to University of Ruse* p. 144 ISBN 978 954 712 756 2

[4] Nikolov M and Kangalov P 2012 Benefits from Maintenance and Repair in Utilization of Resources *Int. Sc. Conf.* **No 1** pp 24-27 ISSN 978-80-7375-625-3

[5] Nikolov M and Kangalov P 2014 Research Methods for Tribological Properties of Restorative and Preventive Coatings in Different Lubricating Media at Sliding Friction *Acta technologica agriculturae* **No 3** pp 70-74 ISSN 1335-2555

[6] Nikolov M and Kangalov P 2018 Full Production Maintenance as Tool for Quality of Repaired Machines *Agricultural Machinery* **No 1** pp 40-42 ISSN 2535-0269

[7] Nikolov M Gospodinova N Pavlov D and Dimitrov M 2009 Tribological Characteristics at Running-in of Repaired Tractor Engines Using Friction Modifiers *Industrial Lubrication and Tribology* **No 4 Vol.61** pp 209-212 ISSN 0036-8792

[8] Nikolov M Kangalov P Kerekov S and Gospodinova N 2016 Investigation of the Tribological Characteristics of Complex Ester of Adipic Acid as a Friction Modifier for Engine Oils on SMC-2 Tribotester *Journal of the Balkan Tribological Association* **No 22 3A-II** pp 3412-3419 ISSN 1310-4772

[9] Todorov I 2019 A Research about Wear Process of Details from Belt Conveyor *Agricultural, forest and transport machinery and technologies* **Vol. VI – Issue 1** pp 5-10 ISSN 2367-5888

IOP Conf. Series: Materials Science and Engineering **977** (2020) 012014 doi:10.1088/1757-899X/977/1/012014

A study of some cars dynamic parameters in urban traffic flow

D Lyubenov and G Kadikyanov

University of Ruse, 8 Studentska Str., 7017 Ruse, Bulgaria

E-mail: dlyubenov@uni-ruse.bg

Abstract. In this paper different methods for determining some dynamic parameters like longitudinal accelerations of cars are presented. The studies were conducted at different road intersections in Ruse, Bulgaria. Data of accelerating cars in real traffic flow with non-contact equipment, VBOX GPS Data Logger measurement system has been acquired. This study can be used by the experts in forensic accident investigation of vehicles. The data presented in this article can improve the calculations accuracy and the quality of expertise during the accident investigation.

1. Introduction

The significant increase in cars over the past years has led to a positive impact on the economy and people's comfort. But at the same time to many problems (pollutions [1], noise [2] etc.), including heavy road accidents. The number of deaths in road accidents in Bulgaria is about three times bigger than some countries in the European Union [3]. In order to clarify the reasons for the occurrence, and the mechanism of the occurrence of these heavy accidents, the police and the court appoint technical expertise investigation to third party independent professionals. Those experts need broad spectrum of knowledge within the transport engineering, mechanics, vehicle dynamics, etc. The main task of the expertise is to determine the pre-collision velocity. And for road accidents occurring at intersections, the velocity can be determined if the vehicle acceleration and the distance traveled to the point of impact are known.

There are various models, methods and studies of vehicles acceleration and traffic flow in safety area (also vehicle fuel consumption and emissions etc.) [4, 5, 6, 7, 8]. However, some vehicle accelerations studies are for specific locations, cities and countries, or specific roads and driving situations [9, 10, 11, 12, 13, 14]. Unfortunately, there are no suitable studies for Bulgaria, tailored to the needs of the expert practice.

The aim of this study is to compare different methods for determining the vehicles accelerations and choose relevant method for accelerations in urban traffic flow.

2. Mathematical model

The vehicle distance in motion at constant acceleration without initial speed can be calculated

$$D = \frac{at^2}{2}, m \tag{1}$$

where a is the vehicle acceleration, ms^{-2}; t – the acceleration time, s.

Using (1) it can be determine the acceleration in such a movement by

$$a = \frac{2D}{t^2}, ms^{-2}. \quad (2)$$

Some of the Bulgarian experts in vehicle accident investigation use dependence (2) to determine the acceleration of the vehicle and then the pre-collision velocity.

The pre-collision velocity can be calculated by

$$V = (2Da)^{1/2}, ms^{-1}. \quad (3)$$

3. Experimental study

In this paper three different methods to determine the vehicle acceleration are used. The objects of the study are randomly selected cars.

3.1. First method

In this method non-contact equipment, VBOX GPS Data Logger measurement system (Racelogic, Ltd) was used [15, 16]. GPS has now become a popular way of measuring due to the very high accuracy. VBOX is a powerful instrument used for non-contact measuring of a moving vehicle and can measure speed, distance, acceleration, lap times, position and much more. The logged data is stored directly onto a compact flash card for easy transfer to a PC.

The test vehicle, equipped with the VBOX Data Logger (Figure 1, position 1), stops in parallel to a vehicle, that stops in front of the intersections stop line (Figure 1, position 2). The test vehicle starts and moves in parallel with the other vehicle (Figure 1, position 3). The VBOX system has a video camera that monitors whether the vehicles' movement is parallel. The acceleration is determined only for vehicles that have moved in parallel for the distance of the study.

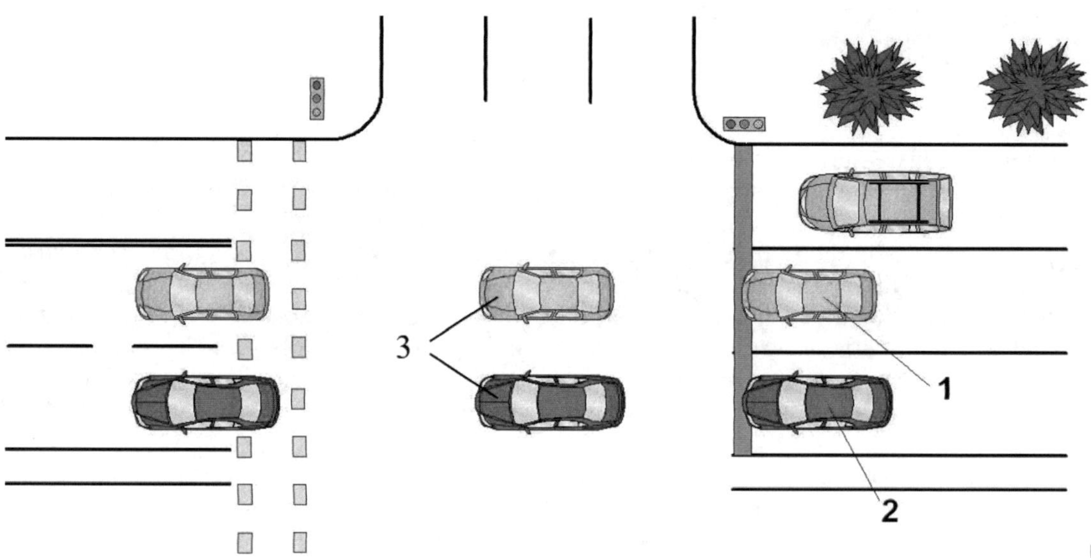

Figure 1. The testing scheme for the first method.

In this way a study of 200 vehicles was carried out. The study was conducted in April in clear and dry weather at intersections of different streets in Ruse. The results of the study were processed with VBOX Tools software [17]. Figure 2 shows the main window of the software and the specified acceleration for a distance of 30 meters, for one of the tests.

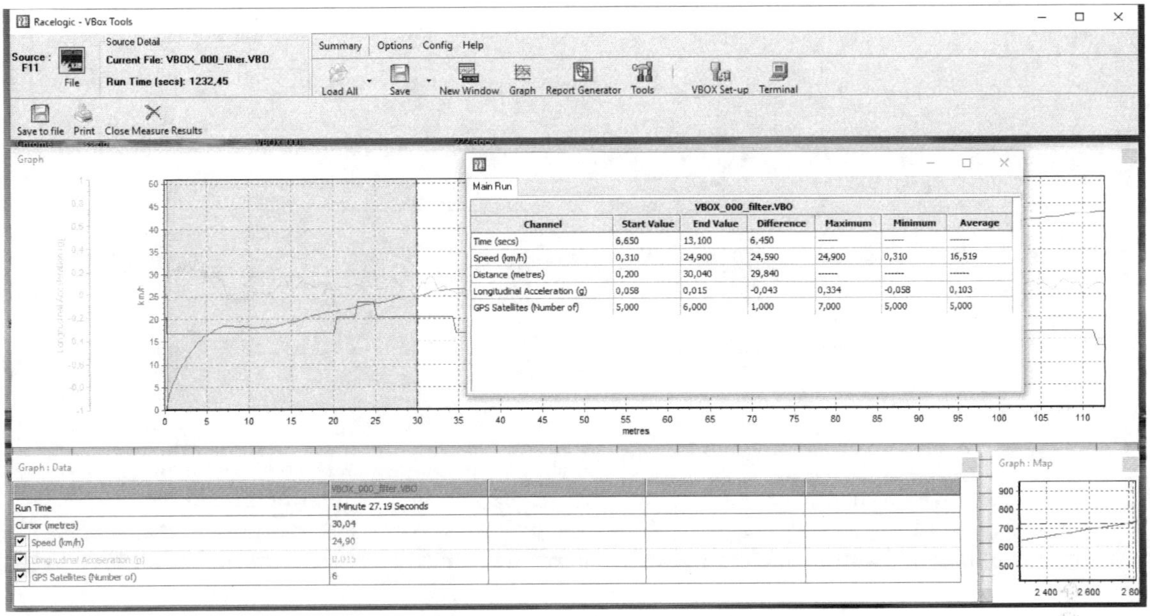

Figure 2. The main window of VBOX Tools software.

The average values of acceleration were determined for the travelled distance (from 10 to 70 meters for each test) with this software.

Figure 3 shows the results for the average values for the acceleration in real traffic flow depending on the travelled distance.

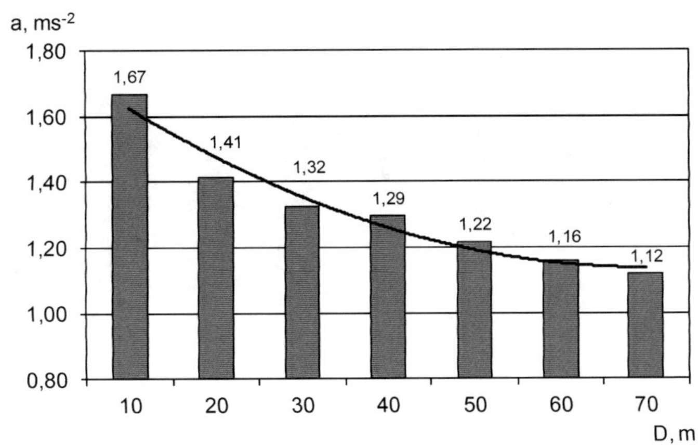

Figure 3. The acceleration versus the travelled distance – average values.

The variation of acceleration according to travelled distance (Figure 3) is

$$a = 0{,}0134D^2 - 0{,}1884D + 1{,}798, ms^{-2} \quad (4)$$

where a is the vehicle acceleration, ms^{-2}; D – the travelled distance, m.

The determination coefficient of equation $R^2=0{,}96$, which indicates that a variance in the acceleration is insignificant. This satisfies the accuracy required by the expert practice.

The Figure 4 shows minimum values of acceleration in real transport traffic flow depending on the travelled distance.

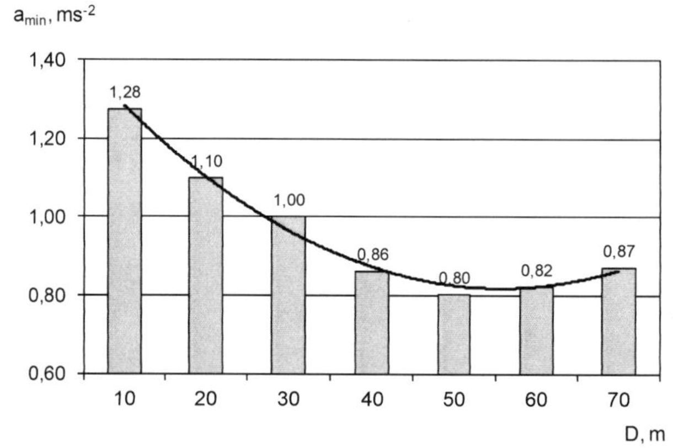

Figure 4. The acceleration versus the travelled distance – minimum values.

For the minimum values, the variation equation describing the acceleration according to the distance is

$$a = 0{,}0223D^2 - 0{,}2482D + 1{,}5093, ms^{-2} \tag{5}$$

The determination coefficient of equation $R^2=0{,}99$, which indicates that 99% of the acceleration variance in the response variable can be explained by the explanatory variables.

The Figure 5 shows the results for maximum values of acceleration, depending on the distance.

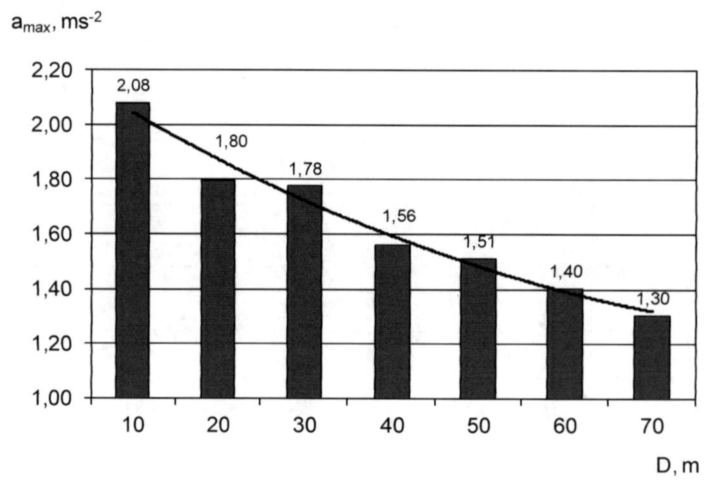

Figure 5 The acceleration versus the travelled distance – maximum values.

In these results (Figure 5) the equation describing the variation of the acceleration according to the distance is

$$a = 0{,}0098D^2 - 0{,}199D + 2{,}2325, ms^{-2} \tag{6}$$

The determination coefficient of equation $R^2=0{,}97$.

The extent R of the statistical scattering is determined by

$$R = a_{max} - a_{min} \tag{7}$$

The Figure 6 presents the extent R of the statistical scattering depending on the distance.

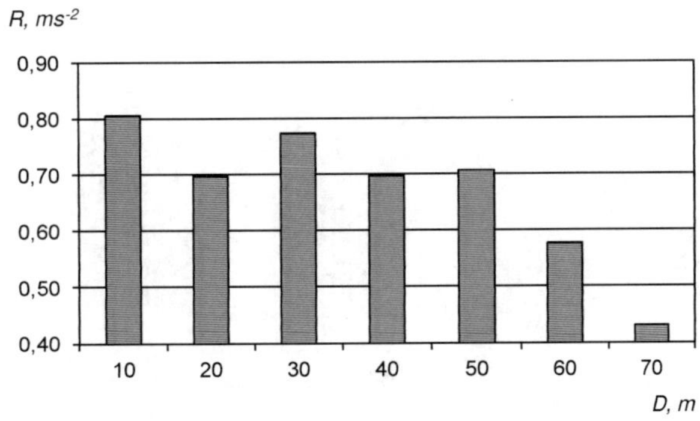

Figure 6. The extent R of the statistical scattering depending on the distance D.

The Figure 6 shows a tendency to decrease R by increasing the distance. At a distance of 70 meters, scattering decreases by 46% compared to a distance of 10 meters. As a result, in the investigation, it is necessary to carefully analyze the evidence on how to accelerate the vehicles involved in the accident.

3.2. Second method

Where there is no possibility to apply expert methods to determine the speed, some Bulgarian experts use the following method. Time measurement carried out with a stopwatch for which the vehicle travels a certain distance. Once the time and distance is known by (1), (2) and (3), the speed of the vehicle can be determined.

Sometimes the same acceleration values are used for different traffic situations, without taking into account the distance traveled by the vehicle. Of the results obtained so far shows that this type of modeling is inaccurate, and leads to poor quality of the expertise.

Further, this method raises certain questions about its accuracy and applicability. In order to answer these questions, a study was conducted to determine the vehicle acceleration. The study was conducted at 20 intersections on different streets in Ruse. And the data was collected from 100 vehicles.

The travel time was measured with a stopwatch at the 20th meter, 30th meter and 40th meter from the stop line. The speed of the vehicles at the beginning of the measurement is zero, i.e. the vehicles have stopped and the drivers are waiting for green light at the traffic signals (Figure 7, position 1).

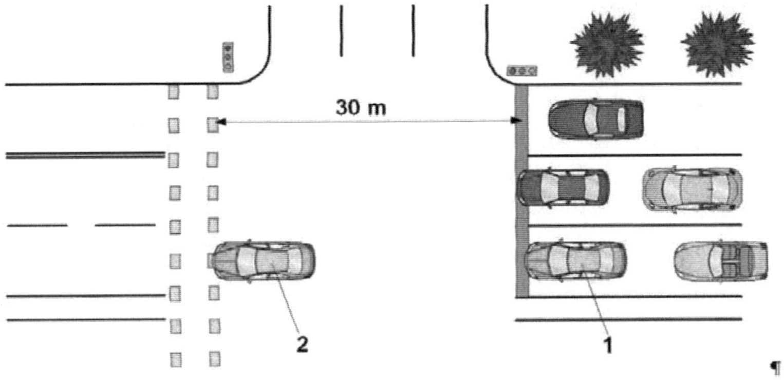

Figure 7. The testing scheme of second method for distance of 30 meters.

The travel time results are obtained at each station (20 m, 30 m and 40 m) and filed in the work sheets (Figure 7, position 2).

Once the time is determined using (2), the vehicle acceleration is calculated. The results of the calculations are presented in Figure 8.

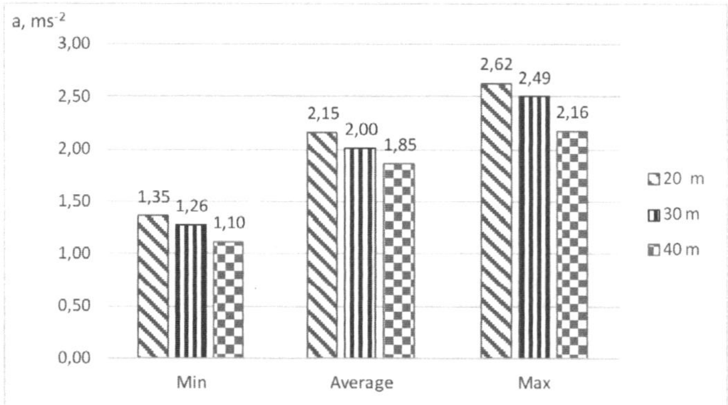

Figure 8. Results of the second method.

The minimum, average and maximum values of the study are presented. The results are for acceleration distances of 20, 30 and 40 meters (Figure 8).

3.3. Third method

The experiment was implemented at a vehicle training ground, i.e. the studies were not conducted within real traffic flow. For the purpose of the study different vehicles ware used, with drivers from 20 to 50 years old. Each driver accelerates with the vehicle in three ways – normal, fast and very fast (with acceleration marks on the road). The VBOX Data Logger registration system [15, 16, 17] was used to determine the accelerations. The results of the study were processed with VBOX Tools software.

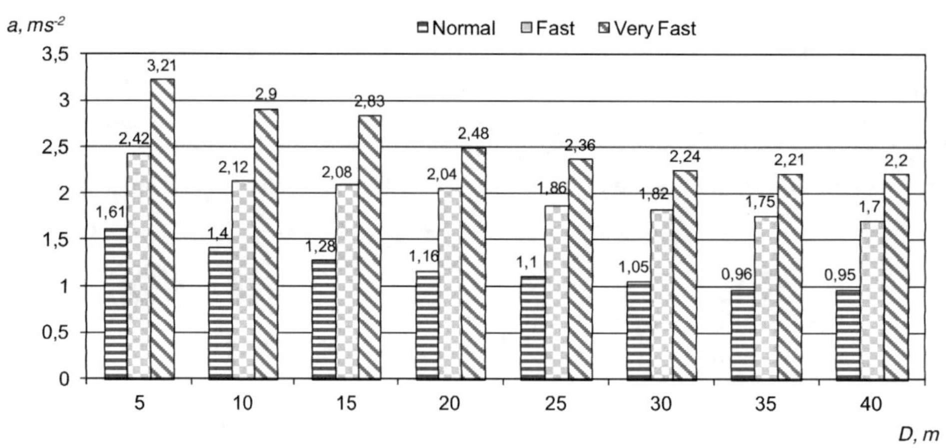

Figure 9. Results of the third method.

From results on Figure 9 it can be concluded that the normal acceleration judged by the drivers differs from the average acceleration values in real traffic flow (Figure 3).

This is very important and these differences need to be carefully analyzed and taken into account by the experts in the investigation of accidents.

4. Conclusions

The data from all the studies presented in this article can be used in expert practice in the investigation of vehicle accident. Using the data can improve the accuracy of the calculations and the quality of the expertise in the investigation of accidents.

For first method longitudinal accelerations are: minimum values from 1,28 to 0,87 ms^{-2}; maximum values from 2,08 to 1,30 ms^{-2}; average values from 1,67 to 1,12 ms^{-2} obtained from predetermined distances from 10 m to 70 m.

Experts using second method must be careful because the mathematical model is for constant acceleration motion. In fact, vehicle does not pass the distances at constant acceleration. This was confirmed by the first method.

As for the third method in real traffic flow the normal acceleration is higher than the normal acceleration on a vehicle training ground. For example in 30 meters distance the average acceleration in real transport flow is 25,7% higher than normal acceleration measured on the vehicle training ground.

References

[1] Hristov R and Stefanov S 2019 Study of the potential for reducing CO2 emissions from road transport in Bulgaria *IOP Conference Series: Materials Science and Engineering* vol 614.

[2] Li Q and Qiao F and Yu L 2015 Testing of in-vehicle noises when driving on freeways with different pavement types *Proceedings of the 108th Air and Waste Management Association (AWMA)* Raleigh North Carolina USA.

[3] European Commission. Transport. Statistical pocketbook 2017. Available at: https://ec.europa.eu/transport/facts-fundings/statistics/pocketbook-2017_en

[4] Goñi-Ros B and Knoop V and Takahashi T and Sakata I 2016 Optimization of traffic flow at freeway sags by controlling the acceleration of vehicles equipped with in-car systems *Transportation Research Part C* vol 71 pp 1–18

[5] Jin W and Laval J 2018 Bounded acceleration traffic flow models: A unified approach *Transportation Research Part B* vol 111 pp 1–18

[6] Leclercq L 2007 Bounded acceleration close to fixed and moving bottlenecks *Transportation Research Part B* vol 41(3) pp 309–319

[7] Wang Y and Bin Y and Keqiang L 2008 Longitudinal Acceleration Tracking Control of Low Speed Heavy-Duty Vehicles *Tsinghua Science & Technology* vol 13 pp 636–643

[8] Zhao H and Liu X and Chen X and Lu J 2018 Cellular automata model for traffic flow at intersections in internet of vehicles *Physica A: Statistical Mechanics and its Applications* vol 494 pp 40–51

[9] Bokare P and Maurya A 2017 Acceleration-Deceleration Behaviour of Various Vehicle Types *Transportation Research Procedia* vol 25 pp 4733–4749

[10] Bokare P and Maurya A 2016 Study of Acceleration Behaviour of Motorized Three Wheeler in India *Transportation Research Procedia* vol 17 pp 244–252

[11] Dey P and Biswas P 2011 Acceleration of queue leaders at signalized intersections *Indian Highways* No 3(1) pp 49–54

[12] Gattis J and Bryant M and Duncan L 2010 Truck acceleration speeds and distances at weigh stations *Journal of the Transportation Research Board* National Academies, Washington

[13] Haas R and Inman V and Dixson A and Warren D 2004 Use of intelligent transportation system data to determine driver deceleration and acceleration behavior *Transportation Research Record* pp 3–10

[14] Wang J and Dixon and Li H and Ogle J 2004 Normal acceleration behavior of passenger vehicles starting from rest at all-way stop-controlled intersections *Transportation Research Record 1883* pp 158–166

[15] Racelogic Company Profile. Available at: http://www.racelogic.co.uk/index.php/en/

[16] Vehicle testing (Video) VBOX data logger. Available at: https://vboxautomotive.co.uk/index.php/en/products/data-loggers

[17] VBOX Tools software. Available at: https://vboxautomotive.co.uk/index.php/en/products/software/vbox-tools

IOP Conf. Series: Materials Science and Engineering **977** (2020) 012015 doi:10.1088/1757-899X/977/1/012015

Sound level study of a hybrid car

G Staneva, K Dimitrova, R Ivanov and G Kadikyanov

University of Ruse, Department of Engines and Vehicles, 8 Studentska Str., 7017 Ruse, Bulgaria

E-mail: glstaneva@uni-ruse.bg

Abstract. One of the main reasons for the urban noise are the cars. When the hybrid cars are moved in low-speed mode, reduced noise is supposed. In the presented paper have been made comparative analysis according to standard methods of the external and internal sound levels of a hybrid car. The experiment at different speeds in two modes – using ICE and using electric traction were carried out. The results of the measurements are visualized in graphs and tables. They show variation of external and internal sound levels at different speeds in separate octave bands. Spectral analysis at each speed is done. The results show changing of sound level vs speed and noise predominant maximums at respective frequencies.

1. Introduction

Noise is one of the main reasons for the deterioration of people's health. It reduces labor productivity, causes the appearance and development of nervous, vascular and other diseases and affects life duration.

According to [1] noise is any unpleasant or unwanted sound. A sound field is the area in which the sound waves reproduce. At each point of the sound field, the pressure and velocity of movement of the air particles change over time. The difference between the instantaneous full pressure value and the average pressure observed in an unburnt environment is called the sound pressure p. The sound pressure measuring unit is pascal [Pa]. The intensity of the sound L is measured with the Bell unit [B], which in the practice is increased 10 times – decibel [dB].

In today's increased use of transport equipment, the problem of combating noise is of great importance. Increasing urban traffic intensity leads to a significant increase in the noise level (sound or sound pressure) in the larger cities.

Vehicle noise as a negative impact on people can be seen from two sides: noise in the car where the driver's and passenger's seats are situated, and noise from the vehicle - the so-called external noise, affecting all residents of the modern city.

In order to limit the continuous increase of the noise level and to preserve the human health in recent years in many countries, norms have been developed and introduced for the permissible noise levels for working, living and relaxation [2, 3]. The work against noise has to be conducted in several directions, but one of the most effective is to reduce its level in the sources that cause it.

Noise sources in the vehicle are separate units, mechanisms and aggregates. For example, according to [4, 5] the internal combustion engine (ICE) is one of main noise and vibration generator. Other types of noise are: the noise caused by the interaction of the vehicle with the air - aerodynamic noise and the noise caused by the interaction of the tires with the road surface. The noise level of the individual vehicle is the sum of the noise energies generated by the separate sources [2, 6, 7].

Usually in hybrid cars, there are two main modes of operation: operating with internal combustion engine (ICE) and with an electric motor. When the hybrid car is running on an electric motor mode, one of the main sources of low-speed noise is reduced. This is an indisputable advantage in urban traffic, but a potential problem under the same conditions. The lack of typical ICE noise makes the vehicle discoverable at shorter distances, and this may be a problem for the elderly people, young children, people with disabilities, and people who are not used to the urban movement.

The noise also depends on the tire profile and the movement of the cars on different road surfaces. According to [8] the reasons for the occurrence of noise are:

- flickering of the tire's pages due to periodically occurring deformations when rolling;
- periodic displacement and suction of air in the profile channels upon movement (air-pumping);
- the greatest share of the sound energy of the rolling radius is the impact on the profile teeth (blocks) when they enter into contact with the road.

The purpose of this study is to make a comparative analysis of the measured maximum intensity of noise in ICE mode and electric motor mode of a hybrid car at different motion speeds. There have been made researches of the external and internal levels of the sound.

2. Method and equipment

The study was conducted with a Toyota Yaris HYBRID (Figure 1) with technical characteristics as shown on Table 1 [9].

The vehicle tires are Dunlop Sport Fastresponse 175/65 R15 84H. The air pressure in the tires is 0,25 MPa. Class of tires C1 corresponds ‘passenger car tires’ in [10].

Table 1. Main technical characteristics of tested vehicle.

Main technical characteristics	Toyota Yaris (P3) Hybrid
Modification	1,5 HSD Hybrid 100 Hp
Max Power / engine speed	55 kW / 4800 min^{-1}
Max torque / engine speed	111 Nm / 4400 min^{-1}
Max power of electric system	45 kW
Max torque of electric system	207 Nm
Capacity of the battery	0,94 kWh
EVRO standard	EURO V
Tires	175/65 R15
Maximal speed	165 km/h
Time for acceleration from 0 to 100 km/h	11,8 s
Empty weight	1 120 kg

Figure 1. Studied vehicle Toyota Yaris HYBRID.

Figure 2. Experimental hard-grained asphalt surface.

IOP Conf. Series: Materials Science and Engineering **977** (2020) 012015 doi:10.1088/1757-899X/977/1/012015

2.1. Measurement of the external sound level

Measurement of the external sound shall be carried out in a free sound field on hard-grained road surface with cracks (Figure 2). The site of testing corresponds to the requirements as described in [11, 12]. The location of the microphones during the tests is shown on Figure 3 as they are located of height 1,2 m over the road.

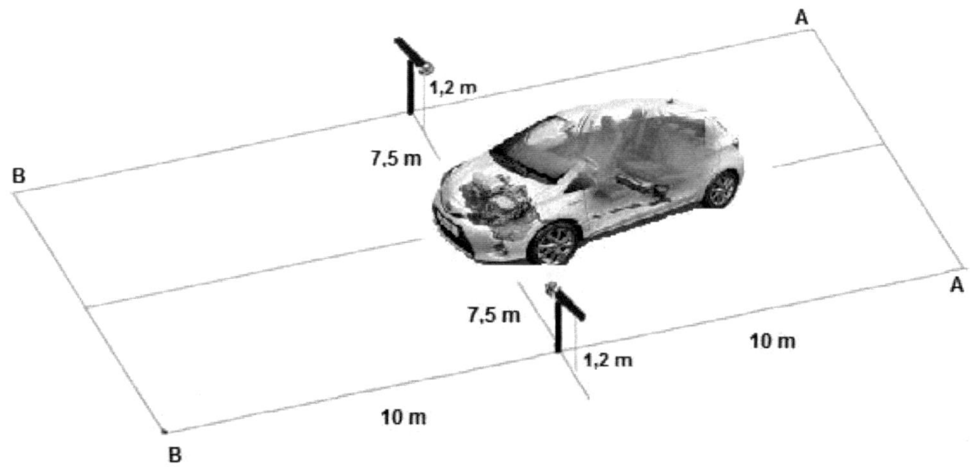

Figure 3. A scheme of the external noise test with distance of the microphone from the middle line of vehicle passing 7,5 m.

The measurements are carried out in the following order:

1. In electric motor mode – the car moves at a steady speed of 20, 30, 40 and 50 km/h in each test. At a speed over 50 km/h, the electric motor mode is impossible.
2. In ICE mode, the car moves steadily at 60 km/h to 120 km/h in each test.

In the tested area shown at Figure 3 a recording is made with the device (Figure 5).

2.2. Measurement of the internal sound level

While measuring the internal sound level, the vehicle is moving on the road as described in the methods for measurement of the external sound level. At a distance less than 25 m near the test site there are no barriers or buildings. At the time of the tests all the doors, the windows and the hatches are closed. The microphone is placed near the ear of the driver (at 700 mm over the seat and at 200 mm aside the longitudinal vertical plain of the seat) oriented forward direction of the moving car (Figure 4) [13].

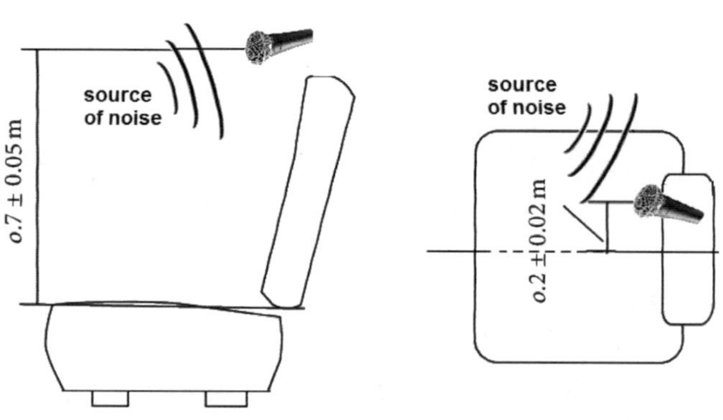

Figure 4. A scheme of the location of the microphone during the measurement of the parameters of the internal sound level.

Experimental studies were performed with a VI-410 Quest Technologies [14]. It is a digital, four-channel device for measurement and analyses of vibrations with channels 1, 2 and 3 using channel 4. VI-410 has internal memory 32 MB for memorizing of the measurement parameters for all the channels for a long time period. The results of the measurements are visualized and processed with the help of software Quest Suite Professional II.

Figure 5. General view of the VI-410 Quest Technologies device.

The device is equipped with sound calibrator having the possibility to emit waves with sound level 94 or 114 dB at frequency 250 or 1000 Hz (Figure 5). Before starting to work with the device VI-410 and after finishing the work, it has to be calibrated (Figure 6) [14].

Figure 6. Sound calibrator QC-10.

Human ear can perceive sounds of different frequencies and levels as equally strong. Differences in the perception of sound depending on the frequency and the level make it necessary to use frequency corrected sound pressure levels. The adjustment is a setting of sound pressure levels and applies to each frequency range.

The sound intensity of the correction curve *A*, expressed in dBA, best suits the subjective perception of sounds at low sound pressure levels [6, 15].

Current sound level measurements are performed using correction filter *A*.

3. Investigation and results

At the end of each measurement, the data is stored in the memory of the VI-410 measuring device.

With its help, the results can be transferred into tables and graphics.

3.1. Results from the external noise level

For each of the tests, the maximum sound level values that correspond to the level of the vehicle passing through the microphone line at a distance of 7,5 m from its central axis (Figure 3) are taken into consideration.

Figure 7 shows, as an example, the change of the sound level in octave bands as well as in the equivalent summary value (1^{-st} left column).

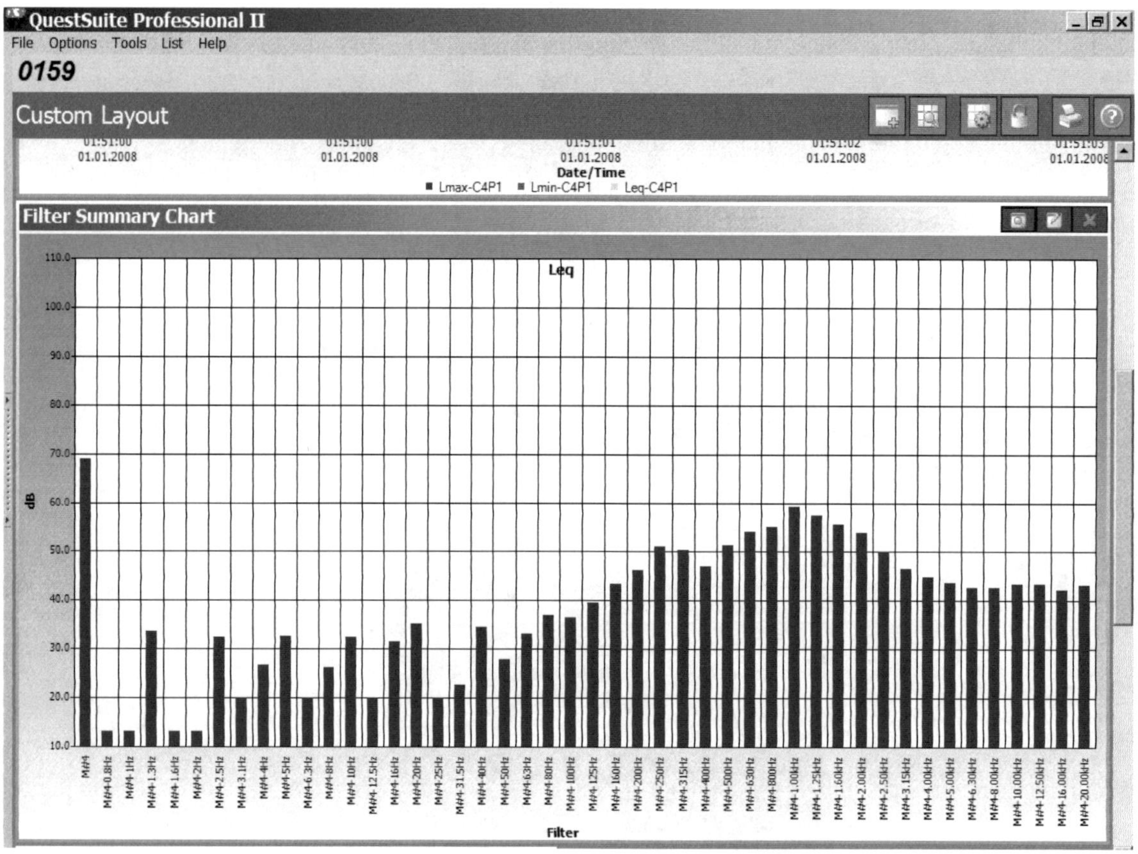

Figure 7. External sound level in third octave bands at speed of 40 km/h.

Figure 8 shows the change of the external sound level (dBA) at different velocity, for the dominant octave band. It is clearly seen that the increase of the car speed leads to increase of the noise level. Slight dispersion of the parameters at 50–60 km/h is seen due to the change of the operational modes of the car – from movement with electric motor mode to ICE mode.

Table 2 shows the summarized data for the change of the external sound level depending on the speed by octave bands at electric motor mode and ICE mode.

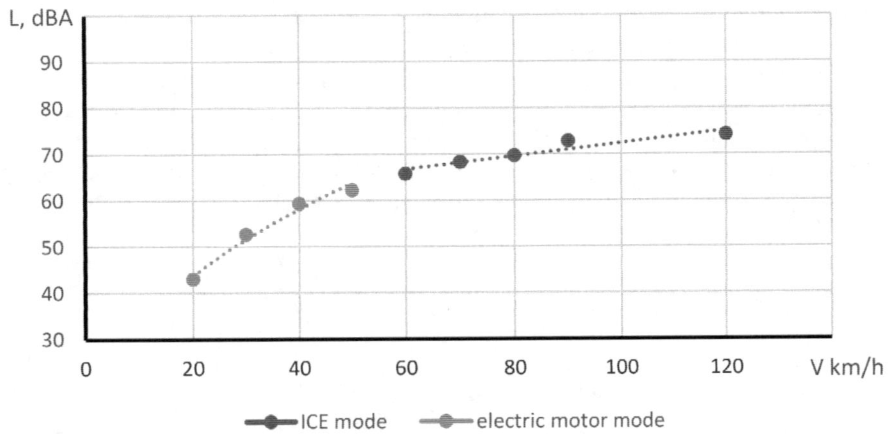

Figure 8. The change in the level of external sound L at 1000 Hz depending on speed V.

Table 2. The measured maximum values of the external sound, dBA.

f, Hz / V, km/h	31,5	63	125	250	500	1000	2000	4000	8000	16000
20	27,5	30,2	33,8	42,3	41,1	43	45,7	39	41,4	42,2
30	27,5	31,7	37,4	47,6	43,6	52,7	47,5	41,4	42,8	43,1
40	23,8	33,4	39,6	51,7	52,4	59,3	54,3	45,5	43,2	42,4
50	27,5	33,7	40	55,1	52,4	62,2	57,5	48,6	42,8	42,1
60	33	33	42,2	56,4	52	65,7	60,5	51,2	46,1	42,5
70	27,4	37,6	49,6	57,4	56,6	68,2	62,1	52,9	47,3	42,3
80	27,5	37,3	48,4	58,7	56,7	69,6	65,4	55,9	48,2	43,7
90	21,1	41,8	48,6	60	62,6	72,7	67,3	57,1	51	44,2
120	21,1	42,7	52,3	62,4	61,1	74,1	70,7	62,2	55,5	45,2

At low frequency up to 125 Hz and low velocity up to 60 km/h, the level of the sound stays close to the background noise, at higher velocity it slightly increases. At frequency 250 Hz and 1000 Hz there are local extremes of the sound level, which increase their values together with the increase of the speed.

Table 2 shows that at ICE mode at speeds higher than 60 km/h and 250 Hz, the change of the sound level is negligible. The main noise source at that working mode is the ICE. Due to the automatic gear box the RPM of engine is changed insignificant.

At frequency 1000Hz more extreme increase of the sound level is observed, which is probably due to the tires, which are the main sources of external noise of the high-speed moving cars.

IOP Conf. Series: Materials Science and Engineering **977** (2020) 012015 doi:10.1088/1757-899X/977/1/012015

3.2. Results of the internal sound level

For the tests of the internal sound level it has been used the same measurement device VI-410. The used methods are described here above. The results are processed by the same specialized software product (Quest Suite Professional II) and are presented and visualized in a way similar to the results for the external sound level.

Figure 9 shows the change of the internal sound level in (dBA) depending on speed, for the dominant octave band. It is seen that speed increase leads to increase of the level of the noise and no substantial deviations are shown upon changing of the working modes from electric motor to ICE.

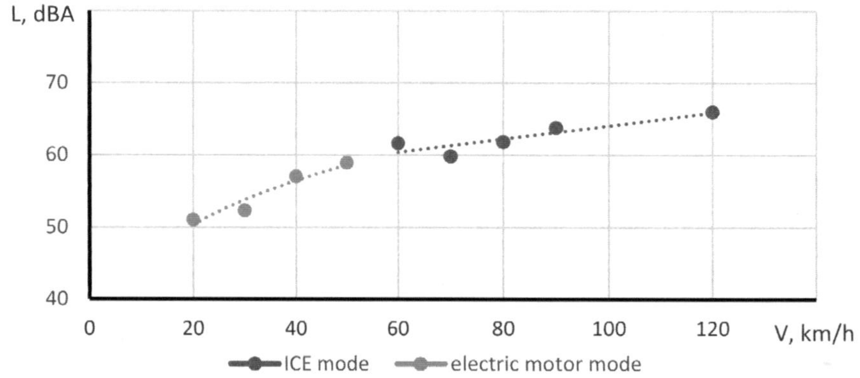

Figure 9. The change of the internal sound level at 125 Hz depending on the different speeds.

Table 3 shows summarized data for the change of the internal sound level depending on the speed in octave bands. It is seen that the level of the sound has a maximum at 125 Hz frequency. This is probably a result from the work of the powertrain.

Table 3. The measured maximum values of the internal sound, dBA.

f, Hz / *V*, km/h	31,5	63	125	250	500	1000	2000	4000	8000	16000
20	39,7	46,6	51	41,7	40,9	37,8	35,6	36,8	42,7	42,2
30	44,9	48,5	52,3	46,9	43,3	39,8	36,5	36,6	40,7	42,3
40	46,6	51,4	57	52,9	49,9	44,6	39,6	38,7	42,5	42,2
50	48,1	52,2	58,9	54,3	51,9	47,1	41,9	37,2	42,6	42,2
60	48,7	52,3	61,6	56,1	54,9	50,1	45,3	38,5	41,8	42,2
70	48,6	51,9	59,8	56,8	56,7	52,8	47,5	39,2	42,4	42,2
80	52,6	51,7	61,8	59	58,1	54,4	49,3	40	42,4	42,3
90	47,6	53,8	63,8	60,4	59,7	61,1	52,1	43,7	42,8	42,3
120	50,6	57,5	66	64,7	64,8	61,1	56,6	49,4	47	42,8

From comparing the external sound level (Table 2) and the internal sound level (Table 3), it is seen that on Table 3 is missing an extreme at frequency higher than 1000 Hz. The reason could be the isolation of the sound from the tires in the hybrid car coupe.

4. Conclusions

A study, concerning sound level of external and internal noise of a hybrid car Toyota Yaris was carried out. The experiments were realized at different speeds in two modes – using ICE and using electric traction. The speed varies from 40 to 120 km/h. The obtained results give possibility to make some conclusions and generalizations.

The general internal and external sound levels increase with increasing of the car speed, following the relation, near to exponent.

There is not clear jump in the curves of internal and external sound levels when the car passes from electric to ICE traction. It is due to the good exhaust system of the car. Significant change in internal noise when changing the mode of power source is not observed due to the good noise isolation of the passenger compartment from the surrounding area. The smooth increase of internal noise can be explained by the dominating acoustic emissions from the vehicle construction vibrations.

When measuring the external noise of the vehicle there are observed two maximum values of external sound levels at 250 Hz and 1 kHz for each speed.

Due to the transforming properties of the construction, the internal noise spectrum has predominant maximums at lower frequencies (125 Hz and 500 Hz).

References

[1] *ISO1999:2014*. Acoustics - Estimation of noise-induced hearing loss

[2] Francis C and Ortega C and Cruz A 2009 Noise pollution changes avian communities and species interactions *Current Biology* vol **19(16)** pp 1415-1419

[3] Golmohammadi R and Abbaspour M and Nassiri P and Mahjub H 2007 Road Traffic Noise Model *Journal of Research and Health Science* vol **7(1)** pp 13-17

[4] Pavlov N 2018 Numerical Simulation on the Vibration of a Vehicle Drivetrain with Dual Mass Flywheel *International Scientific Journal Machines, Technologies, Materials Sofia: Scientific Technical Union of Mechanical Engineering Industry 4.0* Iss 2 pp 49-52

[5] Mihaylov V 2014 Vibroacoustic field of direct injected diesel engine *EKO Varna* **21** pp 423-430

[6] Juis C and Putra A 2011 On a simple technique to measure the airborne noise in a motor vehicle using source substitution *Hari Penyelidikan Fakulti Kejuruteraan Mekanikal* Unviersiti Tecknikal Malaysia Melaka

[7] Li Q and Qiao F and Yu L 2015 Testing of in-vehicle noises when driving on freeways with different pavement types *Proceedings of the 108th Air and Waste Management Association (AWMA)* Raleigh North Carolina USA

[8] Subramani T and Kavitha M and Sivaraj K 2012 Modelling of Traffic Noise Pollution. *Intern. Journal of Engineering Research and Applications (IJERA)* vol **2(3)** pp 3175-3185

[9] Technical data of Toyota Yaris III 1.5 HSD (100 Hp) Hybrid. Available at: https://www.auto-data.net/en/toyota-yaris-iii-1.5-hsd-100hp-hybrid-cvt-19314

[10] *ISO 4000-1:2010*. Passenger car tyres and rims

[11] Arsic D and Becker S An integrated R&D approach for exterior noise development applying contribution analysis and response modification analysis *INTER-NOISE and NOISE-CON Congress and Conference Proceedings, InterNoise16* Hamburg Germany pp 5848-6840

[12] *Regulation (EU) No 540/2014* of the European Parliament and of the Council of 16 April 2014 on the sound level of motor vehicles and of replacement silencing systems

[13] *ISO 5128:1980*. Acoustics - Measurement of noise inside motor vehicles

[14] Quset Technologies *User manual VI – 410*

[15] Ivanov Z 2014 Digital correction filter for noise measurement *EKO Varna* **21** pp 468-475

[16] Gumiela J, Sitnik L and Sztafrowski D 2019 Measuring identification of electromagnetic field emissions in electrical cars Przeglad Elektrotechniczny vol **95 (12)** pp 128-131

[17] Marinov E and Iliev A 2003 Noise maps of urban areas *Scientific works of RU "A. Kanchev "* **40 (8)** pp 22 - 24

Research on the performance characteristics of rapeseed oil for its use as a diesel fuel additive

P Zlateva[1*], S Demirova[2]

[1*]Department of Thermal Engineering, Technical University of Varna, 1 Studentska Str., 9010 Varna, Bulgaria
[2]Department of Industrial Management, Technical University of Varna, 1 Studentska Str., 9010 Varna, Bulgaria

[*] corresponding author: pzlateva@tu-varna.bg

Abstract. Traditionally used fuels can be replaced with one of the possible alternatives such as rapeseed oil. With the development of Industry 4.0 and the high environmental requirements, research into reducing carbon dioxide emissions continues. To this end, research has been carried out on the performance characteristics of rapeseed oil, diesel and mixed fuel - 8% rapeseed oil and 92% diesel. In order to optimize the process, a study and in-depth analysis of the results of the use of diesel with the addition of biodiesel by end-customers has been carried out.

1. Introduction

The rapid pace of development in the fields of artificial intelligence, biotechnology and energy conservation rearranges production and transcends traditional boundaries, creating new opportunities for reducing carbon dioxide emissions. This contributes to the development of technology and is defined as the Fourth Industrial Revolution [1, 2]. It is seen as a model that supports business and society, drives the radical changes and also environmental protection that result from the introduction of new technologies [3, 4].

The reduction of available stocks of traditional fuels in recent decades has led to an increase in the use of renewable sources for the production of biofuels [5, 6]. Unlike other renewable energy sources, biomass can be converted directly into liquid biofuels for transport [7, 8]. The two most common types of biofuels are ethanol and biodiesel [9]. The great advantage of biodiesel over other fuels is that the raw materials it produces are renewable and inexhaustible [10]. When burned, the emissions of exhaust gases, and especially carbon dioxide, are drastically reduced [11]. Its technical structure is similar to that of diesel. It can be used as an additive to diesel to reduce vehicle emissions or in its pure form as biofuel. In recent years, the EU has set itself the goal of reducing harmful emissions from transport, leading to an increase in biofuels produced from food or feed crops [12]. Since the 1990s, the EU has started a course to reduce harmful gases, adopting a directive in 2003 promoting the use of biofuels in transport [13]. Transport emissions are expected to decrease by 20% by 2020 and between 30 and 40% by 2030. The European car fleet is mainly made up of diesel vehicles and this, coupled with the rise in prices of conventional liquid fuels, necessitates an increase of the production of biofuels from vegetable oils. Biofuels can also be mixed with petroleum products [14, 15]. For the production of vegetable oils, it is necessary to use traditional EU-specific raw materials such as rapeseed in Bulgaria. This study

IOP Conf. Series: Materials Science and Engineering **977** (2020) 012016 doi:10.1088/1757-899X/977/1/012016

identifies both the performance of rapeseed oil and consumers' views on the use of biodiesel as an additive to petroleum-based diesel in vehicles.

2. Exposition

The object of this study is rapeseed oil, which can be successfully used as biodiesel. For the purpose of the study, the performance characteristics affecting the quality of rapeseed oil were determined: density, kinematic viscosity, calorific value, flash point and cetane number. A comparative analysis was made between rapeseed oil, diesel and a mixture of diesel and rapeseed oil.

2.1. Raw material for the production of rapeseed oil in Bulgaria

The cultivation of rapeseed in Bulgaria has become increasingly urgent in recent years in connection with its use as a raw material for the production of rapeseed oil. Rapeseed is characterized by a relatively high oil content of 33 to 42%. This makes rapeseed a preferred raw material for oil production mainly for technical purposes. For maximum yield of rapeseed oil and at the same time high quality of the resulting oil, it is necessary to comply with all technological requirements regarding the collection, storage and preparation of the seeds for oil extraction in them. To extract the oil, the seed is pressed, filtered and cooled. Biodiesel is produced from organic oils by chemical reaction with alcohol (ethanol or methanol) in the presence of a catalyst. Ethyl and methyl esters are obtained, respectively. The yield of alcoholic esters (100% biodiesel) is increased when the reaction proceeds with excess alcohol and accelerates with increasing temperature. The process is called trans-esterification and pre-esterification of fats to produce biodiesel and glycerine. The rapeseed oil molecules are transformed into molecules similar to diesel hydrocarbons. Samples for testing rapeseed oil were taken from a production facility. Petroleum-based diesel oil and rapeseed oil were tested individually and then mixed at a ratio of 92% petroleum-based diesel and 8% rapeseed oil. The tests were performed in laboratory conditions and the results are presented as mean.

2.2. Testing Methods for diesel and biodiesel fuels

Diesel, rapeseed oil and blend performance tests have been performed based on European Testing Methods. Table 1 gives the values of the performance characteristics of diesel and biodiesel according to the methodologies and studies of the authors in [16, 17, 18, 19, 20].

Table 1. Testing Methods for diesel and biodiesel fuels.

	Unit	Testing Method	Diesel	Biodiesel
Density at 15°C	kg/m^3	DIN EN ISO 3675, DIN EN ISO 12185	820-845	860 - 900
Viscosity at 40°C	mm^2/s	DIN EN ISO 3104	2 – 4,5	3,5 – 5,0
Calorific Value,	MJ/kg	DIN 51900-3	37-44	33-40
Flash point	°C	DIN EN ISO 2719	min 55	min 101
Cetane number		DIN EN ISO 5165	min 51	min 51

3. Results and discussion

Based on the research performed on rapeseed oil, diesel and mixed fuel - 92% diesel and 8% rapeseed oil, the following results were obtained for determining performance.

3.1. Results and analysis of the results obtained for the performance of rapeseed oil, diesel and mixture fuel.

Table 2 shows the results obtained for the performance characteristics of rapeseed oil, diesel and mixture fuel.

Table 2. Performance of rapeseed oil, diesel and mixture.

Distance (m)	Unit	Rapeseed oil	Diesel	8% Rapeseed oil and 92% Diesel
Density at 15°C	kg/m^3	919	841	858
Viscosity at 40°C	mm^2/s	37.3	4.66	4.79
Calorific Value,	MJ/kg	37.1	42.5	41.8
Flash point	°C	270	56	65
Cetane number		42	49	48.1

The results obtained for the density of the tested fuels indicate that rapeseed oil is higher than that of diesel and a mixture of 8% rapeseed oil and 92% diesel. The viscosity of rapeseed oil is much higher than that of diesel and mixture. The calorific value of rapeseed oil is lower than that of diesel and mixture. The flash point of rapeseed oil is many times higher than that of diesel and mixture. Cetane number for diesel and blend is higher than rapeseed oil. Comparative analysis of the results showed that the resulting values for the mixture of 8% rapeseed oil and 92% diesel were closer to those of diesel than to rapeseed.

3.2. Survey of end-customer attitudes of service stations on the use of a blend of petroleum-based diesel and biodiesel

In connection with the survey, in November 2019 - January 2020, a survey was conducted with the final customers of the gas stations for their opinion on the use of a mixture of diesel oil and biodiesel.

The questionnaire contains eight questions, with semantics affecting the end-customer's opinion about the need to use biodiesel as an additive to diesel (see from figure 1 to figure 8). The synchronization of open and closed questions gives the respondents greater freedom of expression and the opportunity to make specific proposals.

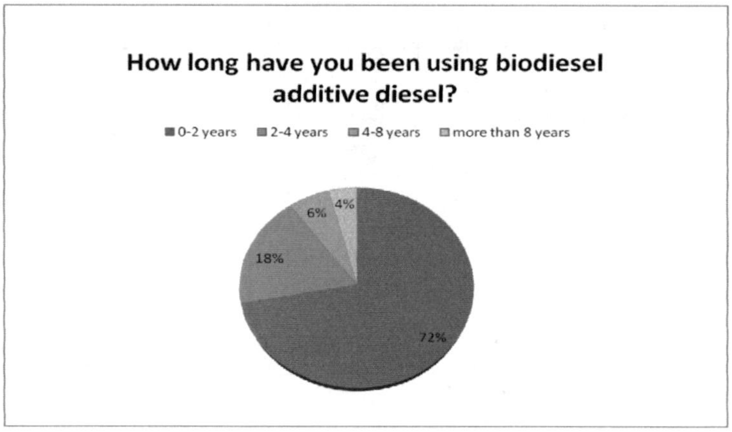

Figure 1. How long have you been using biodiesel additive diesel?

The study shows that over the past two years, biodiesel supplementation has become increasingly popular with customers. This is also due to the fact that the EU has started a course to reduce the harmful emissions of gases into the atmosphere and into diesel fuel so Bulgaria has started adding 6% biodiesel.

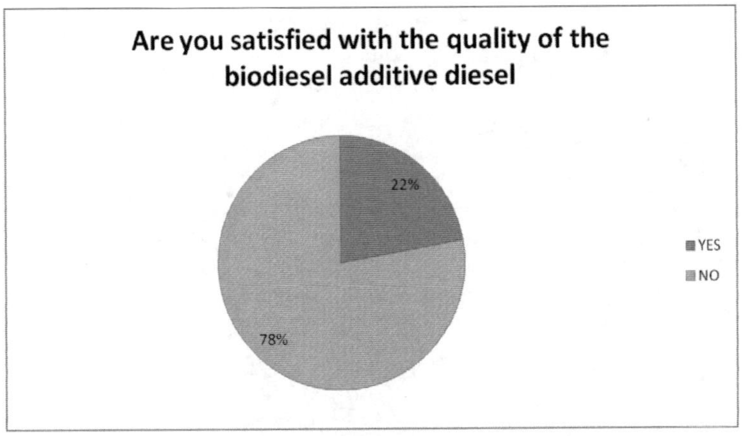

Figure 2. Are you satisfied with the quality of the biodiesel additive diesel?

Customer responses indicate that they are satisfied with the biodiesel supplement. But there will always be sceptics who, for obvious reasons, do not accept the new.

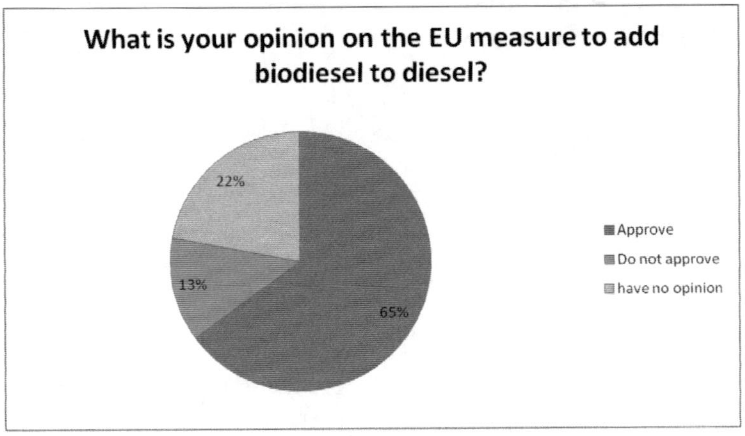

Figure 3. What is your opinion on the EU measure to add biodiesel to diesel?

Data from the answers to this question clearly show that the end customer is following the changes in the EU's environmental programs. This is confirmed by the answers to the following question:

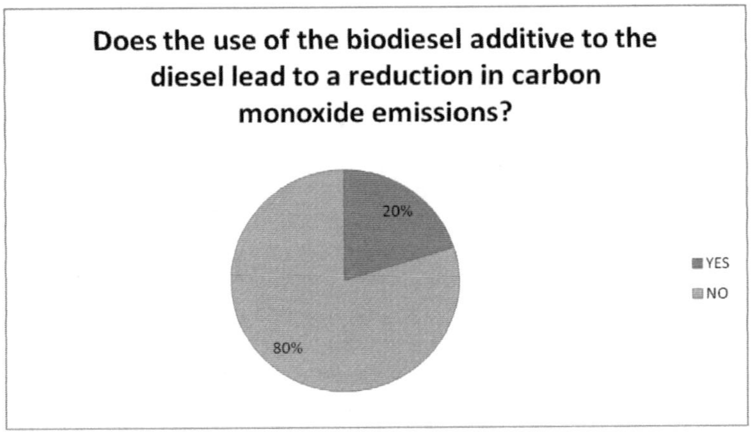

Figure 4. Does the use of the biodiesel additive to diesel lead to a reduction in carbon monoxide emissions?

From the following two questions, it is clear that a large percentage of end users are aware of the main benefits and advantages of using biodiesel as an additive to diesel:

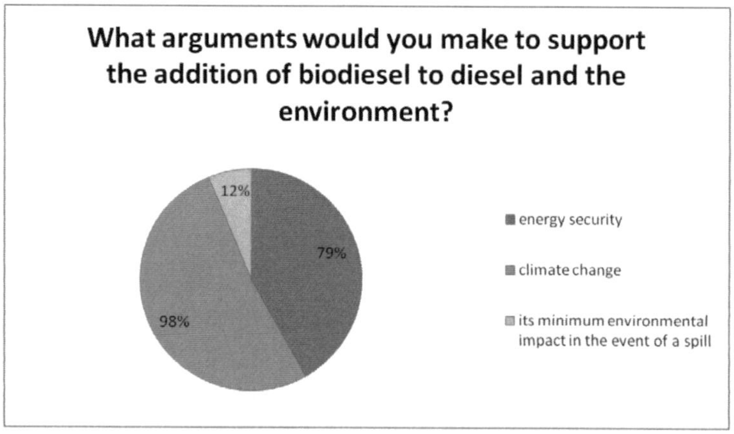

Figure 5. What arguments would you make to support the addition of biodiesel to diesel and the environment?

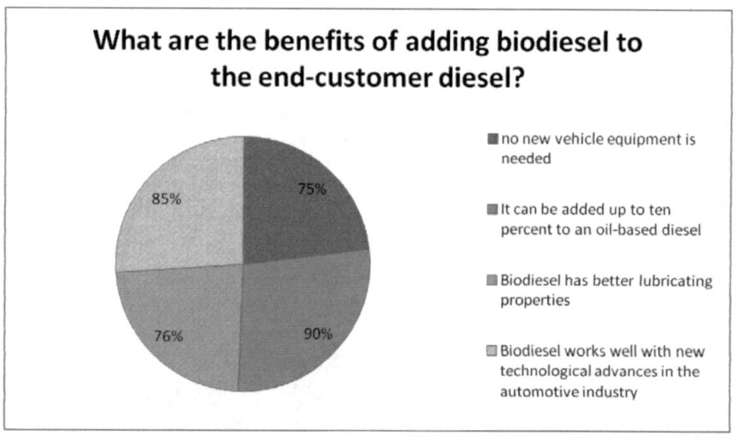

Figure 6. What are the benefits of adding biodiesel to the end-customer diesel?

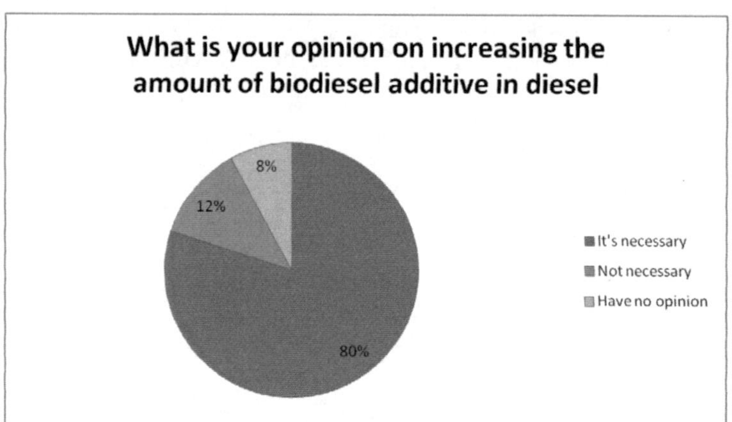

Figure 7. What is your opinion on increasing the amount of biodiesel additive in diesel?

80% of end customers agree that it is necessary to increase the amount of biodiesel additive in diesel. 8% use biodiesel, but do not have an opinion due to the fact that they automatically trust the service station staff and do not seek information from outside sources.

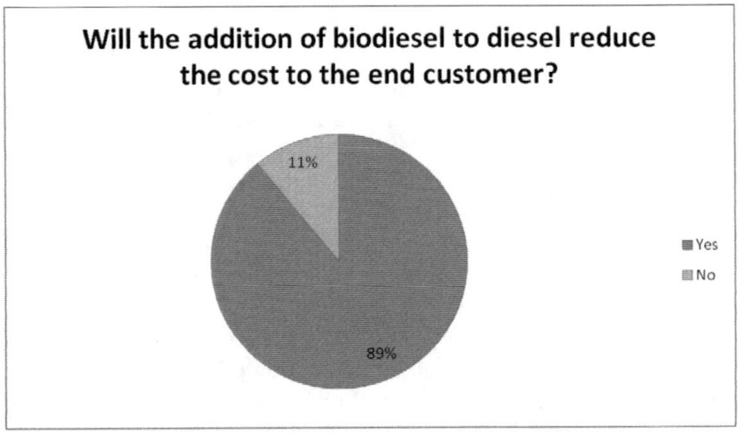

Figure 8. Will the addition of biodiesel to diesel reduce the cost to the end customer?

89% of the respondents strongly hope that the price for them will not increase. They share that everything related to environmental protection should not affect the price for the end customer. However, 11% are absolutely convinced that the end customer always pays the price. This is dictated by the frequent upward shift in fuel prices.

The analysis of the survey shows that the end customer is positive about using biodiesel as an additive to conventional diesel. The personal experience of most of the responders shows that the addition of biodiesel to the diesel does not impair the quality of the diesel engine, and even leads to its improvement. The customer seeks and receives the necessary information through various distribution channels and is inclined to believe that the benefits of adding biodiesel to the diesel are enormous both in reducing diesel engine wear and in protecting the environment. A large percentage of end users rely on the fact that their price will not change. However, there is always one group that is sceptical about pricing questions.

4. Conclusions

The following conclusions can be drawn from the analysis of the rapeseed, pure diesel and 8% rapeseed oil and 92% diesel blend, performance studies and questionnaires for the use of biodiesel as a diesel additive:

1. The rapeseed oil density is 919 kg/m^3 and is higher than the diesel 841 kg/m^3 and the mixture of 8% rapeseed oil and 92% diesel, which is 858 kg/m^3.
2. The viscosity of rapeseed oil is 37.3 mm^2/s and is much higher than that of diesel and the mixture, which is about 4.79 mm^2/s.
3. The calorific value of rapeseed oil is 37.1 MJ/ kg and is lower than that of diesel and the mixture 41.8 MJ/kg.
4. The flash point of rapeseed oil is 270°C and is many times higher than that of diesel and the 65°C mixture.
5. Cetane number for diesel and blend is 48.1 and higher than rapeseed oil 42.
6. Consumers are positive about using biodiesel as an adjunct to petroleum diesel. Experience has shown that the addition of biodiesel to diesel does not impair the quality of the diesel, and even leads to its improvement. They tend to believe that the benefits of adding biodiesel to diesel are enormous in reducing harmful gases into the atmosphere.

The conclusions presented above are grounds for claiming that rapeseed oil can be blended seamlessly and used as an additive to petroleum diesel. The benefits of adding biodiesel to diesel are enormous in reducing harmful gases into the atmosphere.

References

[1] Damianov D 2019 Modernization - an Alternative to Rapid Innovation *Int. Conf. on Creative Business for Smart and Sustainable Growth (CREBUS) IEEE* p 1-5

[2] Demirova S 2019 Turning Knowledge into Innovation and Innovation into an Effective Productn Concept *Int. Conf. on CREBUS IEEE* p 1-4
[3] Bogdanov K and Dimitrov R 2019 Research on the Possibility of Increasing the Power Output of a High-Speed Diesel Engine by Adding Compressed Natural Gas *IOP Conf. Ser.:Mater. Sci.Eng.* vol 614 012007
[4] Morrar R Arman H and Mousa S 2017 The Fourth Industrial Revolution (Industry 4.0): A Social Innovation Perspective *Technology Innovation Management Review* vol 7(11) p 12-20
[5] Owen N A Inderwildi O R and King D A 2010 The status of conventional world oil reserves-hype or cause for concern *Energy Policy* vol 38(8) p 4743–4749
[6] Ward D S and Mahowald N M 2014 Contributions of developed and developing countries to global climate forcing and surface temperature change *Environ.Res.Lett.* vol 9 p 74008– 74017
[7] Iliev S 2016 A Comparison of Ethanol, Methanol and Butanol Blending with Gasoline and Relationship with Engine Performances and Emissions. *Proceedings of the 29th DAAAM International Symposium* p 0505-0514
[8] Kerschbaum S Rinke G and Schubert K 2008 Winterization of biodiesel by micro process engineering *Fuel* vol 87(12) p 2590–2597
[9] Iliev S 2017 Investigation of N-Butanol Blending with Gasoline using a 1-D Engine Model *Proc. of the 3rd Int. Conf. on Vehicle Technology and Intelligent Transport Systems* vol 1 p 385-391
[10] Krumov K S Penkova N Y 2019 Numerical analysis of the transient heat transfer in high temperature chamber furnaces *IOP Conf. Ser.:Mater. Sci. Eng* vol *595 012005*
[11] Hansen J Sato M Kharecha P et al. 2017 Young people's burden: Requirement of negative CO_2 emissions *Earth Syst. Dynam.* vol 8 p 577–616
[12] Dwivedi G Verma P and Sharma M P 2016 Impact of oil and biodiesel on engine operation in cold climatic condition *Journal of Mater. and Environmental Sci.* vol 7(12) p 4540-4555
[13] McDonnell K Ward S Leahy J J and McNulty P 1999 Properties of Rapeseed Oil for Use as a Diesel Fuel Extender *Journal of the American Oil Chemists' Society* vol 76(5) p 539-543
[14] Madiwale S Karthikeyan A Bhojwani V 2017 A Comprehensive Review of Effect of Biodiesel Additives on Properties, Performance, and Emission *IOP Conf. Ser.: Mater. Sci. Eng.* 197
[15] Szabados G Bereczky A 2015 Comparison Tests of Diesel, Biodiesel and TBK-Biodiesel Periodica Polytechnica, *Mechanical Engineering* vol 59(3) p 120-125
[16] Crude petroleum and liquid petroleum products - Laboratory determination of density – method (ISO 3675:1998)
[17] Petroleum products - Transparent and opaque liquids - Determination of kinematic viscosity and calculation of dynamic viscosity (ISO 3104:1994)
[18] Determination of flash point - Pensky-Martens closed cup method (ISO 2719:2016)
[19] Petroleum products - Determination of the ignition quality of diesel fuels - Cetane engine method (ISO 5165:2017)
[20] Testing of solid and liquid fuels - Determination of gross calorific value by the bomb calorimeter and calculation of net calorific value - Part 3: Method using adiabatic jacket (DIN 51900 - 3:2005)

IOP Conf. Series: Materials Science and Engineering **977** (2020) 012017 doi:10.1088/1757-899X/977/1/012017

Investigation upon the operational factors influencing the smoke emissions from diesel engine

I K Moneva, M V Gramenova-Angelova, M S Ivanova and I S Petrov

Technical University in Sofia, Faculty of Engineering and Pedagogy in Sliven

imoneva@abv.bg, mgramenova@abv.bg, ing.mariqna_ivanova@abv.bg, aladina@abv.bg

Abstract. The results of laboratory examinations of the ecological characteristics of diesel internal-combustion engine are represented in this article. The powerful, economical and ecological parameters of a diesel engine investigated in this article are presented by the speed characteristics of effective torque, injected fuel per cycle, and fuel and air consumption per hour, temperature, and smoke emissions of exhaust gases. On the basis of the experimental dependences a mathematical model is designed. Parameters influencing significantly the smoke emissions of exhaust gases are investigated. The factors, injected fuel per cycle and excess-air factor, are examined at different operation regimes of the engine.

1. Introduction

The solution of the ecological problems of our planet is unthinkable without the circumscription of harmful emissions in the environment. Ecological emission standards lay down higher regulations for the concentration of harmful substances in the cars exhaust gases. Therefore, the development of automobile drives and internal-combustion engines is associated significantly with the adherence of the ecological standards during the exploitation of vehicles. The engines working with ignition from compression and using diesel fuel show a considerable inclination in the generation of hard particles due to the character of the running combustible process. Constructions using alternative and combined fuels are developed as well as the optimal concentrations of the substances are investigated by [1] with the goal of minimizing these emissions through the increase of the homogeneous part of diesel combustion process.

Mathematical models for the analytical determination of internal-combustion engines characteristics are designed on the basis of the experimental investigations. The ecological behaviour of the engine at predominant influence of either one or a combination of some factors can be prognosticated by means of such models. Because of the great variety of the constructions the universal models give considerable derivations when they are applied to specific engines. This is the reason why mathematical models, worked out for particular operating processes and applicable for a certain type, to be used more frequently.

The goal of this investigation is the determination of the analytical dependences of diesel engines and the parameters exerting considerable influence on the generation of harmful emissions in the exhaust gases (EG).

2. Presentation

In a diesel engine diffusion combustion of non-homogeneous fuel mixture is taking place. In the combustion zone the composition of fuel mixture is close to the stoichiometric one , $\lambda \ll 0$, and the temperature is high. A section of much enriched mixture is touching to this zone, which creates conditions for the running of fuel pyrolysis with a small quantity of oxygen. Here the generated soot begins immediately to burn at oxygen entrance. The speed of the soot generation exceeds their speed of burning at the beginning of the process, while at the end of the process the speed of soot burning is higher than the speed of soot generation. The soot burning continues through the process of expansion, but around the combustion chamber walls the temperature of the gases is tempered through the whole cycle and due to this the soot fell there is not taking place in the combustion. Soot can be also generated when the fuel flow fall upon considerably cool walls of the cylinder.

The raised amount of soot and smoke emissions of EG sharply increase if the fuel mixture is enriched. This circumstance has a great significance in the conditions of diesel engines operation, as the reasons for this are different. One reason can be the irregular control of the fuel apparatus influencing the amount and the quality of the fuel injection, while another reason can be the irregular choice of the characteristic of the injected fuel per cycle. Moreover, increased smoke emissions can be read not at all but at merely some regimes of operation of the diesel engine. Some investigations in this respect present, that during the vehicles movement in urbanized territories, the engines operate mainly at transitional operating modes.

In such circumstances the working process takes place at different from the optimal thermodynamic parameters of the environment in the combustion space, and as a result the ecological characteristics are made strongly worse [2]. That characteristics of the EG also depend on the type of fuel used - vegetable oils, biodiesel or diesel [3].

Together with the raised smoke emissions due to the enriched fuel mixture the fuel consumption also increases. In order the desired characteristic of the effective torque, M_e, the decreased smoke emissions of exhaust gases, and the decreased fuel consumption to be reached, it is necessary the characteristics of injected fuel per cycle, q_c, and the excess-air factor to be optimized at all speed regimes. The combustion temperature in the combustion chamber, where highly cancerogen and polycyclic aromatic hydrocarbons are not generated, needs to be controlled. Soot contains highly cancerogen and polycyclic aromatic hydrocarbons, which fall in human body by the respiratory system, and are held back in lungs alveoli [4, 5].

An experimental examination for the determination of the smoke emissions of EG, the characteristic of injected fuel per cycle, and the powerful and ecological characteristics of EG is carried out without catalytic treatment by means of test stand. The test stand is equipped with diesel engine, dynamometer, smoke-emissions-meter Hartridge, fuel and air flow-consumption-meter, sensors for measuring the EG-temperature in the release collector, and etc.

The smoke emissions of EG is determined with dynamometer Hartridge by measuring the opacity (optical density) of EG in relative units for transparency from 0 to 100. The relation between the smoke emissions according Hartridge, K, and the coefficient of light absorption by smoke is represented in [6], and is of the form

$$k = -\frac{1}{L} \cdot ln\left(1 - \frac{K}{100}\right),$$

where k is the coefficient of light absorption, m^{-1}; L is the measuring length of dynamometer chamber, m; K is the smoke emissions according Hartridge, %.

The diesel engine investigated has a power $N_e = 142\ kW$ at $n_{cs} = 2100\ min^{-1}$; maximal torque $M_{max} = 697\ Nm$ at $n_{cs} = 1300\ min^{-1}$, with direct injection.

The external and partial speed characteristics are taken for the determining the analytical dependences of the effective torque. The changes in the external (curve 1) and the partial characteristics (curves 2, 3, and 4) of the effective torque as a function of the crankshaft speed are represented in Figure 1, while the corresponding characteristics changes of the injected fuel per cycle are shown in Figure 2.

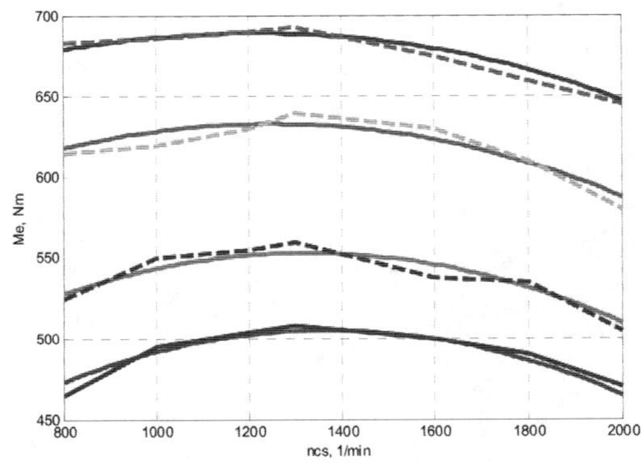

Figure 1. A change of the effective torque M_e at different crankshaft speed n_{cs}.
Experimental – - - -; Analytical – ——.

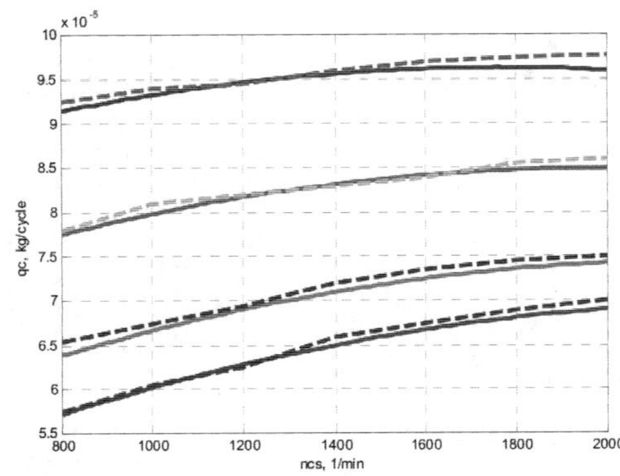

Figure 2. A change of injected fuel per cycle q_c at different crankshaft speed n_{cs}.
Experimental – - - -; Analytical – ——.

During the process of operation the automobiles work with variable load and variable speed regimes. For each speed regime the engine can undergo changeable load and develop the adequate torque. In the diesel engines the load change is achieved by change of the injected fuel per cycle. We assume boundaries of the investigated field, which includes the speed regimes at crankshaft speed from $800\ min^{-1}$ to $2000\ min^{-1}$, and load regimes at injected fuel per cycle from $40\ mg/cycle$ to $100\ mg/cycle$.

The approximation equations of the mathematical model are valid for these boundaries.

The experimental investigations are carried out by repeated measurements of the parameters (the investigated characteristics) of the multi-factorial experiment. A mathematical model of two-factorial experiment is chosen during the experiment planning. The uniform plan of the two-factorial experiment is used for the determining the mathematical dependence of the effective torque on both injected fuel per cycle and the crankshaft speed. After the design of the planning matrix the regression coefficients are defined, and the following formula is got

$$M_e = \begin{bmatrix} a_{e_1} + a_{e_2} \cdot q_c + a_{e_3} \cdot n_{cs} + \\ +a_{e_4} \cdot q_c^2 + a_{e_5} \cdot n_{cs}^2 + a_{e_6} \cdot q_c \cdot n_{cs} - \\ -(a_{n1} + a_{n2} \cdot n_{cs} + a_{n3} \cdot n_{cs}^2) \end{bmatrix}, \quad (1)$$

where M_e is the effective torque; $a_{e_1} \div a_{e_6}$ and $a_{n1} \div a_{n3}$ are the regression coefficients; q_c is the injected fuel per cycle, and n_{cs} is the crankshaft speed.

The matrix, containing experimental data is supplemented with the corresponding analytical magnitudes of the effective torque.

The external and the partial characteristics of the injected fuel per cycle as a function of the rotation frequency of fuel injection pump (FIP) and the rack position are presented with the polynomial

$$q_c = b_1 + b_2 \cdot n_p + b_3 \cdot h_p + b_4 \cdot n_p^2 + b_5 \cdot h_p^2 + b_6 \cdot n_p \cdot h_p, \quad (2)$$

where $b_1 \div b_6$ are the regression coefficients; n_p is the rotation frequency of FIP, and h_p is the rack position of FIP.

The results from the analytical formula (2) are presented in Figure 2 as thick curves.

For the determination of the analytical dependence of smoke emissions of engine EG are constructed the experimental characteristics, $K(n_{cs}, q_c)$, presented in Figure 3, where curve 1 is the external characteristic, while curves 2, 3, and 4 are partial (at different position of the accelerator pedal).

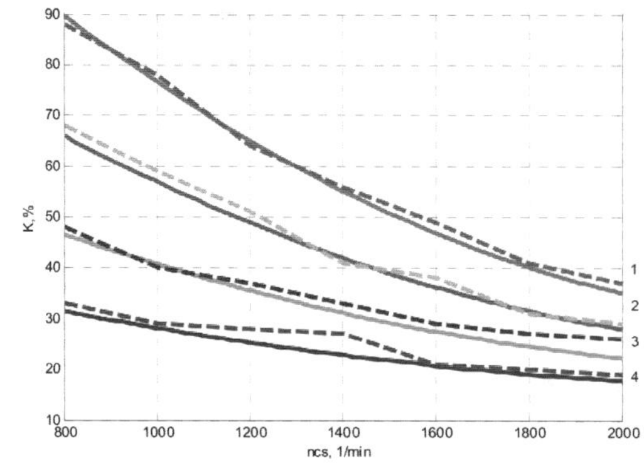

Figure 3. A change of smoke emissions of EG of diesel engine, K, at different crankshaft speed n_{cs}: curve1 is the external characteristic; curves 2, 3, and 4 are the partial characteristics. Experimental – - - -; Analytical – ——.

The adequacy of the mathematical model and the magnitudes of the coefficients for all investigated parameters are defined by using mathematical statistics methods and criteria. For this purpose, the sums of the numbers squared of the derivations of the experimental values from the calculated ones of the investigated parameters, the degree of freedom of the experiment, the remaining dispersion, and the dispersion of reproduction are determined.

The adequacy is checked and proved by F-test.

A commonly accepted theory for the soot formation is not available, but the numerous experimental data, found in many references, allow the more important factors, represented as decisive, one of which is the excess-air factor, λ, to be examined.

The excess-air factor in this article is determined by indirect measurement, by experimental determining the fuel consumption, G_f, and the air consumption according the following formula

$$\lambda = \frac{G_a}{G_f \cdot L_0}, \qquad (3)$$

where λ is the excess-air factor; G_a is the air consumption, kg/h; G_f is the fuel consumption, kg/h; L_0 is the theoretically necessary air amount for the combustion of one unit quantity of fuel.

The experimental curves of the changes in air consumption, G_a, and fuel consumption, G_f, as a function of the crankshaft speed, n_{cs}, are shown Figure 4 and Figure 5, respectively, where curve 1 is the external characteristic, and curves 2, 3, and 4 are the partial characteristic of G_f.

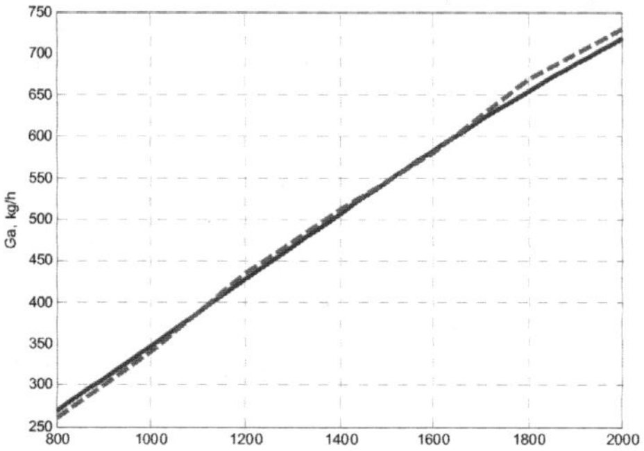

Figure 4. A change of air consumption, G_a, at different crankshaft speed n_{cs}.
Experimental – - - -; Analytical – ——.

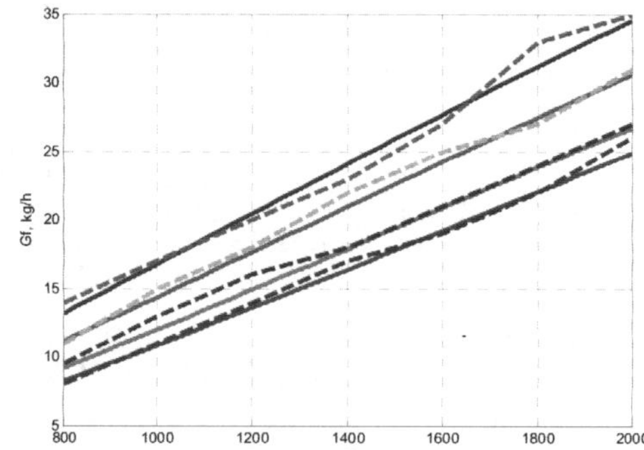

Figure 5. A change of fuel consumption, G_f, at different crankshaft speed n_{cs}.
Experimental – - - -; Analytical – ——.

The approximation equation describing the change of fuel consumption is of the form

$$G_f = n_{cs} \cdot i \cdot i_{dp} \cdot q_c, \quad (4)$$

where G_f is the fuel consumption, kg/h; n_{cs} is the crankshaft speed; i is the number of cylinders; i_{dp} is the driving ratio between the crankshaft and the shaft of fuel injection pump; q_c is the amount of injected fuel per cycle. After the mathematical processing of equations (2) and (3), the well-known formula from [7] $\lambda = \frac{V_{en} \cdot \eta_f \cdot \rho_a}{q_c \cdot L_0}$, and the experimental data presented in Figure 4, are obtained the following formulae for

- air consumption G_a

$$G_a = V_{en} \cdot i \cdot \eta_f \cdot \rho_a \cdot \frac{n_{cs}}{\tau}, \quad (5)$$

where G_a is the air consumption, kg/h; V_{en} is the engine working volume, m3; i is the number of cylinders; ρ_a is the air density and η_f is the volumetric efficiency; τ is the engine stroke.

- volumetric efficiency η_f

$$\eta_f = a_{\eta_1} + a_{\eta_2} \cdot n_{cs} + a_{\eta_3} \cdot n_{cs}^3, \quad (6)$$

where η_f is the volumetric efficiency; $a_{\eta_1} - a_{\eta_3}$ are regression coefficients; n_{cs} is the crankshaft speed.

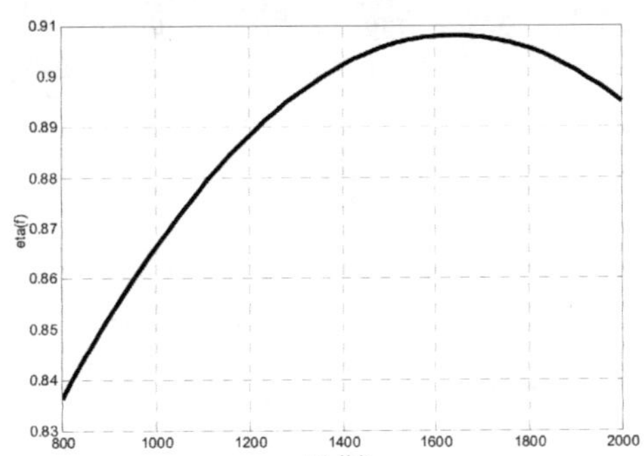

Figure 6. An analytical curve of change of the volumetric efficiency η_f at different crankshaft speed n_{cs}.

Then the change of the excess-air factor λ as a function of crankshaft speed n_{cs} acquires the form presented in Figure 7.

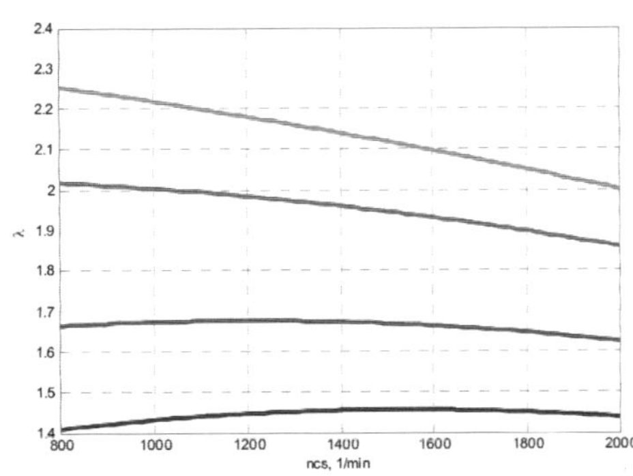

Figure 7. A change of the excess-air factor λ at different crankshaft speed n_{cs}.

One of the most widespread methods for the decrease of smoke emissions and improvement of fuel economy is the recirculation of cooled exhaust gases [8, 9]. The degree of cooling of EG influences the smoke emissions and the wearing of engine details. The higher EG temperature leads to higher smoke emissions and greater wearing of engine details. This makes the experimental investigation of the temperature of exhaust gases T_{eg} actual. The experiments carried out prove the dependence of EG temperature on the crankshaft speed and the injected fuel per cycle. The experimental characteristics are presented in Figure 8.

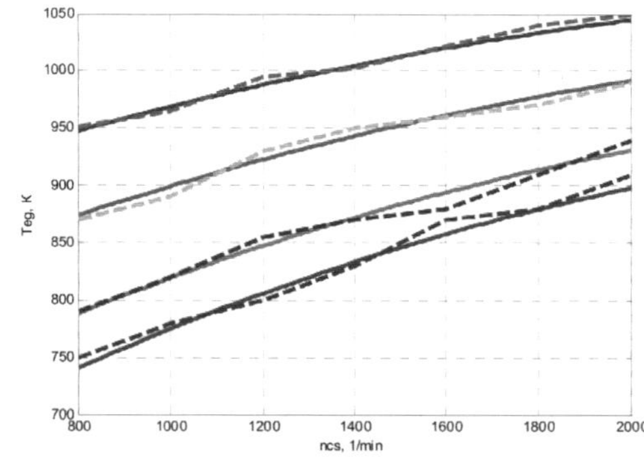

Figure 8. External and partial speed characteristics of EG temperature T_{eg}.

The regression coefficients of the approximation polynomial are got by using the uniform plan of the two-factorial experiment. The obtained polynomial of second degree is of the form

$$T_{eg} = a_{T_1} + a_{T_2} \cdot n_{cs} + a_{T_3} \cdot q_c + a_{T_4} \cdot n_{cs}^2 + a_{T_5} \cdot q_c^2 + a_{T_6} \cdot n_{cs} \cdot q_c, \tag{7}$$

where T_{eg} is the temperature of exhaust gases, K; $a_{T_1} \div a_{T_6}$ are regression coefficients; n_{cs} is the crankshaft speed; q_c is the amount of injected fuel per cycle.

During the experimental examination is established that the smoke emissions of EG vastly depends on two parameters – excess-air factor λ, and crankshaft speed n_{cs}.

The experimental curves of smoke emissions, K, depending on the excess-air factor, λ, and the crankshaft speed, n_{cs}, are constructed at constant values of crankshaft speed, and are presented in Figure 9. For the values $n_{cs} = 600\ min^{-1}$ and $n_{cs} = 800\ min^{-1}$ the curves are constructed by the method of extrapolation. The change of smoke emissions K at constant magnitudes of excess-air factor $\lambda = 1{,}5;\ 2{,}0;\ 2{,}5;\ 3{,}0$ is shown in Figure 10. These dependences are used for the extrapolation of curves at $n_{cs} < 1000\ min^{-1}$.

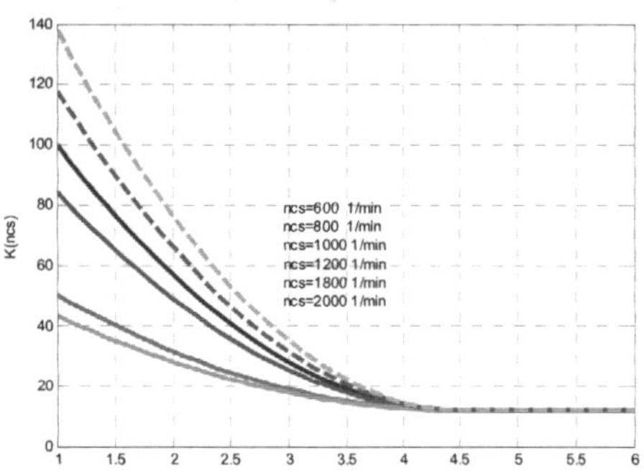

Figure 9. A change of smoke emissions of EG, K, as a function of excess-air factor λ, at different but constant crankshaft speed.

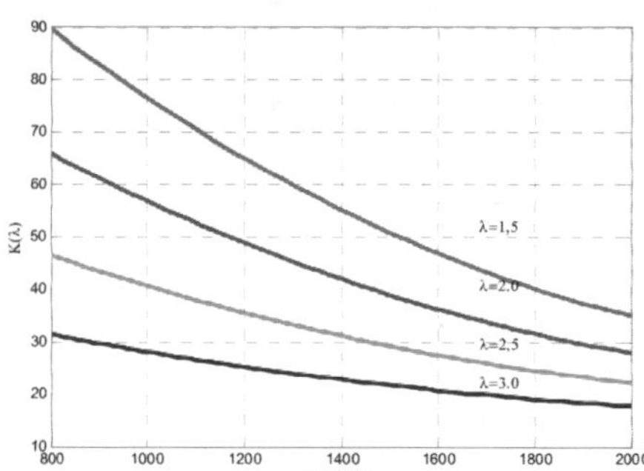

Figure 10. A change of smoke emissions of EG, K, as a function of the crankshaft speed at constant excess-air factor $\lambda = const$.

As it is seen in Figure 9 the curves of smoke emissions K are parabolas originated from one and the same point. A new coordinative system „$K - x$", which origin of coordinates coincides with the beginning of parabolas, and axis „x"coinciding with λ and a connection between „x" and „ λ" , as $x =$

$(b_{K_4} - b_{K_5} \cdot \lambda)$, is introduced for determining the equation of these curves. It is settled that the parabolas can be described by the following equation

$$K = A.x^2, \tag{8}$$

where K is the smoke emissions, %; A a coefficient depending on the crankshaft speed, n_{cs}.

The dependence of coefficient A on crankshaft speed, obtained by the method of least squares, is depicted with the following polynomial

$$A = (b_{K_1} \cdot n_{cs}^2 + b_{K_2} \cdot n_{cs} + b_{K_3}). \tag{9}$$

After a mathematical processing of equations (8) and (9), and a return to the coordinative system "$K - \lambda$", the following equation is got

$$K = (b_{K_1} \cdot n_{cs}^2 + b_{K_2} \cdot n_{cs} + b_{K_3}) \cdot (b_{K_4} - b_{K_5} \cdot \lambda)^2 + b_{K_6}, \tag{10}$$

where $b_{K_1} \div b_{K_3}$ are approximation coefficients for both equations (9) and (10); $b_{K_4} and\ b_{K_5}$ are coefficients of the equation of the connection between both coordinate systems; b_{K_6} is the magnitude of smoke emissions independent on λ; n_{cs} is the crankshaft speed.

This equation is valid for λ quantities in the limits $\lambda_{min} \leq \lambda \leq \lambda_{max}$, (in this case1,3 $\leq \lambda \leq 4,5$). In Figure 9 is clear that the smoke emissions of EG at $\lambda > \lambda_{max}(\lambda > 4,5)$ in practice does not depend on λ, and gets magnitude $K = b_{K_6}$. Experimental data for the smoke emissions behavior K at $\lambda < \lambda_{min}$ ($\lambda < 1,3$) are not generated, but presumably with the decrease of λ, the smoke emissions of EG is sharply raised.

The investigation carried out and the comparative analysis between the theoretical and experimental dependences for the smoke emissions of EG show that with the slowing down of crankshaft speed the smoke of EG is increased, and becomes greater than the allowed practical standard. The most probable reason for the increased smoke emissions of EG at low values of crankshaft speed could be the raised injected fuel per cycle q_c and the reduced excess-air factor λ.

In order the influence of both parameters q_c and λ to be determined, theoretical investigations with the analytical dependences obtained at conditionally assumed maximal quantity of smoke emissions $K = 45$ % ($k = 1,2\ m^{-1}$) are carried out. The investigations pointed out in references [2, 3] have proved that the simultaneous reduction of the basic harmful emissions from diesel engines, i. e. NO_x and soot, is impossible. During the present investigations it is assumed the smoke emissions to be limited up to $K = 45$ %, which does not lead to an increased amount of NO_x.

The results from the theoretical investigation show, that in order the smoke emissions of EG to be reduced at low values of crankshaft speed, it is necessary the injected fuel per cycle, q_c, to be decreased and the excess-air factor to be increased, as is shown in Figure 11 and Figure 12, respectively (see in both cases dot lines).

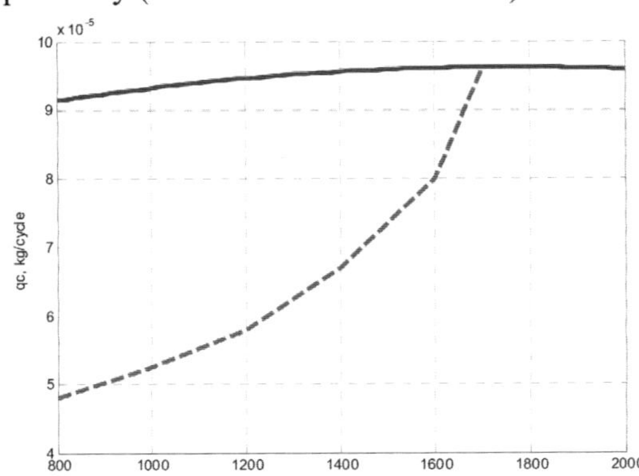

Figure 11. A change of injected fuel per cycle q_c at different crankshaft speed n_{cs}.

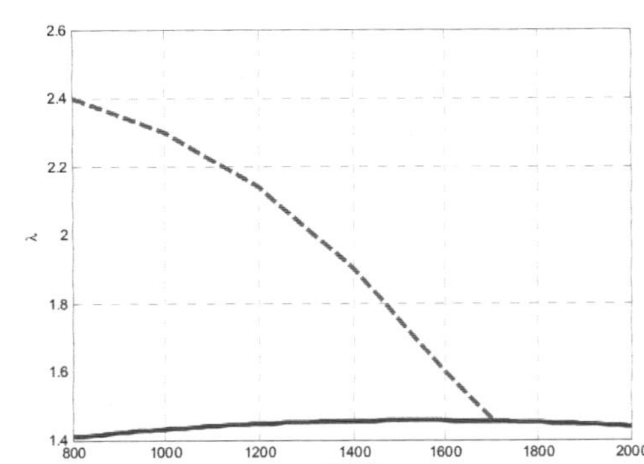

Figure 12. A change of excess-air factor λ at different crankshaft speed n_{cs}.

3. Conclusion

During the investigation is settled that the smoke emissions of EG is raised at low crankshaft speed due to worse quality of fuel injection and slow down of incoming air, which leads to oxygen shortage. At higher crankshaft speeds the diffuse combustion is disturbed because of the greater fuel amount.

The developed mathematical model refers to trucks, building machines and buses driven by diesel engines. The model points out the dependence of powerful and ecological parameters on crankshaft speed and engine loading. The factors injected fuel per cycle and excess-air factor influencing the smoke emissions of exhaust gases are examined at different operation regimes of the engine.

For the engine investigated the increase of smoke emissions of exhaust gases leaving directly the engine is greatest at crankshaft speed below $1200\, min^{-1}$ and excess-air factor less than 2,5. The characteristics pronounced to exert influence on the excess-air factor by changing the injected fuel per cycle could be applied in the systems for control of such a type of engines with the aim of optimizing the regeneration processes of filters for hard substances.

Acknowledgements

The authors would like to thank the Research and Development Sector at the Technical University of Sofia for the financial support.

References

[1] Dimitrov R, Ivanov Z, Zlateva P and Mihaylov V 2019 Optimization of biogas composition in experimental studies *E3S Web Conf.* **112** 02007

[2] Ivanov Z and Mihaylov V 2019 Transient operation of a direct injection diesel engine, IOP Conf. Ser.: Mater. Sci. Eng. **614** 012008

[3] Hristov R, Dimitrov A and Bogdanov K 2008 Comparative tests of RABA MAN D2156 HM 6U engine with diesel fuel, biodiesel and vegetable oils, *XI Int. Sc.– Pr.Conf. "Fundamental and Applied Problems of Piston Engines Improvement"* (Vladimir, Russia) pp 36-40

[4] Lloyd A and Cackette T 2001 Diesel Engines: Environmental Impact and Control *J. Air & Waste Management Association* **51** pp 809-847

[5] Lyons K 2003 Diesel Particulate Matter Emission Reduction Measures, Assessment of Potential Strategies to Reduce Emissions from Diesel Engines in Washington State, (Washington State Department of Ecology) 05-02-005

[6] Ivanov Z, Mikhailov V and Kolev A 2012 Investigation System for Determining the Ecological Characteristics of Automobile Engines EKO Varna **19** pp 502-10

[7] Andryusenko P, Burtzev O and Gutarevich Y 1978 Characteristics of Vehicles and Tractors Engines Higher School (Viischaya shkola) Kiev

[8] Burzev K and Stankov E 2017 *Ecological Problems in Transport* (Rousse)

[9] Gureev V, Khairullin A, Varlamov F, Gumerov I and Khafizov R 2016 Influence of exhaust gas recirculation on technical, economic, and environmental performance of high-uprated diesel aircraft engine *Russian Aeronautics* **59** pp 554–8

IOP Conf. Series: Materials Science and Engineering **977** (2020) 012018 doi:10.1088/1757-899X/977/1/012018

An integrated model for designing of technical systems: another step to coherence between theory and practice

B Angelov[1], **V Haralanova**[2] **and A Ahmed**[3]

[1,3] Department of Transport, University of Ruse „Angel Kanchev", Ruse, Bulgaria
[2] Department of Mechanical Engineering, Linnaeus University, Växjo, Sweden

bangelov@uni-ruse.bg[1], valentina.haralanova@lnu.se[2], aahmed@uni-ruse.bg[3]

Abstract - A number of authors recommend selecting the systems design methods to be used depending on the practical tasks to be solved. At the same time, results of studies aiming at clarifying and systematizing practical tasks are very rare. In the frames of the research performed, it was revealed that in the companies four major tasks use to be solved - adapting an existing solution to work in new conditions, refining an existing solution, combining it with previously developed elements and creating a new solution. Resultantly, in the research an integrated model for design of technical systems was proposed. This model was obtained by combining (overlaying) the methods for solving the four main tasks defined and subsequently the content of the different stages in the proposed model was clarified.

1. Introduction

In the second half of the last and the beginning of this century, scientific publications emerged that outlined the theoretical foundations of modern science for development of technical systems. A number of authors, summarizing the experience gained so far and adding their own personal contribution, presented completed studies that are justifiably considered fundamental from a methodological point of view (Birkhofer ed., 2011; Cross,2004; Hubka,2004; Hubka and Eder, 1992; Hubka and Eder, 1996; Koller, 1985; Pahl and Beitz, 1977; Pahl et al.,2007; Roth, 1982 and Ullman, 2010). The scientific results found their place in the training of students and generations of engineers gained methodological knowledge about the design work content and the tools available for implementation in the different stages of development of the new technical systems´.

Despite the results achieved and the students` training completed, it has been found that methods of systematic product development are used with limits in practice (Birkhofer ed., 2011; Lopez-Mesa and Thompson, 2002).The cited authors explain this phenomenon as follows:

- There is not enough time to embed the new methods into practice;
- Misuse of various methods leads to disappointing results;
- Some methods are used according to their popularity, not depending on the task to be solved;
- Most of the methods are supported in the theoretical sources by inappropriate examples;
- The number of scientific publications is increasing in an avalanche and most of them claim that the published methods are universal;
- There is a discrepancy between the methods offered and the needs of practice.

In his analysis of problems in transferring design methods into practice, K. Wallace notes that "no-one's territory" has been formed between theory and practice (Birkhofer ed., 2011). The author cites the results of years long research done by a group, led by Prof. H. Birkhofer, according to which, in order to adopt a theoretical method from the practice, it is necessary to fulfil the following conditions - the method must be simple, adaptable, well-publicized, and the necessary training for its use must be carried out. In this context, K. Wallace proposes to be created interdisciplinary groups of specialists who would turn theoretical knowledge into reliable and simple tools for solving practical problems. According to other researchers, methods of systems development should be systematized into computer-based catalogues to assist practitioners in selecting the appropriate method for solving any particular, practical problem (Birkhofer ed., 2011; Lopez-Mesa and Thompson, 2003).

In the presented material, an attempt has been made to clarify and systematize the main tasks being solved at the companies, and on this basis, an integrated model for development of technical systems has been proposed.

2. Analysis of existing studies

Results of studies, aiming at clarifying and systematizing the problems to be solved in the practice, are relatively rare in the technical literature. When creating his theory for solving inventive problems (TRIZ), G. Altshuller assumes that practical problems have five levels of complexity (Altshuller, 1998). The first level refers to tasks solved by using existing solutions. As the level of complexity increases, one moves on to the tasks of refining known solutions, combining pre-existing solution fragments, or creating a new technical solution. The fifth level refers to the tasks in which, based on the new technical solution, a new technical system is to be developed. The tasks solved in the practice at the companies, according to the analysis done by K. Roth's, can be attributed to one of the following five types of design - technological, subjective, functionally specific, functionally abstract and original design (Roth, 1982). The author proposes a flexible work plan where, depending on the complexity of the task, the solution can start from a different level.

According to V. Hubka and W. Eder, one of the following four major tasks is to be solved in the development process - creating a new solution, refining an existing solution, adapting a known solution to new conditions, or creating a modified series of technical systems (Hubka and Eder, 1996). In their analysis on the different types of design and the different tasks of the design engineers in the process, G. Pahl and W. Beitz define three types of design - variant, adaptive and new (Pahl and Beitz, 1977). Routine, creative and innovative product development are the essence of J. Gero's design method using design prototypes (Gero, 1990). Design prototypes defined by the author are models (schemes) for solving the types of design tasks to which each new problem should relate. In the proposed general design theory (Yoshikawa, 1981) and later in the extended general design theory (Tomyama and Yoshikawa, 1987), H. Yoshikawa and T. Tomiyama examine the design process in the context of the axiomatic set theory. T. Tomyama proposes that if, at the beginning of the process, there is no suitable existing solution, then one of the following three options can be used - to refine an existing solution, to combine pre-existing elements or to create a new solution (Tomyama et al., 2009).

The scientific publications contain research results clarifying specific issues of the product development process such as: different features of the product design in small and medium-sized enterprises; the impact of production volume and complexity of the designed technical system on the activities performed by the design engineers throughout the life cycle of the technical system, etc. (Ernzer and Birkhofer, 2002; Kristina, 2005; Lopez-Mesa et al., 2002; Löfqvist, 2009).

3. An integrated model for developing technical systems based on company practice

Based on the analysis of the research published and the direct observation of the processes from the practice, it can be concluded that in the companies the following four main tasks are solved - adaptation of an existing solution to work in new conditions, refinement of an existing solution, combining pre-developed elementary solutions and creating a new solution. The empirical information used for the purposes of the current research was collected by observing and analysing the development process at two companies TISEM-Ltd and METAREM-Ltd. As a result of the investigation, an integrated model for development of technical systems was proposed. The model was obtained by combining (overlaying) the methods for solving the four main tasks defined (Figure 1). The model is part of previously done studies on the downward hierarchical link between the life cycle of the technical system and the stages in it such as developing the product to be produced; developing the production (Angelov et al., 2013). From this perspective, the model has a separate entrance (input) and exit (output), which is becoming input to the next stage developing the production (Angelov, 2012). In accordance with the model presented in Fig. 1, the design process begins by clarifying and defining the problem. There is no consensus in the research literature on the content of this stage. Proposals range from a general presentation of the problem to a list of requirements and constraints, according to which the solution should be sought.

An objective basis for clarifying and defining the task can be obtained if a technical system is considered as an element of a higher-level system for the conversion of matter, energy and information (Angelov and Haralanova, 2012). Within this system, the technical system interacts with humans, with the object under change and with the environment. Through this interaction, the main function (purpose) of the technical system is realized – the TS is the means by which the desired change of the properties of the object is accomplished. The pre-planned change of the properties of the object is realized by technology (recipe, algorithm), which remains unchanged throughout the whole process or changes depending on the intermediate results obtained. It is clear that clarifying and defining the problem at this level requires systematizing the available information about:

- the object under change and its states - initial, intermediate and final;
- the technology, which prescribes how the properties of the object are modified during the interaction between the technical system and the object;
- the interaction between the technical system and the human;
- the interaction between the technical system and the environment.

Thus, the first stage regulates only the functional connection between the technical system developed and the elements of the system for the conversion of matter, energy and information, and does not set any restrictions on the design engineers in the execution of the next stages of development process.

Conducting feasibility studies is another important prerequisite for successful completion of the task. They reduce the cost and time of design and guarantee the up-to-date level of the technical system developed. The purpose of the feasibility studies is to analyse existing solutions or theoretical research results regarding the structure and processes in the systems of the three hierarchical levels as follows: the system for the transformation of matter, energy and information on the first (highest) level; the technical system designed - on the second level; the executing organs of the TS (subsystems) on the third level (Angelov, 2012). Based on the feasibility studies, the following two questions should be answered:

- which are the requirements and constraints, defining the space of possible design solutions;
- which of the following four alternative strategies should be applied for solving the problem - adapting and implementing an existing solution into new conditions, refining an existing solution, combining partial solution elements, or creating a new solution.

The requirements and constraints shall be systematized in the specification. The content and the mode of formulating the specification are most often regulated within the company. In the practice, there are examples of standardization of this document on a national level. The particular importance of the specification is determined by the fact that at a later stage it is used for evaluating the quality of the design. For this purpose, a prototype is made and tested following the design. The test results are weighed against the requirements and limitations in the specification and on this basis, a decision is made on the further development (Angelov, 2012). In the first three stages of the proposed integrated model, the content and volume of work depend mainly on the strategy chosen for solving the problem. However, there is an overlap of activities, which is a sufficient reason for their combined presentation in Figure 1.

The main difference between the four main tasks defined is the selection or synthesis of a technical solution. If there is a suitable existing solution, the task is to adapt this solution to the new conditions. In this case, the technical solution is preserved or it is only slightly modified. The technical solution changes more if there is an existing solution that needs to be refined. Although there is a perception that refinements solve relatively simpler tasks, changes can cover the entire development process. For example, if the technology in the system for conversion of matter, energy and information is improved, changes at the functional level will follow, and as a consequence, the technical solution will change.

The problem is solved similarly, while combining existing partial solutions. In this case, additional specific tasks arise, such as evaluating alternatives, evaluating the functional and structural compatibility between the elements combined, or the need to include additional matching or joining elements.

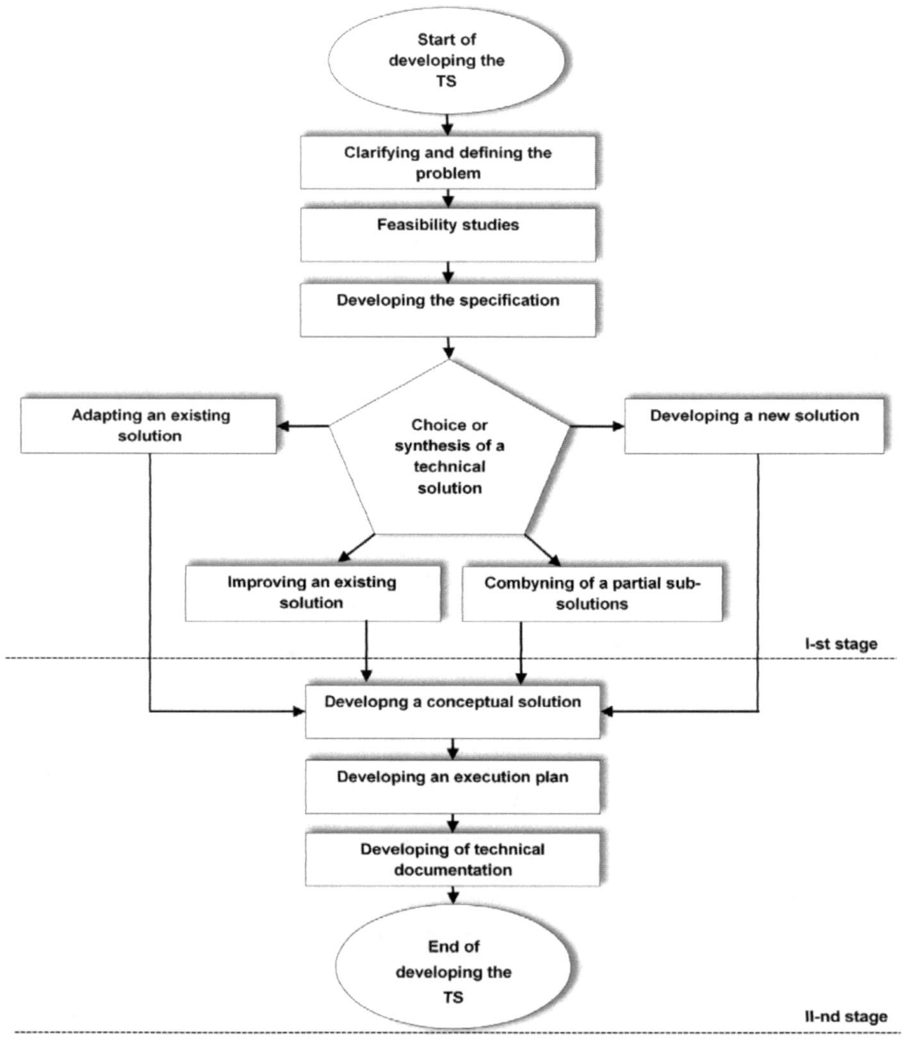

Figure 1. Integrated model for the practical methods of developing technical systems

The step of synthesizing a technical solution is developed to full extend only when a new solution is created. After determining the initial, intermediate and final states of the object under change, the technology in the system for transformation of matter, energy and information is compiled (Angelov and Haralanova, 2012). Then the technology is presented as a collection of sequential or parallel elementary operations. Operations, in their turn, are presented as physical processes. For the execution of those processes, appropriate physical structures (executive organs) are sought. The compilation of the executive organs (sub-systems) results in a new technical solution.

By selecting or synthesizing a technical proposal, the first design phase - the functional phase - is completed (Figure 1). The development of the technical solution to the level of a detailed execution plan and on this basis, development of complete technical documentation are carried out in the second phase of the product development - the subject phase. The activities of the second phase, regardless of the

IOP Conf. Series: Materials Science and Engineering **977** (2020) 012018 doi:10.1088/1757-899X/977/1/012018

chosen strategy for solving the problem, can also be unified - the process of developing returns to the basic algorithm (Figure 1).

In the practice it can be seen that despite the overlapping of the steps and the repeated return to the technical solution and the requirements and limitations of the specification, the following three stages can be distinguished in the second phase - development of a conceptual design, development of an execution plan and development of technical documentation. Separation of the second phase into three separate stages creates an opportunity for better planning and management of the process and for carrying out intermediate control at the end of each stage.

By developing a conceptual solution, a step is made from the functional phase to the objective design phase. The technical solution is developed to the level of an overall preliminary project. For this purpose, sub-systems for interaction with humans and the environment are added to the conceptual solution. Then, the interaction between the subsystems, the interfaces and their relative positioning within the technical system are clarified. The structure and internal processes within the individual subsystems are also clarified to the necessary extent. The proposed solution is appropriately visualized and the fulfilment of the requirements and limitations specified in the specification are evaluated.

In the next stage, the preliminary design is developed to the level of a detailed execution plan. Based on the clarified interactions and the mutual arrangement between the executive organs, the parallel design of the components at the different hierarchical levels of the new technical system is carried out. At the core of this complex process is the structural and functional dependence between the components at the different levels. The development of a complete working project is achieved after repeated solving of strength-dimensioning problems and development of structural and functional linkages between the individual components. The working project shall be submitted in accordance with the principles of the technical documentation. The execution of the project shall be evaluated taking in consideration the fulfilment of the requirements and limitations set out in the technical assignment.

The development of the new technical system ends with the development of complete technical documentation. The components of the different levels are presented graphically following the rules of 2D technical documentation or module based documentation. The next stage after the development of a new technical system is the stage of developing the production process. The technical documentation must contain the information needed to organize the production, in accordance with the applicable company and national standards. An important condition for the smooth transition between development of a new technical system and production planning is the compliance of the working documentation with the planned production volume and technological capabilities of the manufacturing company.

4. Conclusion

As a result of the analysis of the recent scientific publications in the area and the direct observation of the processes at the companies, it was found that in the practice the following four main tasks are solved: adaptation of an existing solution for application in new conditions; refinement of an existing solution; combining existing sub-solutions and creating a new solution.

Based on these findings, an integrated model for designing a technical systems has been proposed. It is obtained by combining (overlaying) the methods for solving the four main tasks defined. The design of a new technical system is considered to be an integral part of the first stage of the technical system's life cycle - production planning. During this stage, the design and development of a new technical system is accomplished and the production plan and technological support of the production are carried out. The analysis of the processes in their interconnection makes it possible to accomplish the "whole picture" of the production set out and to make the content of this process an element of the training of the engineering design students.

The content of the developed model, its correspondence to the development process in the companies and the availability of tools for implementation at each stage of the model have been discussed repeatedly with specialists from universities and companies in Bulgaria and Sweden. The main results of the study are used in the educational process in the field of technical design and development, with students from the University of Ruse "Angel Kanchev" (Bulgaria) and Linnaeus University, Växjö (Sweden).

References

[1] Angelov B, Haralanova V and Mateev S 2013 Preparations for production - the first stage of the life cycle of the new technical system *Jurnal Mechanical engineering* (Varna) No.18 ISSN 1312-8612 pp 19-22

[2] Altshuller G 1998 *Creativity As An Exact Science: The Theory of the Solution of Inventive Problems* Book 5 Gordon and Breach Science Publishers Inc. (Netherlands) ISBN-10:0677212305

[3] Angelov B 2012 About the Contents of Design Work in the Preparation Phase for the Production of New Technical System *International Scientific Conference MendelTech* (Brno, Czech Republic) ISBN 978-80-7375-625-3

[4] Angelov B 2012 About the Hierarchical Relationship between the System for the Conversion of Matter, Energy and Information, the Technical System and the Executive Organs of the Technical System *International Scientific Conference MendelTech* (Brno, Czech Republic) ISBN 978-80-7375-625-3

[5] Angelov B and Haralanova V 2012 The System for the Conversion of Matter, Energy and Information - the Missing Element in the Theory of Technical Systems *The Seventh International Symposium KOD 2012 – Machine and Industrial Design in Mechanical Engineering* (Balatonfüred, Hungary) ISBN 978-86-7892-399-9 pp 65-70

[6] Birkhofer H (Editor) 2011 The Future of Design Methodology *Springer-Verlag* (London Limited) ISBN 978-0-85729-614-6

[7] Cross N 2004 Engineering design methods: strategies for product design *John Wiley & Sons Ltd* (England) ISBN 10:0-471-87250-4

[8] Ernzer M and Birkhofer H 2002 Selecting methods for life cycle design based on the needs of a company". *IDC Design 2002* (Dubrovnik) pp. 1305-1310

[9] Gero J S 1990 Design Prototypes: A Knowledge Representation Schema for Design *AI Magazine* Vol. 11 Number 4

[10] Hubka V 1984 Theorie Technischer Systeme: Grundlagen einer wissenschaftlichen Konstruktionslehre *Springer-Verlag* (Berlin) ISBN 9876-3-662-10446-0

[11] Hubka V and Eder W E 1992 Einführung in die Konstruktionswissenschaft: Übersicht, Modell, Ableitungen *Springer-Verlag* (Berlin) ISBN 987-3-642-95673-7

[12] Hubka V and Eder W E 1996 Design Science: Introduction to the Needs, Scope and Organization of Engineering Design Knowledge *Springer-Verlag* (London Limited) ISBN 13:978-3-540-19997-7

[13] Koller R 1985 Konstruktionslehre für den Maschinenbau. Grundlagen des methodischen Konstruirens *Springer-Verlag* (Berlin Heidelberg) ISBN 3-540-15369-1

[14] Kristina L 2005 Job design for good design practice *Elsevier Design Studies* Vol. 26 (2) pp 191-213

[15] López-Mesa B, Thompson G and Willianderm 2002 Managing Uncertainty in the Design and Development process by Appropriate Methods *IDC Design 2002* (Dubrovnik) pp 829-830

[16] López-Mesa B and Thompson G 2002 The application of the 4Ps model to the management of creativity and innovation in product development *The 9th IPDM Conference* May 27-28 (Sophia Antipolis: European Institute for Advanced studies in Management) pp 587-601

[17] López-Mesa B and Thompson G 2003 Exploring the need for an interactive software tool for the appropriate selection of design methods *Proceedings of ICED 03, the 14th International Conference on Engineering Design* pp 627-628 (exec. sum.), full paper no. DS31_1118FPB

[18] Löfqvist L 2009 Design Processes and Novelty in Small Companies: a Multiple Case Study *Proceedings of ICED 09, the 17th International Conference on Engineering Design* 24-27 August pp 265-278

[19] Pahl G and Beitz W 1977 Konstruktionslehre: Handbuch für Studium und Praxis *Springer-Verlag* (Berlin Heidelberg GmbH) ISBN 978-3-662-02289-4

[20] Pahl, G, Beitz W, Feldhusen J, and Grote K H 2007 Konstruktionslehre: Grundlagen erfolgreicher Rroduktentwicklung Methoden in Anwendung *Springer-Verlag* (Berlin Heidelberg) ISBN 10:3-540-34060-2

[21] Roth K 1982 Konstruieren mit Konstruktionskatalogen: Systematisierung und zweckmäßige Aufbereitung technischer Sachverhalte für das methodische Konstruieren *Springer-Verlag* (Berlin Heidelberg New York) ISBN 3-540-09815-1

[22] Ullman D G 2010 The Mechanical Design Process *Fourth Ed. McGraw-Hill*, ISBN 978-0-07-297574-1

[23] Tomiyama T and Yoshikawa H 1987 Extended General Design Theory *In: H. Yoshikawa and E.A. Warman (eds.) Amsterdam Design Theory for CAD* (Amsterdam North-Holland) pp 95-130

[24] Tomiyama T, Gul P, Jin Y, Lutters D, Kind C and Kimura F 2009 Design Methodologies: Industrial and Educational Applications *CIRP annals: manufacturing technology* 58(2) pp 543-565

[25] Yoshikawa H 1981 General Design Theory and a CAD System *In T. Sata and E.A. Warman (eds.), Man-Machine Communication in CAD/CAM* (Amsterdam: North-Holland) pp 35-58

Acknowledgments

The present document was supported with the financial assistance of the Project 2020-RU-03.

IOP Conf. Series: Materials Science and Engineering **977** (2020) 012019 doi:10.1088/1757-899X/977/1/012019

Developing of automatic lap counting system for electric vehicle at Shell Eco-marathon competition

E Mitev[1], S Iliev[1] and D Gunev[2]

[1]University of Ruse, Department "Engines & vehicles", 8 Studentska Str., Ruse, Bulgaria
[2]University of Ruse, Department "Industrial Design", 8 Studentska Str., Ruse, Bulgaria

e-mail: dgunev@uni-ruse.bg; emitev@uni-ruse.bg; spi@uni-ruse.bg

Abstract. The article shows the development of a closed circuit track counting system made from a prototype electric car that was specifically developed for the Shell Eco Marathon. The system records the laps made by transmitting and receiving radio waves (RF). The entire electronic system was built on the Arduino microcontroller platform and one-way radio frequency transceiver modules. The design of the system was developed using Catia V5 3D modeling software. Additive technology was used to reduce the mass of the system.

1. Introduction

Shell Eco-Marathon is a unique competition that challenges students from all over the world to design, construct, and compete with the most energy-efficient cars. The competition develops the innovative thinking of students to solve problems and tasks related to energy efficiency and environmental problems. The initial project was based on two working purples: the design, manufacturing of prototype battery electric vehicle (PBEVs) and the participation in a few of the competitions held annually in Europe [1, 2]. Both purples are closely connected. First, the design of the vehicle is carried out according to the regulations of these competitions, which limit many of its features such as physical dimensions, engine power, battery type, etc. and they also have very strict specifications and regulations regarding safety such as the characteristics of the braking system, safety switches, etc [3]. On the other hand, the involvement in races is an ideal benchmark to evaluate the performance of the designed and manufactured vehicles.

The experience gained in these races is used to identify problems and needs and to seek solutions for them, which will be tested and analyzed again in future races. In the Shell Eco-marathon, the teams are used to record the number of laps on the track, which is positioned in a certain part of the track determined by the race organizers. This method of reporting laps predisposes to errors due to various factors (such as distraction of the persons involved in the reporting, poor visibility from the place of reporting, engaging the pilot's attention during reporting, etc.). The construction of a system for counting the laps is necessary, from the point of view of eliminating the human factor. Improper lap counting can lead to pilot leaving the runway earlier or later. These mistakes lead to disregard for the laps made according to the rules of the competition. In the lap counting system, the pilot receives information from the system for the exact time of his last lap. This information allows the pilot to adjust his finishing time

IOP Conf. Series: Materials Science and Engineering **977** (2020) 012019 doi:10.1088/1757-899X/977/1/012019

by reducing or increasing his speed on the runway. All of this has an impact on energy consumption at laps, which is the other important factor requiring the design of the lap counting system.

2. Construction of the electronic part of the lap counting system

The system consists of two separate parts, transmitting and receiving. The emissive part of the system is placed near the start-finish line of the track, from where the transmitter sends a constant signal, which is intercepted by the receiving part when the electric vehicle is passing, where the receiving part of the system is located.

The SYN115 / SYN480R transceiver pair was chosen for one-way communication (Table 1) due to its low cost, small size and relatively low frequency of radiation, which allows the waves to spread more smoothly in space and also low energy consumption. In addition, broadcasting of radio waves is in the range of 15-20 meters from the transmitter to the receiver. This allows the counting of the laps to take place only in the area of the start-finish line (Figure 1).

Table 1. Technical characteristics of the transceiver chip modules

On-board single-chip module	**SYN115** (transmitter)	**SYN480R** (receiver)
Module transmitting frequency	433MHz	433MHz
Module output power:	<10dBm	-107dBm
The module transmission rate	<10kbps	2.5kbps (SWP), 10kbps (FIXED)
The module supply voltage	1.8-3.6V	3.3-5.5V
PCB size:	15*11.8(mm)	17.2*11.8(mm)

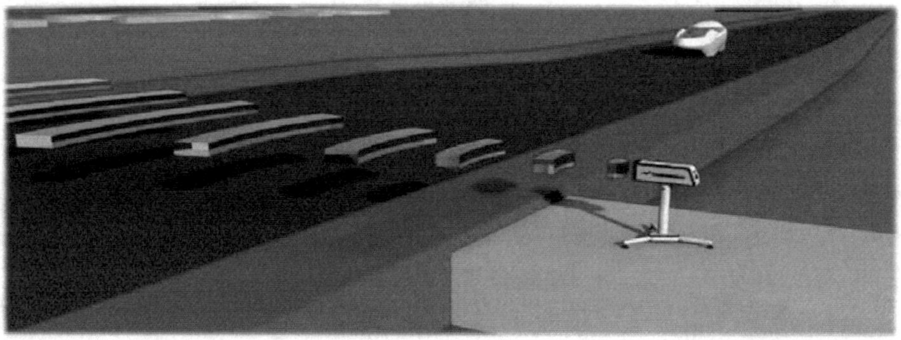

Figure 1. Positioning of the emitter in relation of the runway

(a) (b) (c) (d)

Figure 2. Components of the electrical part of the emitter

a- Arduino microcontroller; b- 9V power supply battery; c- transmitter chip module SYN115 / F115; d- antenna

The emitting part consists of four main components: Arduino microcontroller (Figure. 2, a), 9V power supply battery (Figure. 2, b), SYN115 / F115 emitting chip module (Figure. 2, c) and antenna (Figure. 2, d).

The Arduino Nano microcontroller control requires a supply voltage of 5V via its USB port, which means that it can also be powered via a portable powerbank. It is also possible to supply from 6 to 12V, when connecting the positive terminal of the power supply, in this case using a 9V battery, to the VIN pin of the board (red wire).

Figure 3 shows the connection diagram of the signal emitting part of the system. The signal emitting chip requires a power supply of 1.8 to 3.6V, for this reason it is connected to the 3.3V pin on the board (orange wire), and the ground of the battery (black wire) is used as ground. The generated signal is transmitted from the TX1 pin of the microcontroller to the emitting chip (green wire), the signal is being transmitted by radio waves. As an antenna, a copper wire with a cross-sectional diameter of 0.6 mm is used, which is connected directly to the emitting chip (blue wire). It is encased in an aluminium shell in order to direct the distribution of radio waves.

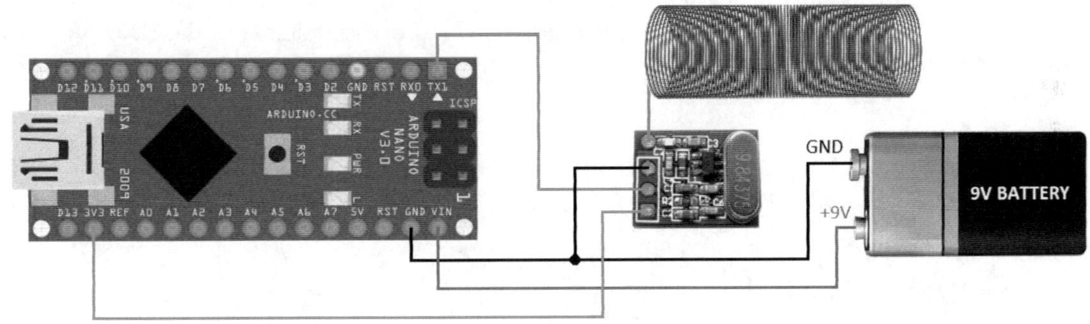

Figure 3. Wiring diagram of the transmitting part of the system

The receiving part of the system is located in the steering wheel housing. This is done in order to reduce the length of the wires and to avoid unwanted electromagnetic interference resulting from the operation of the electric motor, which is located in the rear of the electric vehicle. This part of the automatic lap counting system consists of three main components: Arduino Due microcontroller (Figure 4, a), SYN480R chip receiving module (Figure 4, b) and antenna (Figure 4, c).

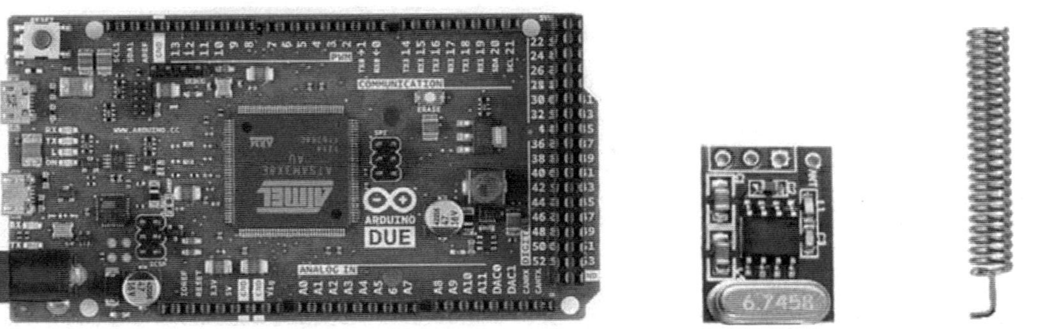

(a) (b) (c)

Figure 4. Components of the electrical part of the receiver

a- Arduino Due microcontroller; b- receiving chip module SYN480R; c- antenna

Arduino Due is used in this section of the microcontroller system. This is the first Arduino board based on a 32-bit ARM core (Atmel SAM3X8E ARM Cortex-M3 CPU) with 54 digital input / output and 12 analog pins [4]. It is chosen because of the many calculations it has to make based on the input signals. This board provides faster performance than other boards, which increases the number of data processing cycles. It performs the function of receiving a signal for reading the laps, but also for receiving and processing data based on the set data of the route, as well as the path, speed and time travelled by the electric vehicle. As after processing this data, they are visualized in an easily perceptible form on a 3.5 "TFT LCD display with a 16-bit parallel port.

The Arduino Due microcontroller can be powered by its USB port (5V), which means that it can be powered via a portable powerbank. The control is also equipped with a power socket, which allows a voltage of 6 to 12V. In this case, a voltage transformer is used because the power supply of the electric car, which has a voltage of 48V, is used [5].

Figure 5 shows the connection diagram of the receiving part of the system. GND pins on the board (black wire) are used for the table. The requirements of the technical characteristics of the receiving chip (SYN480R) - a voltage between 3.3 and 5.5V, which allows it to be connected directly to the 5V pin of the board (red wire). The transmitted signal is received by means of a small antenna, because the whole chip module and antenna are placed directly in the steering wheel, and the antenna is connected directly to the receiving chip module. The signal is received by the antenna, and the signal processing is performed by the receiving chip module, which pass on the processed signal to the RX0 pins of the Arduino Due board (green wire).

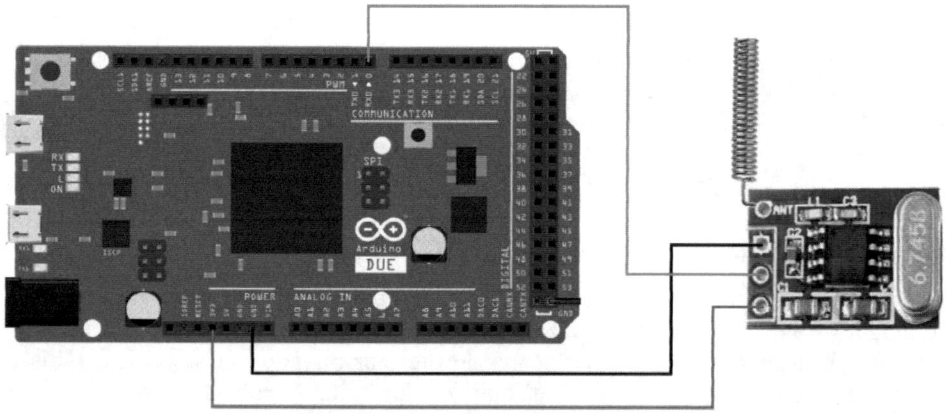

Figure 5. Wiring diagram of the receiving part of the system

3. Construction of a housing for the radiating part of the system

The production of individual components in the automotive industry is obstructed by a many factor. These factors are monitored from the initial design, during their construction and verification of their functionality until the moment they enter into exploitation. Due to the complexity of their work, some elements require additional upgrades, which cost a lot of resources and time for the manufacturers [6, 7,8]. This necessitates the use of software product with a high degree of flexibility [9]. The French company Dassault Systems provides the exact CAD software. The architecture of CATIA V5 is based on a modular principle and it includes a set of modules, which correspond to specific engineering tasks or processes [10]. Because of the common user interface on all platforms, the cost is minimized and the efficiency of introducing new opportunities for expanding the scope of use is maximized. To build the body of the radiating part of the system in a virtual environment, the 3D modeling software CATIA V5 is used [11]. In the construction of such a housing, the dimensions of the components, such as the antenna, the microcontroller control, the radiating chip module and the battery, are of key importance, as well as the possibility of the transmitter housing being easy to disassemble, making it easily portable. Another important aspect in the modeling of the case is the possibility to build it using FDM technology for 3D printing, i.e. to be consistent with the technical capabilities of the 3D printer.

As is well known, a system of three points in space describes a planar surface. The housing stand itself is in contact with the ground at three fulcrums (Figure 6), which ensure the stability of the housing at all times.

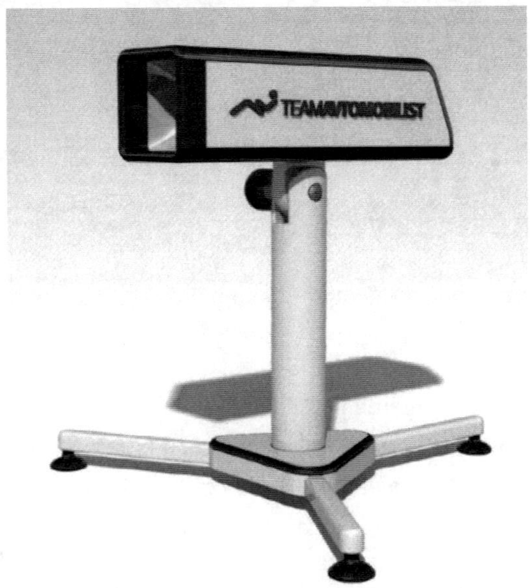

Figure 6. Overall view of the designed housing of the transmitting part

To activate the signal transmission, it is necessary to turn on the circuit breaker (Figure 7, b), which connects the electric battery (9V) with the Arduino Nano microcontroller. To confirm the operation of the system, the controller sends signals to an RGB LED (Figure 7, a), which is tested by simultaneous transmission in the three spectra (red, blue and green), emitting visible white light for a period of 1 second, after which the emission of light ceases. As it is permissible for the control to work smoothly from 6V to 12V, when the battery voltage drops below 7V, the LED indicates by flashing a bright red light. To avoid the need to replace the battery during operation, a small hole is made that provides access to the mini USB port of the Arduino Nano (Figure 7, c). Thus, by means of an external battery (powerbank), the transmitter can continue to operate smoothly.

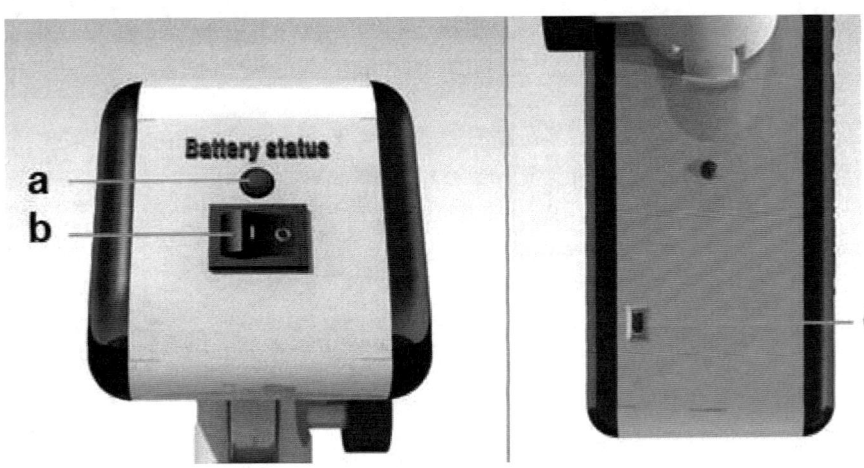

Figure 7. The body of the radiating part viewed from behind and below

In the main body of the transmitter are placed all the components necessary for the signal broadcasting (Figure 8).

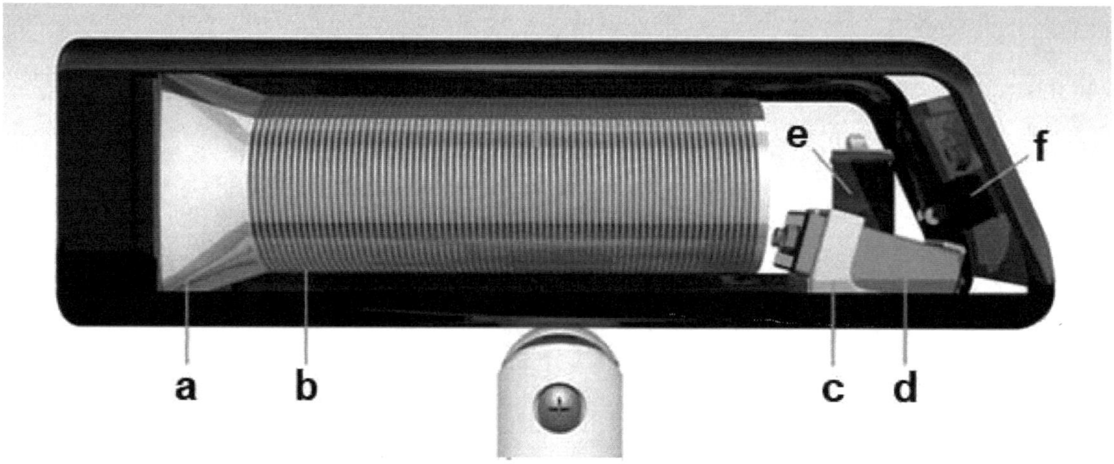

Figure 8. Arrangement of the components inside the body

a- Antenna support; b- antenna winding; c- battery holder; d- battery; e- Arduino Nano microcontroller control; f- switch

As mentioned, the design of the case allows it to be disassembled for more compact and easy carrying. The housing is divided into several main parts, which can be assembled and disassembled in a few seconds. The main body is connected to the stem by means of a bolt assembly with a nut, which allows to choose an inclination under which to fix the main body. And the stem is connected to the base to which 3 legs are mounted by pressure (Figure 9).

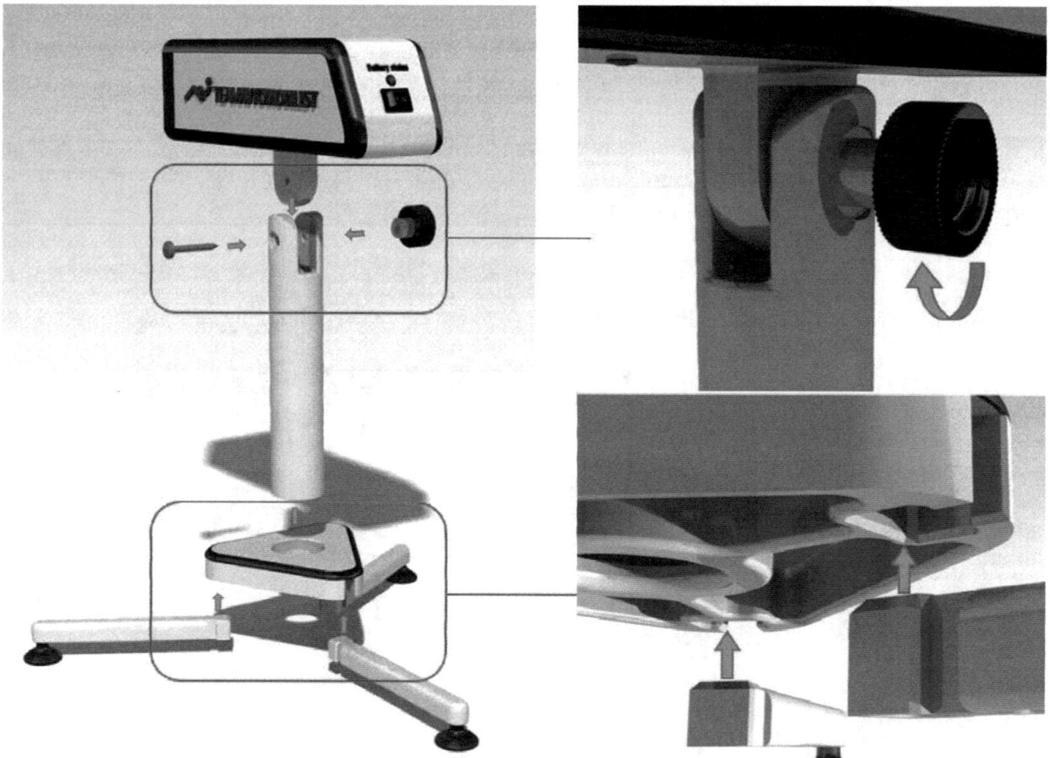

Figure 9. Representation of the way of assembling all the main components

IOP Conf. Series: Materials Science and Engineering **977** (2020) 012019 doi:10.1088/1757-899X/977/1/012019

4. Conclusion

The system was implemented in the latest prototype electric car DTT-3 by team "Team Avtomobilist". Experiments have shown that:

1. The system uses a minimum amount of energy for its operation;
2. Thanks to the system, the pilot concentrates his full attention on the runway, because of the fact that it is not necessary for him to monitor the laps made;
3. The system provides up-to-date information to the pilot about the laps he has made and the exact time of the last lap;
4. The design of the housing of the transmitting part allowed its easy transportation in disassembled form in a box with dimensions 250x100x120mm.
5. The design allows assembly to take place within 30 seconds.

Acknowledgments
The present document was supported with the financial assistance of the Project 2020-RU-03.

References

[1] Iliev S, Gunev D and Dobrev V 2017 Improving engineering education through the design and manufacture of electric car for the Shell Eco-marathon competition, ERS Spectrum (Educational Research Service) Vol. 29 No. 2
[2] Dobrev I, Massouh F, Danlos A, Todorov M and Punov P 2017 Experimental and numerical study of the flow field around a small car, MATEC Web of Conferences 133 02004
[3] Shell eco-marathon, 2020 official rules chapter I
[4] Margolis M 2011 Arduino Cookbook (O'Reilly Media, Inc.)
[5] Hristov R and Grozev G 2014 Information system for monitoring of the electricity consumed by KVAZI *EKO Varna* **21** pp 42-59
[6] Angelov B 2012 About the Hierarchical Relationship between the System for the Conversion of Matter, Energy and Information, the Technical System and the Executive Organs of the Technical System Int. Sc. Conf. MendelTech, ISBN 978-80-7375-625-3
[7] Angelov B 2012 About the Contents of Design Work in the Preparation Phase for the Production of New Technical System Int. Sc. Conf. MendelTech, ISBN 978-80-7375-625-3
[8] Antonova D and Stoycheva B 2018 Approved model of factors, influencing the management process in developing new products IMES 2018 Czech Republic, Vysoká škola ekonomická v Praze Nakladatelství Oeconomica – Praha Nakladatelství Oeconomica pp. 38-54
[9] Petrova Tz 2016 Guide for engineers and students. Part 1: CATIA V5. Design of machine-building details. Plovdiv: "Haycad Infotech Publishing". ISBN 978-619-7365-01-6.;
[10] Prof. Univ, S. T. P, CATIA V5-6R2015 for Designers, 13th Edition. CADCIM Technologies; 13 ed. ISBN-13: 978-1942689218, 2016;
[11] Weaver, J. M., Zamani, N. G. (2012). CATIA V5 Tutorials Mechanism Design & Animation Release 21. SDC Publications. ISBN: 978-1-63056-056-0;

IOP Conf. Series: Materials Science and Engineering **977** (2020) 012020 doi:10.1088/1757-899X/977/1/012020

Vehicle's energy efficiency via pilot work's efficiency in Shell Eco-marathon competition

D Gunev [1], S Iliev[2] and E Mitev [2]

[1] University of Ruse, Department "Industrial Design", 8 Studentska Str., Ruse, Bulgaria
[2] University of Ruse, Department "Engines & vehicles", 8 Studentska Str., Ruse, Bulgaria

e-mail: dgunev@uni-ruse.bg; spi@uni-ruse.bg; emitev@uni-ruse.bg

Abstract. Shell Eco-marathon is a competition about energy efficiency and the first thing one have to do, is to design and build lightweight, aerodynamic vehicle with fine mechanical transmission and tires with low rolling resistance. Depending on power unit type, there is a need of perfect way of transforming chemical or electrical energy to mechanical one. In addition, almost every team develops their own driving strategy, applied by the pilot in the competition. Doing this, in the best way they can do, most competitors decide before-mentioned is the maximum, but if one can look to the vehicle plus the pilot like a system, there is more possibilities for improving the efficiency of that system. This paper includes analysis of vehicle control functions and specific organizational distribution and design of vehicle's controls, in accordance with the capabilities of the pilot.

1. Introduction

In the history of mankind all types of vehicles: ground, marine and aerial were design and build to multiply human abilities in transportation of materials, goods, people etc. There were, there are and there will be extreme examples of that vehicles with no utilitarian but specific function: scientific, research, sport competition, which helps people to extend their knowledge about the nature and the man as part of it.

Shell Eco-marathon is competition with extreme, efficient vehicles, which are built with and use for their movement minimum natural resources.

2. Analysis of vehicle control functions

Every vehicle is a mechanical system, which main function is: movement and main sub functions are: steering – giving direction of moving; acceleration – getting from state of resting to state of moving or getting to move rapidly; deceleration (braking) – getting from state of moving to state of resting or getting to move slowly. Without control of those sub functions, state of movement will be impossible or chaotic. While the steering must be applied permanently without interruption through the whole process of movement, acceleration and deceleration have to be applied successively without overlapping and discretely.

2.1. Steering

Steering is the sub function which needs most of all precision. That's why almost every vehicle is steered by hands, or by hands and foot in addition.

There are exotic examples of vehicle steering with arms, legs or weight of the driver. Sledge is one of them.

To steer the sled, the slider uses his or her calves to apply pressure to one of the runners, deforming it slightly in turn direction, or shifts their weight sideways using their shoulders (Figure 1).

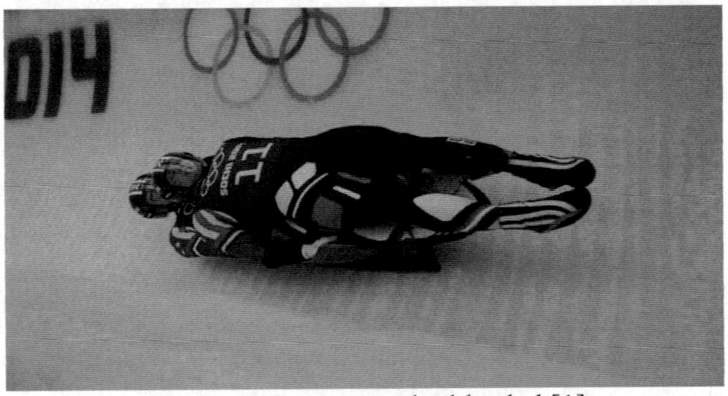

Figure 1. Lugers on double sled [1]

Historical sight in watercraft and ground vehicles will show that first steering control was simply lever called tiller. It was used in the first and not big in dimensions watercrafts and ground vehicles. Tiller provides leverage in the form of torque for the helmsman (Figure 2) or the driver (Figure 3) to turn the device that changes the direction of the vehicle, such as a rudder on a watercraft and the wheels on cars and two or three wheeled vehicles. Bicycles and motorcycles use some kind of double tiller called handlebar (Figure 4).

Figure 2. Boat tiller [1]

Figure 3. Tiller on a car [1]

Figure 4. Bicycle's handlebar [1]

In bigger vehicles the force to operate the tiller is bigger too and at one point it is impossible for human being alone to steer the vehicle. The solution is to use intermediate leverage or reducer between the control and steering device. In this case the force to steer the vehicle becomes smaller but the movement of the control becomes bigger and as a rule there is not enough space for this. Using wheel (boat's wheel (Figure 5) or steering wheel (Figure 6)) as a control solves this problem.

Figure 5. Boat's wheel. [1]

Figure 6. Steering wheel of a car [1]

In modern vehicles the use of power steering or drive by wire systems, with hydraulic or electric actuators, size of the wheel becomes smaller or even is replaced by joystick (Figure 7).

Figure 7. Joystick in a car [2]

Analysis of human-machine interaction [3] and steering sub function of a vehicle in detail can be made and one can see the following:

The grasp of standard car and truck steering wheels is variable and has three main variants:

- Basic - the steering wheel is hold with the whole hand - all fingers and palm are involved.
- Grasp for operating the controls behind the wheel - the thumb and palm are constantly involved in steering and the other fingers are used either to work with the controls or for additional holding, to help the palm.
- Grasp for operation with the controls on the front of the steering wheel - the palm and four fingers are involved in steering without the thumb, which works with the controls.

The grasp of motorcycle and bicycle steering controls is variable, and there are three main variants too:

- Basic - the steering wheel is hold with the whole hand - all fingers and palm are involved.
- Grasp for operating the controls in front of the handle - the thumb and palm are constantly involved, and the other fingers are used either to work with the controls or for additional holding, to help the palm.
- Grasp for operating the controls on the inner side of the handle - involves the palm and four fingers without the thumb, which works with the controls.

Acceleration and Deceleration are sub functions which need precision only in ground vehicles, because of the character of the environment of moving.

2.2. *Acceleration*

In common case sub function - acceleration is linked with process of active moving when the driver have intention to move faster or to keep moving with constant speed (when acceleration equals to zero). This means that handling the controls of acceleration is constant thru the process of moving such like the handling of the control of steering and consequently controls of acceleration is possible to operate with the same limb only if the operator have constant contact with both controls – steering and acceleration. This requirement is fulfilled in motorcycles and other light vehicles where the steering is conducted by handlebar where in the process of steering motorcyclist as a general rule have to keep his /her hands on the handlebar permanently, instead car driver have to grasp the steering wheel alternatively with right and left hand during bigger turn. These are the reasons, we have the control of acceleration – in form of a pedal, operated by foot in the cars and twisting handle or small lever operated by hand or thumb in the motorcycles.

2.3. *Deceleration (Braking)*
This sub function refers to passive part of moving where the vehicle decreases its speed when no engine power is supplied due to different drag forces or when active brake force is applied to the vehicle. This function is randomly used and in most of the cases when the vehicle is moving straight forward (not in a turn). This means that one can use any limb, doesn't matter it is used in steering or not, to operate this function in sense of overlapping actions. Another important moment in the process of choosing how to operate brake controls is the force needed to. That's why in bigger ground vehicles like cars and trucks (especially before the invention of servo systems) logical choice was leg to operate brake controls and in small two to four wheeled vehicles hand or hand and foot because of the combinations with other controls.

3. Analysis of Shell Eco-marathon prototype class vehicle control functions

3.1. *Shell Eco-marathon rules*
In Shell Eco-marathon 2019 Official Rules, Chapter I, one can read:

"ARTICLE 42: TURNING RADIUS AND STEERING

a) Only front wheel steering is permitted. If the Organisers are not satisfied with the effectiveness and/or control of a vehicles steering system, this vehicle will be removed from the competition.

b) The turning radius must be 8 m or less. The turning radius is the distance between the centre of the circle and the external wheel of the vehicle. The external wheel of the vehicle must be able to follow a 90° arc of 8 m radius in both directions. The steering system must be designed to prevent any contact between tyre and body or chassis.

c) Electrically operated indirect steering systems are permitted providing they are operated by a steering wheel or similar (rotary potentiometer), joystick operation is not permitted. If electronic steering systems are used, in the event of system failure, the vehicle must be equipped with manual steering override.

...

ARTICLE 43: BRAKING

a) Vehicles must have with two independently activated braking systems; each system must comprise of a single command control (lever or foot pedal), command transmission (cables or hoses), and activators (callipers or shoes). Brakes that act on the tyres are not permitted.

i. One system must act on the front wheels, the other must act on the rear wheel(s). For the front wheels, there must be one activator (calliper or shoe) for each individual wheel, with both activators commanded by the same command control. In addition, the right and left brakes must be properly balanced.

ii. The rear system must act on each wheel, unless they are connected by a common shaft in which case, they can have a single system.

iii. It must be possible to activate the two brake systems at the same time without taking either hand off the steering system.

...

d) A single foot-operated hydraulic brake is recommended for the front wheels. For 2020, foot-operated front brakes will be required." [4].

3.2. Battery - electric prototype specific features
Battery electric prototype propulsion system usually consist of electric motor, one-way clutch, transmission and driving wheel. In this case there is need of only one control – accelerator and there is no need of clutch control.

The vehicle has as a rule three wheels two in front and one at the rear. Tyres' plane of rotation should be strictly perpendicular to the ground thru the movement for minimal rolling resistance.

For the DTT-3 battery – electric prototype of University of Ruse, team "Avtomobilist" geometric parameters are:

wheelbase - 1450 mm
front track – 510 mm.

To fulfil Article 42, “a” and “b”, turning angle of the front wheels must be 9.96 deg for the outer wheel and 10.60 deg for the inner one.

3.3. Requirements for the sub-functions performance
Because of the small turning angles needed for front wheels and relatively small mass of the vehicle and the pilot (between 70 and 85 kg) steering control could be something like handlebar rather than steering wheel. Accelerator control could be operated with hand or foot. Front brakes foot operation is mandatory (Article 43, “d”) and rear brake control could be operated with hand or foot.

Proper organization and distribution of these controls is the base for fast and precise, in other words effective way of driving the vehicle [5].

4. Concept for the controls

4.1. Distribution
To eliminate errors in the processes of acceleration and deceleration, which results in loss of energy or can provoke accidents, one can decide to separate these two controls between different limbs. And once the front brakes are mandatory to be operated with foot, it’s logical and natural to operate rear brake with foot too. This is the reason why in DTT-2 and DTT-3 vehicles front and rear brake controls, are in form of a foot pedal.

Steering and acceleration which needed highest possible precision was decided to be operated with hands and fingers.

4.2. Organization
Taking in account that fact that more people are right-handed (and pilots in the “Avtombilist” team too) front brake control is operated with right foot for bigger precision and sensitivity and rear brake control is operated with left foot like emergency brake used in addition to the front brake or separately.

For the same reason acceleration control is oriented to the right hand and fingers.

There are additional sub functions that are very important for safety reasons – horn, for example which can be operated with fingers.

4.3. Construction
Compared to the motorcycle and car steering controls in the new steering wheel, the *main concept* is that the grasp has only one variant: *ulnar-palmar grasp* (Figure 8) and is performed only by the ring finger and the little finger (in which the ulnar nerve ends) and the palm of the hand. The thumb, forefinger and middle fingers operate on the controls or complement the grasp.

Figure 8. Ulnar-palmar grasp of the steering control

The steering wheel consists of three main elements: left and right handles and a main body. The handles lie on opposite sectors of the thought circle passing through them. The three elements are connected into whole one by means of transversely located on each handle a bridge connecting the middle of the handle with the upper end of the main body. In this way, the handles acquire a T-shape, turned horizontally. In the lower part the zones for realization of the ulnar-palmar grasp are formed, and in the upper part - zones for placement of controls (buttons) and surfaces for supplementing the grasp (Figure 9).

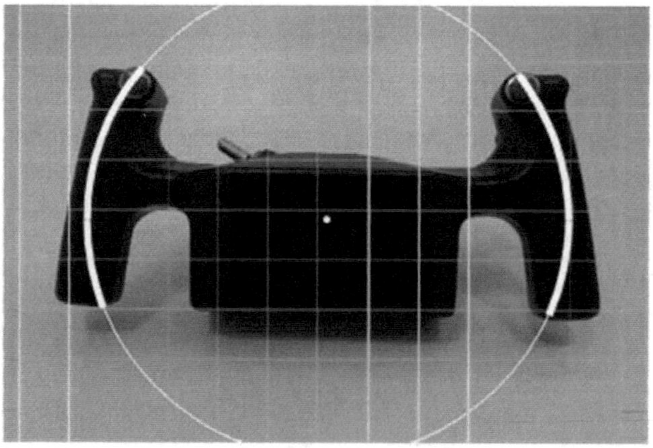

Figure 9. New steering wheel

According to scientific studies [6] [7] [8], control and display locations must be determined very carefully for both accuracy and safety reasons. In this extreme but not complex case, controls and display were arranged in a way of avoiding action overlap in a process of controls use by analysis of timing of action.

On the front wall of the main body there is a display, located at an angle to the axis of the steering shaft for better visibility. In the central part of the rear wall of the housing there is a flange element with oval (slot) holes arranged in a concentric circle, designed for fine adjustment of the position of the steering wheel relative to the steering shaft (Figure 10).

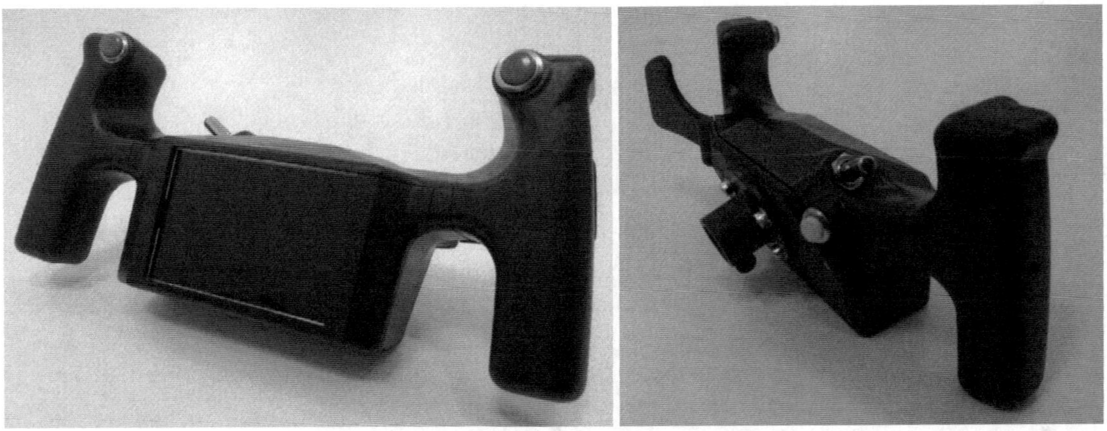

Figure 10. Steering wheel - features

Behind the right handle, a finger pedal is positioned for controlling the speed and acceleration of the vehicle, and it is operated with forefinger and middle finger simultaneously (Figure 11) [9].

Size and all dimensions of the steering wheel are consistent with ergonomic requirements. [10]

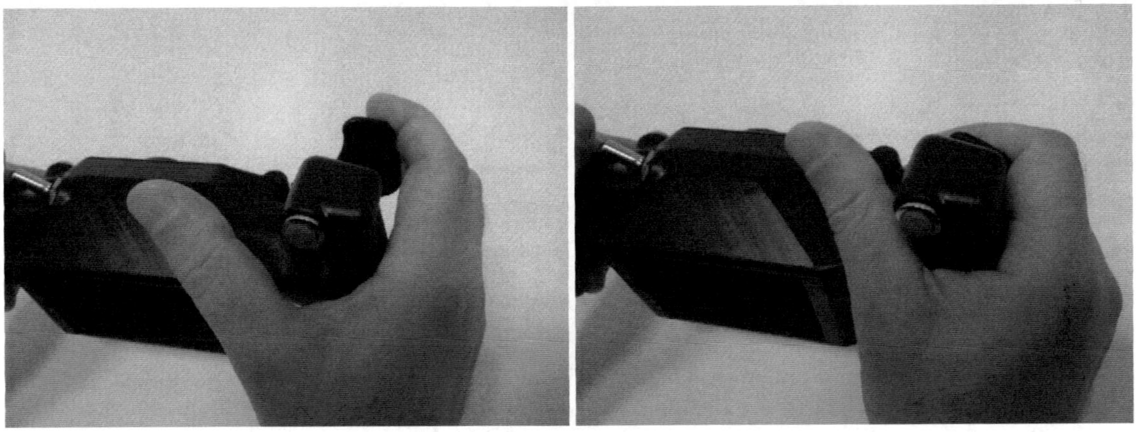

Figure 11. Accelerator control operation

5. Conclusion

The steering wheel is designed for precise and safe driving of the vehicle by:

- All operations with the controls can be performed without leaving the steering wheel grasp ensuring that it can be gripped continuously with both hands without the need to release it when performing other basic or additional activities in the steering process;
- The controls are positioned to require minimal effort and movement from parts of the human body, which reduces fatigue and therefore improves driving efficiency and safety.
- Every control is operated with different limb or finger, or combination of them that leads to situation of coding control commands by place, by executable part of the human body and by it specific movement during the operation which is base for a faultless and efficient driving.

Acknowledgments
The present document was supported with the financial assistance of the Project 2020-RU-03.

IOP Conf. Series: Materials Science and Engineering **977** (2020) 012020 doi:10.1088/1757-899X/977/1/012020

References

[1] https://en.wikipedia.org
[2] https://www.saabplanet.com/saab-9000-drive-by-wire-1992
[3] Bridger R.S. (2003), Introduction to Ergonomics. *Taylor & Francis*
[4] The Shell Eco-marathon Global Rules 2019, Chapter I
[5] Hristov R and Petrov P 2019 Exploring different car suspension solutions for the shell eco-marathon competition, *proc.of university of Ruse* vol.58 book 4 pp 30-33
[6] Angelov B 2012 About the Hierarchical Relationship between the System for the Conversion of Matter, Energy and Information, the Technical System and the Executive Organs of the Technical System Int. Sc. Conf. MendelTech, ISBN 978-80-7375-625-3
[7] Green, P., Kerst, J., Ottens, D., Goldstein, S., & Adams, S. (1987). Driver Preferences for Secondary Controls. Ann Arbor: *The University of Michigan Transportation Research Institute.*
[8] Murata, A., Moriwaka, M. (2005). Ergonomics of steering wheel mounted switch - how number and arrangement of steering wheel mounted switches interactively affects performance. *International Journal of Industrial Ergonomics*, 35 (2005), 1011-1020
[9] Angelov B 2012 About the Contents of Design Work in the Preparation Phase for the Production of New Technical System Int. Sc. Conf. MendelTech, ISBN 978-80-7375-625-3
[10] Tilley A. R., Henry Dreyfuss Associates (1993). The Measure of Man and Woman: Human Factors in Design. *Whitney Library of Design.*

Precision vibration measurement systems

S Stefanov*, R Hristov and P Toteva

Faculty of Mechanical Engineering and Technologies, Technical University of Varna, 1 Studentska Street, Varna 2010, Bulgaria

* corresponding author: stefanov_48@abv.bg

Abstract. This report examines the design of a vibrostand for precision vibration measurements by comparing the parameters of the transducer with that of a reference transducer. The report gives the methodology for designing a stand for precise vibration measurements. Based on the theoretical calculation, a stand for precise vibration measurements was designed and manufactured. The manufactured stand has the lowest possible natural frequency, a characteristic of the suspension approaching the ideal, the possibility to adjust the stiffness of the suspension and a large load capacity.

1. Introduction

Engineering requires that the values of different parameters be measured. The measured parameters' accuracy depends on the objectives of the tasks. For this purpose, the companies producing different types of measuring instruments produce also transducers converting measurement parameters of different accuracy classes. The measured parameters have heterogeneous nature, such as pressures, temperatures, displacement, velocity, acceleration, etc. These instruments and transducers may, over time, register some deviations from the set parameters [1,2]. Each country has its metrological control service, which verifies the parameters and issues a certificate of serviceability and suitability of measuring instruments. Precise measurements need very accurate equipment because measurements with less accuracy may be incorrect [3,4,5]. Vibration measurements at low frequencies are used in several applications, as for example: environmental monitoring, seismic measurement, human-body vibration assessment, mechanical tests. A comparison calibration system requires a calibrated reference standard to quantify the measured with high accuracy [6].

This report discusses the design of vibrostand for precision vibration measurements, comparing the parameters of the transducer with those of a reference transducer.

2. Explanation

The main objective of project described at this paper is:

1.To design and develop vibration isolation system for precision vibration measurements of piezoelectric and inductive vibration transducers measuring vibration acceleration, vibration velocity and vibration displacement.

2.1. Operating Principle

The system must ensure maximum isolation of the tested vibration transducer from external influences.

After studying similar systems and structures, pneumatic vibration isolation scheme was selected.

IOP Conf. Series: Materials Science and Engineering **977** (2020) 012021 doi:10.1088/1757-899X/977/1/012021

The main indicators that the system must meet are: low natural frequency; characteristic of the suspension must approach the perfect one, adjustability of the suspension stiffness, high load capacity and relatively low cost [7].

The general view of the system is shown in figure 1. The two units (left and right) are absolutely identical and can work both together and independently of each other. The pumping of vibration isolators 5 is performed by compressor 2, with nozzle 11 being placed in one of the eight valves. The compressor is switched on by switch 9. The pressure in vibration isolators (0,08 - 0,1) [MPa] is read by pressure gauge 10.

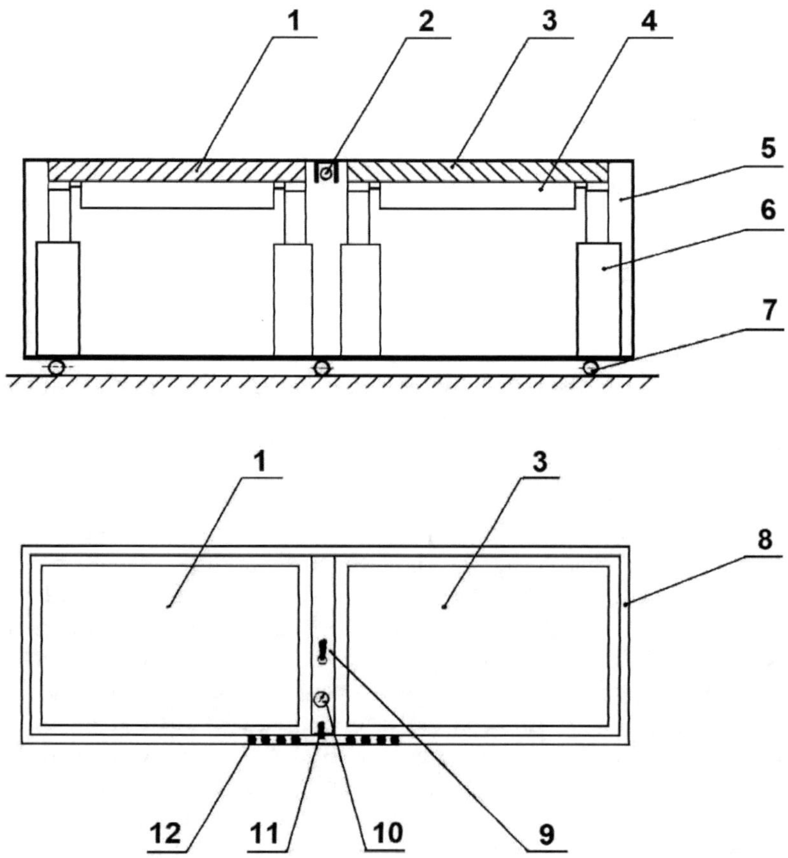

Figure 1. General view of the system

2.2. Theoretical Statement:

Figure 2 shows the vibration insulated plate, which is considered as a solid body with six degrees of freedom. Two coordinate systems are introduced: movable - with the main inertial axes of the plate (x, y, z), which is fixed and coincides with the movable plate at its static equilibrium (X, Y, Z). The plate displacements during oscillations are insignificant. For this reason, the rotation of movable, compared to the fixed coordinate system is set with the elementary angles - φ, ᴪ, æ.

The foundation free oscillations are described by a system of linear second order differential equations with constant coefficients:

$$m.\ddot{X} + a_{11}.X + a_{12}.Y + a_{13}.Z + a_{14}.\varphi + a_{15}.\psi + a_{16}.æ = 0$$
$$m.\ddot{Y} + a_{21}.X + a_{22}.Y + a_{23}.Z + a_{24}.\varphi + a_{25}.\psi + a_{26}.æ = 0$$
$$m.\ddot{Z} + a_{31}.X + a_{32}.Y + a_{33}.Z + a_{34}.\varphi + a_{35}.\psi + a_{36}.æ = 0$$

$$m.\rho_x^2.\varphi + a_{41}.X + a_{42}.Y + a_{43}.Z + a_{44}.\varphi + a_{45}.\Psi + a_{46}.æ = 0 \quad (1)$$
$$m.\rho_y^2.\varphi + a_{51}.X + a_{52}.Y + a_{53}.Z + a_{54}.\varphi + a_{55}.\Psi + a_{56}.æ = 0$$
$$m.\rho_z^2.\varphi + a_{61}.X + a_{62}.Y + a_{63}.Z + a_{64}.\varphi + a_{65}.\Psi + a_{66}.æ = 0$$

where:
m is the foundation mass
X, Y, Z is the foundation plate displacement compared to the fixed coordinate system
φ, Ψ, æ are angles of rotation of the plate relative to X, Y, Z
aij are constant coefficients reflecting the linear, linear-angular and angular stiffness of suspension. The matrix of these coefficients is symmetric - aij = aji.
X, Y, Z are suspension points' coordinates in the movable coordinate system
ρx, ρy, ρz are inertial radii of mass relative to x, y and z axes.
α, β, γ are guidance vectors of the vibration isolators' stiffness axes.

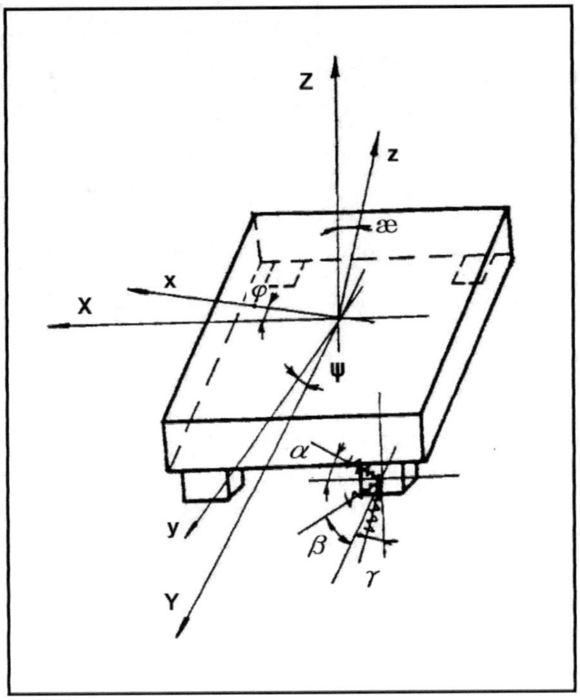

Figure 2. Vibration insulated plate

The analysis of the equation system shows that the action of force or torque on the system causes oscillations in the six main coordinates and that the foundation will perform interconnected oscillations. The theory of oscillations shows that, other things being equal, their Interconnection leads to expansion of the spectrum of free oscillations, and that with its increase the difference between lower and higher frequencies of free oscillations increases. Removing oscillations' interconnection is equal to separating oscillating motions into directions along and around the X, Y, and Z axes. Complete separation can be obtained if the matrix of coefficients aijis diagonal (aij = 0 for i other than j).

2.3. Design of Precision Vibration Measurement System

The scheme of the system is shown in Figure 3. The axes of fixed and movable coordinate systems coincide with the main inertial axes of the foundation plate. The vibration isolators are located symmetrically to XZ and YZ planes and lie in XY plane. Their main stiffness axes are parallel to the coordinate axes. After removing the zero terms from the differential equation system, the natural oscillations of this suspension system are described by the following expressions:

$$\begin{aligned} & m.\ddot{X} + a11.X = 0 \qquad && Jx.\varphi + a44.\varphi = 0 \\ & m.\ddot{Y} + a22.Y = 0 \qquad && Jy.\varphi + a55.\psi = 0 \\ & m.\ddot{Z} + a33.Z = 0 \qquad && Jz.\varphi + a66.æ = 0 \end{aligned} \tag{2}$$

where:

Jx, Jy, Jz are mass inertial torques of the plate relative to the coordinate axes

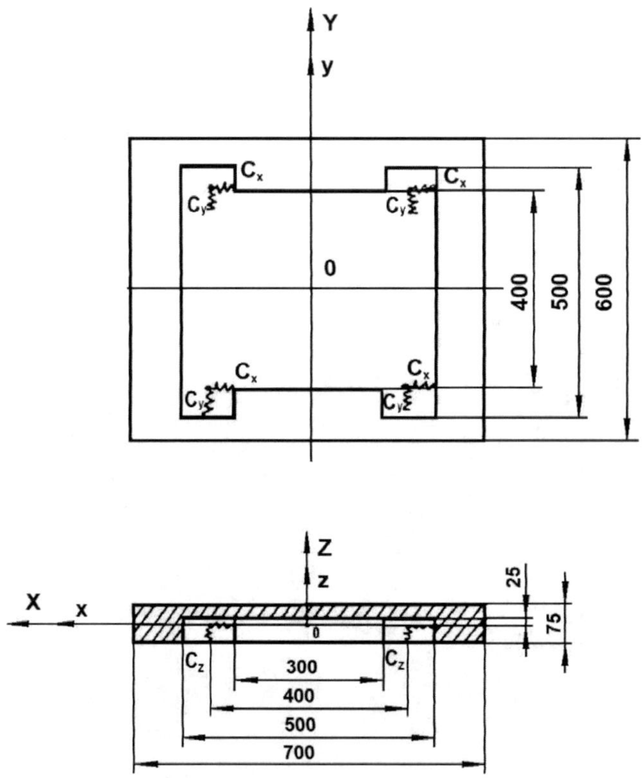

Figure 3. Scheme of the system

The calculations are made by optimising the parameters in order to obtain compactness of the structure.

1. Number of vibration isolators on one stand – N = 4
2. Mass of one stand, together with the vibrotable and sensor, at the time of measurement - m = 180 [kg].

3. Working absolute pressure in the vibration isolators [8]

$$p = \frac{m.g}{4.F} + 0{,}981.10^5 = 1{,}86.10^5 [N/m^2] \tag{3}$$

where: g = 9.81 [m / s2] - gravity variations;

$F = \frac{\pi.D_a^2}{4}$ - active area of a vibration isolator

Da = 0,1 [m] - vibration isolator diameter

4. Volume of one vibration isolator.

$$V_a = \frac{\pi . D_a^2}{4} . h = 1{,}414 . 10^{-3} [m^3] \qquad (4)$$

where: h = 0,18 [m] - vibration isolator height

5. Volume of the receiver to the vibration isolator.

$$V_p = 4 . V_a = 5{,}655 . 10^{-3} [m^3] \qquad (5)$$

$$V_p = \frac{\pi . D_p^2}{4} . h_p = 5{,}67 . 10^{-3} [m^3] - \text{ volume, at } D_p = 0{,}16\ [m] \text{ and hight } h_p = 0{,}282\ [m]$$

6. System natural frequency.

$$\omega_n = \sqrt{\frac{k.p.F^2}{m.V_p}} = 5{,}087 [s^{-1}] \qquad f_n = \frac{\omega_n}{2.\pi} = 0{,}80\ [Hz] \qquad (6)$$

7. System resonant frequency at optimal damping.

$$\omega_c = \omega_n \sqrt{\frac{2.N}{N+2}} = 5{,}87 [s^{-1}] \qquad f_c = \frac{\omega_c}{2.\pi} = 0{,}93\ [Hz]$$

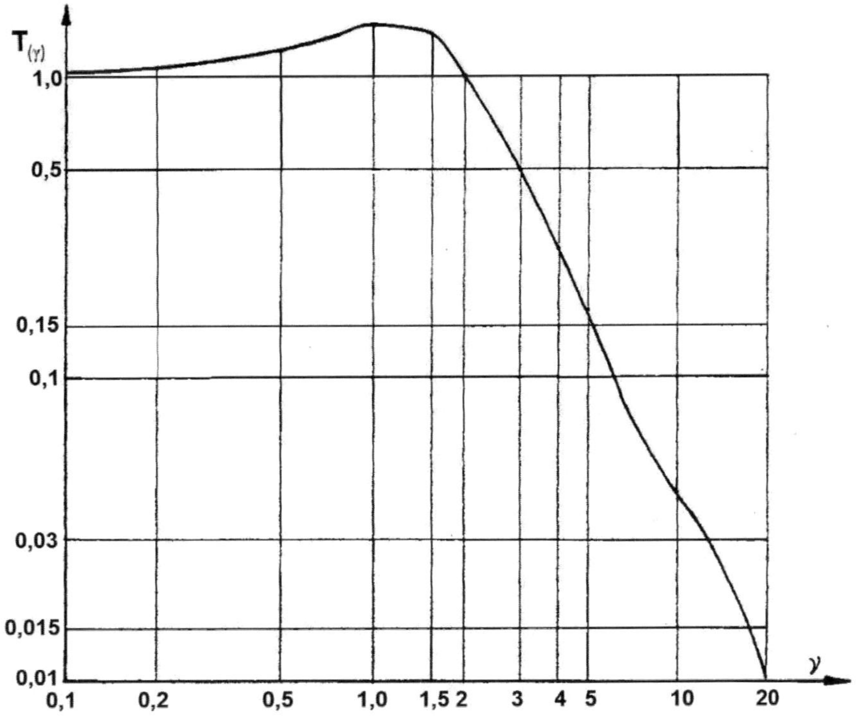

Figure 4. Characteristic of this system

8. Dimensions of the throttle capillary tube connecting the vibration isolator and receiver.

$$d = \left[\frac{128.\mu_g.L.F^2}{2.\pi.\omega_n.D_a.m}\right]^{\frac{1}{4}} \qquad \mu_g = 1{,}81.10^{-5}\left[\frac{kg}{s.m}\right] \tag{7}$$

d = 1.7.10-3 [m] at L = 0.05 [m]

The characteristic of this system - $T_z(\gamma)$ is given in figure 4 and is subject to the following dependence:

$$T_z(\gamma) = \sqrt{\frac{1+4.D_0^2.\gamma^2}{4.D_0^2.\gamma^2.\left(1-\frac{\gamma^2}{N}\right)^2+\left(1-\frac{1+N}{N}.\gamma^2\right)^2}} \tag{8}$$

$$\text{where: } \gamma = \frac{\omega}{\omega_c};\ D_0 = 0{,}968$$

Conclusions

Based on research that has been done, the following conclusions can be made:

- An appropriate methodology has been selected and the precision vibration measurement system has been calculated.
- According to the obtained results, the precision vibration measurement system is made.
- It is used successfully in Varna Quality Centre.

References

[1] T Feist and U Kaulbars 2017 Measuring or estimating vibration dose: Comparison of new systems with standard-conforming measurements *Technische Sicherheit* 7(10) pp 20-24
[2] J Tuo and Z Deng 2017 A New Type of Sensor System with Torsional Quasi-Zero Stiffness for Angular Vibration Measurement *Journal of Xi'an Jiaotong University* **51 (8)** pp. 90-95
[3] Y Gao, W He and Q Liu 2020 A novel vibration velocity measurement method for electromagnetic vibrators *Meas.Sci. Techn.* **31** 105008
[4] D P Towers, C H Buckberry, B C Stockley and M P Jones 1995 Measurement of complex vibrational modes and surface form-a combined system *Meas. Sci. Techn.* **6 (9)** pp 1242-49
[5] D Wang et al 2018 IOP Conf. Ser.: Mater. Sci. Eng.**452** 042166
[6] C D Ferreira et al 2015 J. Phys.: IOP Conf. Ser. **575** 012003
[7] Y Yamakawa and T Yamazaki *2016* Compensation method and measurement accuracy to floor vibration in electronic balance system *IEEE SENSORS Orlando* pp. 1-3
[8] E Marinov, M Zhechev and E Ruseva 1985 Vibration isolator testing system *Scientific-practical conference "Car and the natural environment"* (Razgrad) pp. 130 - 134.

IOP Conf. Series: Materials Science and Engineering **977** (2020) 012022 doi:10.1088/1757-899X/977/1/012022

Investigation of the electromagnetic field in electric and hybrid cars

R Hristov*, S Stefanov and P Kostov

Faculty of Mechanical Engineering and Technologies, Technical University of Varna, 1 Studentska Street, Varna 2010, Bulgaria

* corresponding author: rosen.hristov@tu-varna.bg

Abstract. Electric and magnetic fields are present everywhere electricity flows. In electric vehicles, passengers sit very close to electrical system with significant power, usually for significant period of time. The relatively high currents achieved in these systems and the short distance between power devices and passengers mean that the latter are exposed to the relevant magnetic fields. In view of this, it is important to study the values of electromagnetic field parameters on board electric vehicles to ensure compatibility with international standards. The measurements were performed inside the four conventional vehicles under the standard condition. Data of international and Bulgarian standards for safe levels of electromagnetic fields are given. After the measurements, it was established that these levels were not exceeded.

1. Introduction

This report will present a study of electric and magnetic fields (EMF) exposure inside electric vehicles and hybrid vehicles. The purpose of the report is to present the parameters of electromagnetic environment inside different vehicles and to compare them with the requirements of international and Bulgarian standards. The task of the study is to determine whether the vehicles in question have negative effect on the human body through their non-ionizing radiation. It is assumed that the new cars are in the norm, this should be checked by their manufacturers. We decided to measure on cars that have been in operation for several years to see if there is any significant increase in electromagnetic radiation.

The definition of a kind "What is an electromagnetic field" can also be found in the science world. For instance, Einstein defined a field as "regular, geometrical curvature of space" [1]. Electromagnetic fields can be either time constant or time varying. If the sources which produce electromagnetic fields are not varying in time, the electromagnetic fields themselves are also time constant. To reduce the magnetic field inside the compartment, some car manufacturers choose to integrate the inverter into electric motor and the power management system into the battery system [2].

There have been other publications on EMF and noise research in recent years [3,4,5], but we are expanding the number of measured points in the car. Fields in non-hybrid cars were higher at the front seats, while in hybrid cars they were higher at the back seats [6].

2. Experimental setup

2.1. General Design of Electric Vehicle

Unlike the complex and low-efficiency design of vehicles with internal combustion engines, electric vehicles have extremely simple system, consisting mainly of drive and power supply module. Control is much simpler, and the natural features of electric motors effectively replace the expensive and

IOP Conf. Series: Materials Science and Engineering **977** (2020) 012022 doi:10.1088/1757-899X/977/1/012022

complex systems, such as: ABS braking system, off-road drive, speed differentiation when cornering, etc. Moreover, the reversing mode of operation of electric motors allows the use of unnecessary inertia to generate energy and to charge the batteries.

Electric vehicle efficiency depends mainly on the right choice and design of its electrical system. If we have to compare individual components of electric vehicle and those of conventional one, we may say that the batteries play the role of a tank. They store energy, which will be converted later on into some mechanical motion. The high voltage battery, which is the heaviest electrical component of the vehicle, is located very low below the passengers. Most electric vehicles have front-axle drive and the battery modules are located in the floor between the front and rear axles. This configuration provides a lot of free space for passengers, large trunk, and very low centre of gravity, which improves the overall stability of the vehicle (Fig. 1)[7].

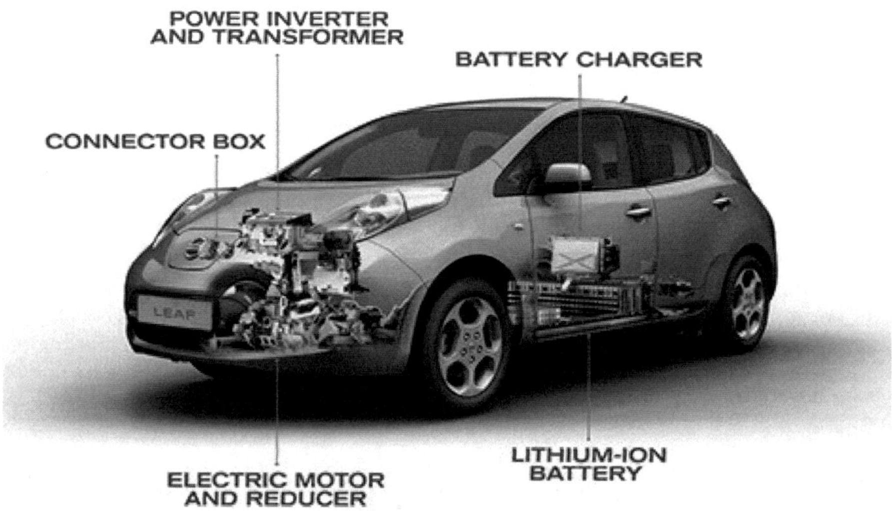

Figure 1. General design of electric vehicle.

Battery is the key component of electric vehicle for the following reasons:

- vehicle take-off run depends almost entirely on the battery;
- this is the heaviest electrical component;
- this is the most expensive electrical component.

The torque is provided by electric motor. Two types of electric motors, which together form a third type of electric motor, are mainly used in production of passenger cars:

- permanent magnet motors;
- induction motors;
- motors with variable magnetic resistance.

It is more appropriate in this case to call them electric machines instead of motors, since they can generate electricity when the vehicle is not moving. This mechanism is called energy recovery/ regeneration or regenerative mode (recuperation).

The controller is used to control the electric motor. It determines the speed at which the electric motor will rotate. Its purpose is the same as that of the carburettor. The controller doses the fuel supplied to it, depending on the speed, which we want to move with. In our case, the supplied fuel is voltage from the batteries. The power electronics control module has several subsystems and each of them has control function. When the vehicle is charged by the mains - 230 [V], the rectifier converts alternating current

(AC) into direct current (DC), which is fed into the high voltage battery. DC-DC converter is responsible for lowering high voltage (e. g. 400 [V] of a three-phase grid) to low voltage of 12 [V].

The inverter controls electric machine speed and torque, converting direct current from the battery into AC three-phase current for the electric machine. When the vehicle is in the energy recovery (braking) phase, the inverter performs reverse conversion, from 3-phase AC to DC.

2.2. Methods used to measure electromagnetic field (EMFs)

Emissions from electric vehicles are completely different from the exposures of power lines, which arise greatest concern in society. EMFs in electric vehicles are characterised by complex and wide frequency spectrum and specific wave patterns. The information obtained from the measurements in this work will allow determining the specific characteristics of electromagnetic field that could be harmful to human body. The careful determination of electromagnetic fields inside the vehicle under different operating conditions, obtained from their field measurements, can provide information, which is important to reduce the exposure. Based on the presented location of electrical components and to measure and evaluate the exposure to EMF, the following methodology was used. Each passenger's and driver's seat in the vehicle's body was measured at three points: at the level of lower limbs (feet), at the level of torso (seat pan) and at the level of head. Points have been selected at the level of the limbs due to the close location of the batteries in the floor of the electric vehicle interior (See: figure 1); the same applies to the points selected at torso levels.

Schematically selected measurement points are presented in figure 2. The presented diagram shows that the measured EMF exposure values are personal for each passengers and driver seat, thus locating the areas with the highest values, which will be indicated and compared with the standards for EMF exposure limit values.

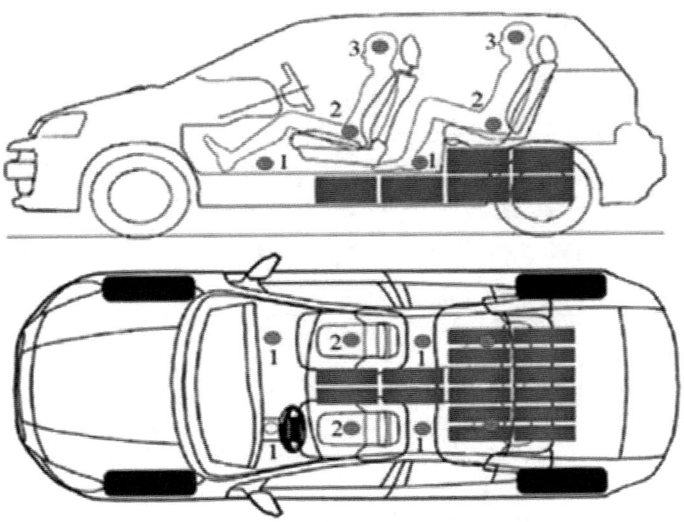

Figure 2. Diagram of the points where measurements were made [8]. The diagram is additionally processed.

2.3. Instruments Used in the Study. Measured Parameters

EMF 450 device was used for this purpose (see figure 3) This device measures simultaneously and displays magnetic field, electric field and radio frequency force and has three orthogonal sensors. The parameters and types of measurements are expressed in units of electric and magnetic field and power density. This device is very suitable for measurements of power lines, electrical appliances, industrial devices, mobile phones, base stations and microwave ovens. The device is delivered fully tested and calibrated and, when used properly, provides reliable results.

Figure 3. EMF 450

The following physical characteristics were measured to describe the exposure to electromagnetic fields:

- Electric field strength (E); [Vm-1] (Volt per meter).
- Magnetic flux density (B), measured in micro tesla, [μT].

Standards for EMF exposure levels and ordinances existing in Bulgaria. Limit values. (ICNIRP, IEEE, Ordinance No RD-07-5 of 15 November 2016)

For the purpose of this study, the approximate frequency range where the electromagnetic fields generated by electric vehicles irradiate should be determined. Most electrical systems, including electric vehicles, are characterised by time-varying electrical quantities. Many magnetic field generators are located in these vehicles. The elements of electronic control, inverters, switches, etc. supplement traction drive and battery. According to the modern technologies for production of electric vehicles, electromagnetic fields generated by electric vehicles represent relatively wide frequency spectrum, from 0 [Hz] (at rest) to about 1 [MHz]. For this reason, standard limit values in the range of these frequencies need to be set.

On a world scale, there are standards for the levels of EMF exposure to which the human body can be exposed. World organisations develop these standards in accordance with research and conclusions drawn from these developments. Two of these organisations are presented here - these are ICNIRP and IEEE.

ICNIRP (International Commission on Non-Ionizing Radiation Protection). It provides scientific advice and guidance on the impact of non-ionizing radiation on human health and its impact on environment, and develops and proposes measures to protect people and environment from harmful exposure to the aforementioned radiation.

ICNIRP has established the following exposure limit values for the entire body in Guidance on Exposure Limits for Variable Electric and Magnetic Fields (100 [kHz] to 300 [GHz])". [8]

For frequency range 0,1-30 [MHz] - E (electric field strength [V / m]) $660/f^{0.7}$. H (magnetic field strength [A / m]) 4.9 / f. As f - frequency in [Hz].

IEEE (Institute of Electrical and Electronics Engineers)

The institute approves the limit values in IEEE Standard for Safe Human Exposure Levels for Electric, Magnetic and Electromagnetic Fields from 0 to 300 [GHz]" (IEEE Std C95.1™-2019) [9].

For frequency range 0.1-1.34 [MHz] E (electric field strength) - 614 [V / m]. H (magnetic field strength [A / m]) 16,3/f. As f - frequency in [Hz].

In Bulgaria, similar standards have been approved and entered in force with Ordinance No RD-07-5 of 15 November 2016 on the Minimum Requirements for ensuring Health and Safety of Workers at Risks related to Exposure to Electromagnetic Fields [10]. It is issued by Minister of Labour and Social Policy and Minister of Health, published in State Gazette No 95 of 29 November 2016, and is in force from 1 March 2017. The ordinance is based on Directive 2013/35 / of the European Union. Due to the low and extremely low frequencies at which the electric vehicle is irradiated, the lowest values of magnetic flux density - B and electric field strength - E are taken into account for frequency range from 1 [Hz] to 1 [MHz] by the above-mentioned Ordinance. The values referred to in this Ordinance have been used in this study.

As regards magnetic flux density, the Ordinance has set the following exposure limit values:

- In the range from 1 to 8 [Hz] - 200,000 to 3125 [μT];
- In the range from 8 to 25 [Hz] - 3125 to 1000 [μT];
- In the range from 25 to 300 [Hz] - 1000 [μT];
- In the range from 300 [Hz] to 3 [kHz] - 1000 to 100 [μT];
- In the range from 3 [kHz] to 1 [MHz] - 100 [μT].

As regards electric field strength, the Ordinance has set the following exposure limit values:

- In the range from 1 to 25 [Hz] - 20,000 [V / m];
- In the range from 25 to 50 [Hz] - 20,000 to 10,000 [V / m];
- In the range from 50 to 1.64 [kHz] - 10,000 to 305 [V / m];
- In the range from 1.64 [kHz] to 3 [kHz] - 305 to 167 [V / m];
- In the range from 3 [kHz] to 1 [MHz] - 170 [V / m].

2.4. Tested Vehicles

For the purpose of the study, the electromagnetic fields in three electric vehicles and a hybrid vehicle were measured. The vehicles tested are of the following brands and models and have the following technical characteristics (Table 1).

Table 1. Test vehicles measured in the study.

No	Brand and model	Vehicle type	Drive	Power	Working voltage
1	Kia Soul	electric vehicle	front-wheel	81 [kW]	375 [V] DC
2	Nissan Leaf	electric vehicle	front-wheel	80 [kW]	400 [V] DC
3	Smart	electric vehicle	rear-wheel	55 [kW]	375 [V] DC
4	Toyota C-HR	hybrid vehicle	front-wheel	60 [kW]	201,6 - 600 [V] DC

3. Experimental Results from EMF Measurement in Electric Vehicles and Hybrid Vehicles

Spot measurements were made while the vehicle is at rest (is not moving), with the engine running and idling in a standby mode in an insignificant background field (at rest), and at different vehicle speeds - 10, 25, 35, 50 [km / h] and in reverse mode (in motion) of the engine. EMF values are measured at each of the four seats indicated in figure 2 - a total of 12 points. At each seat, the instrument was stationary at all specified points, so as to measure the levels of the electromagnetic fields and the location of their highest values.

The exposure to electromagnetic fields in the studied electric vehicles and hybrid vehicles has the following highest values according to the tests:

1. Magnetic flux density – B (figure 4).

- Vehicle 4 moving with 35 [km / h] - at the rear right seat in point 1 (at floor level) - magnetic flux density reaches a value of 2 [μT];

- Vehicle 1 (in reverse mode, at the driver's seat in point 1) - magnetic flux density reaches a value of 1.27 [μT];
- Vehicle 3 (moving with 25 [km / h] in point 3) - magnetic flux density reaches a value of 0.9 [μT]

The average mean of the highest readings Bmax (trend lines) do not exceed 1 [μT].

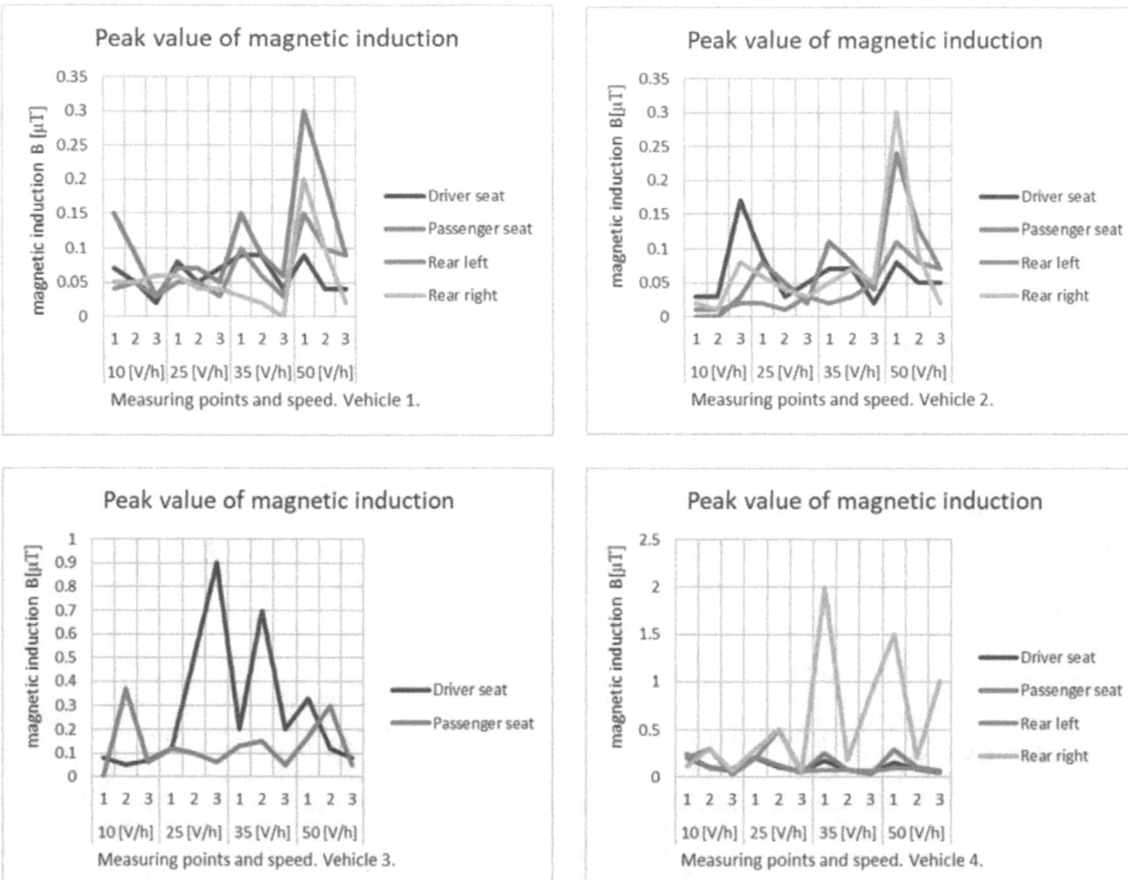

Figure 4. Peak value of electromagnetic induction

2. Electric field strength – E (figure 5).

- Vehicle 3 in a standby mode in point (2) at the driver's seat - electromagnetic induction reaches a value of 12 [V / m];
- Vehicle 4 moving with 25 [km / h] in points 1 and 2 - with 35 [km / h] in point (1) and with 50 [km / h] in point (1) - electromagnetic induction reaches values of 10 [V / m] at the rear right seat;
- Vehicle 3 moving with 35 [km / h] in point 2 at the driver's seat - electric field strength reaches a value of 9 [V / m].

All other readings have values of 7 [V / m] or smaller ones. The trend lines (average means) of the highest readings Emax do not exceed 8 [V / m].

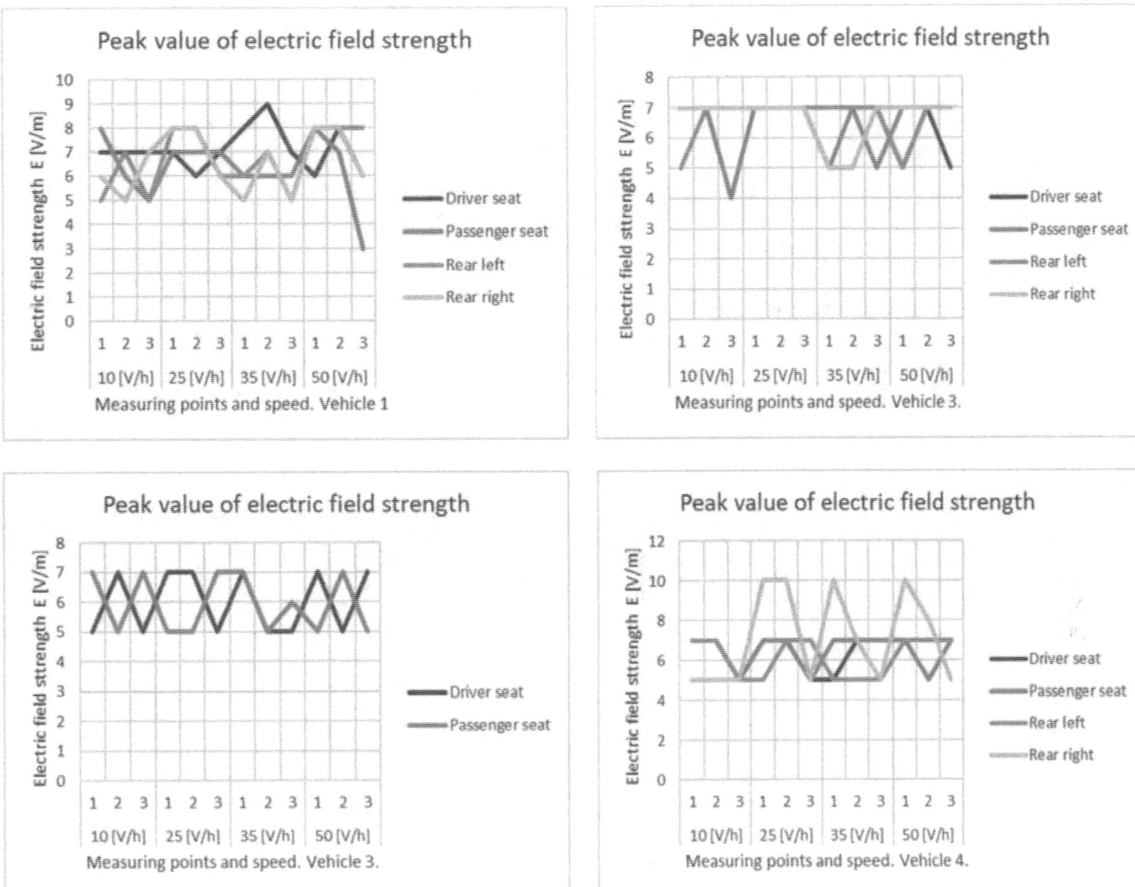

Figure 5. Peak value of electromagnetic field strength

Conclusions

The results of the study are compared with those referred to in the standards recommended in Ordinance No RD-07-5 of 15 November 2016 on the Minimum Requirements for ensuring Health and Safety of Workers at Risks related to Exposure to Electromagnetic Fields.

The study indicates the measured readings of EMF exposure. They range from 0 to 2 [μT] (Vehicle 4) for magnetic flux density ***B***, with average values not exceeding 1 [μT]. Compared to the limit values set in Ordinance No RD-07-5 of 15 November 2016 - 100 [μT], they are only 2% of the lowest reading required by standard. As regards electric field strength ***E***, the measured values vary between 3 and 12 [V / m] in different modes and in motion, with the average values of the peaks not exceeding 8 [V / m]. In this case, the Vehicle 3 showed the highest results. Compared to the lowest possible limit values specified in the above-mentioned Ordinance, namely 167 [V / m], the measured peak readings are 7.19% of the lowest values specified in the standard.

Acknowledgments

This work received funding from the Technical University of Varna under project KD3 with a project leader Assoc. Prof. PhD R. Hristov.

References

[1] Prša M and Kasaš-Lažeti K 2018 *IOP Conf. Ser.: Mater. Sci. Eng.* **294** 012001
[2] Ionescu V, Săpunaru A, Popescu M and Popescu C 2019 Magnetic field constraints in the

passenger compartment of electric vehicles *IEEE ISGT-Europe* (Bucharest, Romania) pp 1-5

[3] Gumiela J, Sitnik L and Sztafrowski D 2019 Measuring identification of electromagnetic field emissions in electrical cars *Przeglad Elektrotechniczny* vol 95 issue 12 pp 128-131

[4] Gombarska D, Smetana M and Janousek L 2019 high-frequency electromagnetic field measurement inside personal vehicle within urban environment *12th International Conference on Measurement* (Smolenice, Slovakia) pp 223-6

[5] Kang Q, Gu P, Gong C and Zuo S 2019 Test and analysis of electromagnetic noise of an electric motor in a pure electric car *SAE Technical Paper* 2019-01-1492

[6] Hareuveny R, Sudan M, Halgamuge M, Yaffe Y, Tzabari Y, Namir D and Kheifets L 2015 Characterization of extremely low frequency magnetic fields from diesel, gasoline and hybrid cars under controlled conditions *Int. J. Environ. Res. Public Health* **12** 1651-66

[7] Group 1 Nissan Retrieved from https://www.group1nissan.co.za/nissan-leaf-electric-vehicle/

[8] Moreno-Torres P, Lafoz M, Blanco M and Arribas J 2016 *Passenger Exposure to Magnetic Fields in Electric Vehicles* INTECH pp 47-52

[9] ICNIRP Statements-Guidelines for limiting exposure to time-varying electric and magnetic fields (1 Hz to 100 kHz) 2020 Health Physics 118(5) pp 483-524

[10] IEEE Standard for Safety Levels with Respect to Human Exposure to Electric, Magnetic, and Electromagnetic Fields, 0 Hz to 300 GHz in IEEE Std C95.1-2019 (Revision of IEEE Std C95.1-2005/ Incorporates IEEE Std C95.1-2019/Cor 1-2019) pp 1-312

IOP Conf. Series: Materials Science and Engineering **977** (2020) 012023 doi:10.1088/1757-899X/977/1/012023

Determination of the stages of the injection process for Common Rail injectors using vibration pulses

Sergey Belchev[1,2], **Zdravko Ivanov**[1], **Trifon Uzuntonev**[1] **and Veselin Mihaylov**[1]

[1] Department of Automotive Engineering and Technologies, Technical University - Varna, 1 Studentska Str., city of Varna, Bulgaria

[2] E-mail: sergtu@abv.bg

Abstract. The study and modeling of the combustion process of diesel engines require knowledge of the injection characteristics. The information on the exact start, duration and end of the injection is required. For Common Rail injectors with electromagnetic control, the beginning and duration of the electronic control pulse are precisely known. The actual start and duration of the injection are not the same as those of the impulse and they are related to the movement of the nozzle needle. The experimental determination of the movement of the nozzle needle requires specially prepared experimental installations and expensive measuring instruments. In addition, the experimental injector must be prepared with design modifications allowing the incorporation of sensors. The article presents a method for determining the start and duration of injection of Common Rail injectors using the registration and analysis of vibration pulses generated in the injector body. The method is based on the fact that the nozzle needle causes vibrations in the injector body when lifting and seating.

1. Introduction

In modern Common Rail systems, fuel control is realized by changing the start and duration of the electronic pulse that triggers the electromagnetic injector. This, combined with the different injection pressures, allows precise fuel metering. The start of the actual injection is considered to be the moment of raising the nozzle needle, and the end is considered the moment of its closing [1]. The start and end of the control pulse do not coincide with the start and end of the injection. This is a result of the principle of operation of the electromagnetic injectors and the processes accompanying the injection [2]. Determining the delay with which the injection starts and ends relative to the control pulse is important both for the design stages in the development of fuel systems for new engines and in determining the impact of wear of the injector elements during their operation. The changed delay of injection as a result of operation compared to the beginning of the control pulse changes the advance of injection and significantly affects the working process of the engine. This is the reason for the large number of scientific studies in this field.

Different methods can be used to determine the needle lift, with custom build sensors - Hall effect or optical sensors [3]. Common to these methods is the need to make changes to the construction of the test injector and the need for specialized equipment. This makes it impossible to study a large number of different injectors, including ones with long-term operation.

To study the lift of the needle for gasoline [4] and gas [3] engines, a method based on the recording of vibration pulses of injector body was used. When the valve needle with solenoid armature move and especially at reaching its end positions, distinct vibration pulses are generated, which can be registered.

The purpose of the presented work is to show to what extent the recorded vibration pulses in the injector housing can be used to determine the beginning and end of lifting of the nozzle needle.

2. Experimental Studies

The experiments were performed on a standard injector test bench Common Rail CRI-NTR816E. The elements of the system delivering, measuring and regulating the pressure belong to the company "Bosch". The measurement is performed by the "Cobolt" system, with a measurement accuracy of 0.1 mm^3. The tests were performed with two BOSCH injectors with numbers 0445110259 and 0445110044. In addition to the stand, additional accelerometers and pressure transducer were used, figure 1.

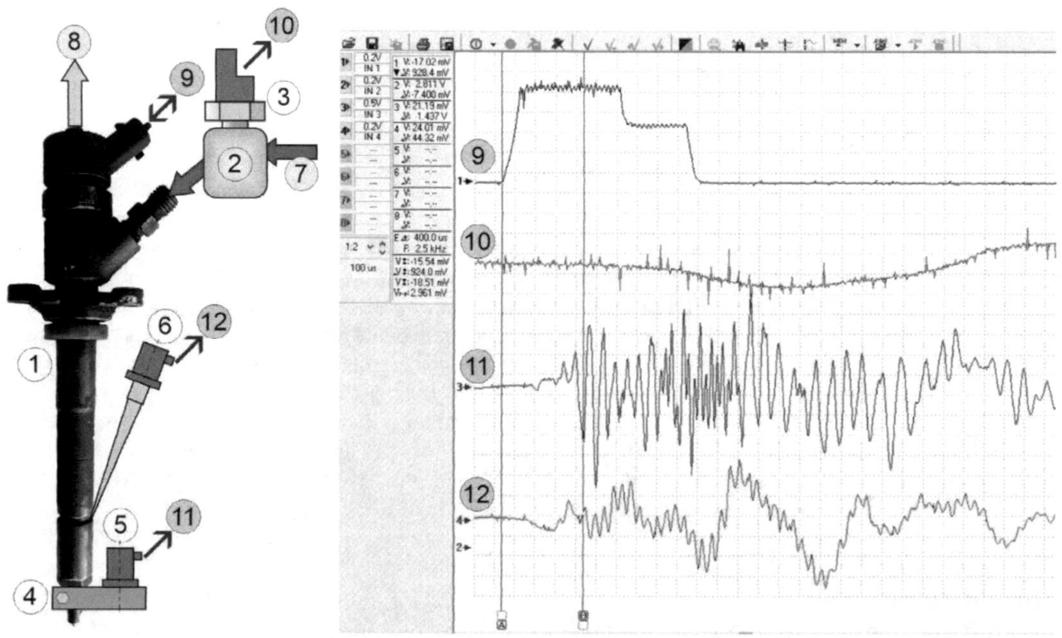

Figure 1. Scheme of the experimental setup.

To measure the pressure in front of the injector (1), a body (2) with a minimum volume is used, on which is mounted a standard car fuel pressure sensor (3). An accelerometer (5), type RFT KD35, is firmly mounted in the front part of the nozzle housing to read the vibration pulses by means of a plate (4). In addition to accelerometer (5), a movable one (6) of the same type is used to read the vibration pulses at the top part of the nozzle housing. The contact between the accelerometer and the injector is made by means of a rod, which is held by hand. Fuel (7) is supplied from the stand via high pressure tubes, and the excess fuel (8) is sent to the flow meter of the stand. A control pulse (9) is generated by the stand, which is recorded simultaneously with the signal (10) from the pressure sensor, with the signal (11) from the vibration sensor (5) and the signal (12) from the vibration sensor (6). Figure 1 shows recorded signals using a digital oscilloscope. At each attempt, the amount of injected fuel is measured.

The experiment was performed with two different injectors, figure 2. They differ significantly in length. The mass of the nozzle needle and the armature of the electromagnet of the two injectors are approximately equal, but the length and mass of the control piston (valve rod) of injector 1 are approximately twice as large as those of the injector two. This implies greater inertia and longer injection delays in this injector.

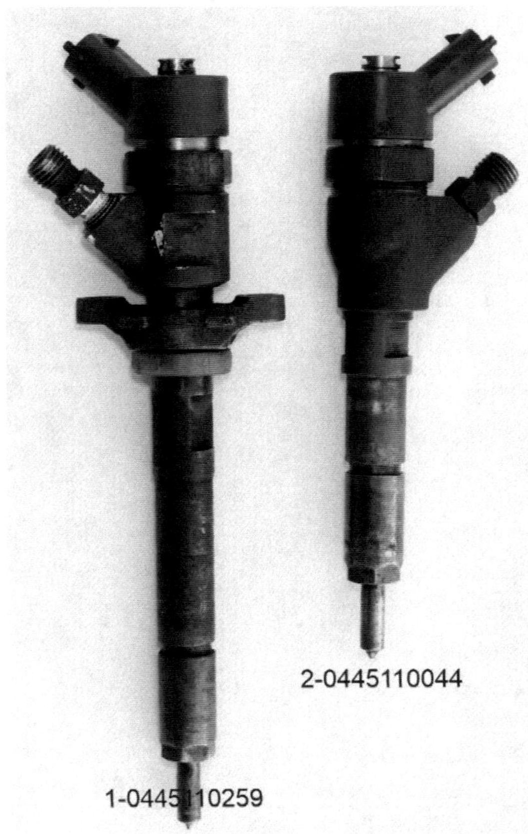

Figure 2. Tested Bosch injectors

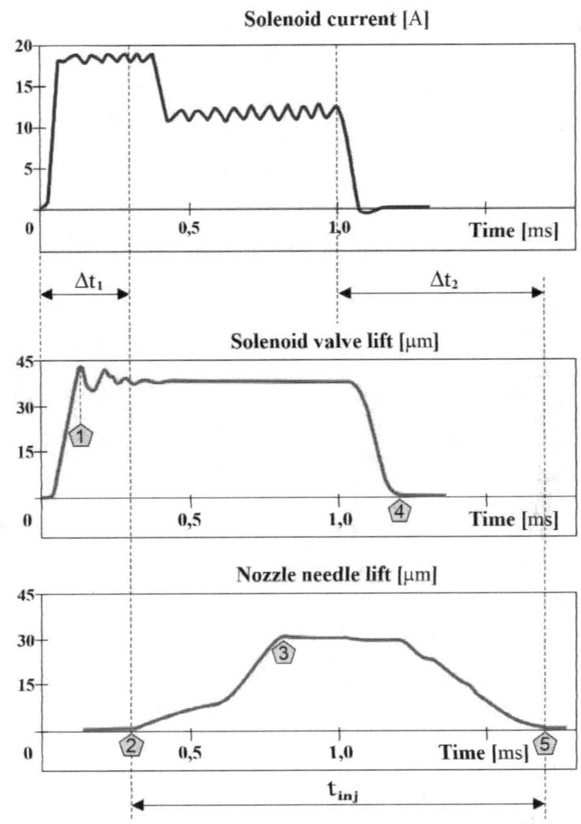

Figure 3. Current, armature and nozzle needle lift

During the preliminary analysis of the vibration signals, a clear repeatability of the signal change was found. This allows five points to be determined during the operation of the injector, figure 3: (1) - the impact of the armature of the electromagnet at the stop at maximum lift; (2) -the beginning of the lifting of the nozzle needle, the housing of the injector shakes and moves back from the movement of the needle; (3) - the impact of the control piston at the stop, the end of the lifting of the needle; (4) - the impact of the anchor of the electromagnet when closing; (5) - the impact of the needle during its squatting, end of the injection.

For each of the injectors, experiments were performed at three different pressure levels in the fuel accumulator - 25MPa, 80MPa and 135MPa. For each pressure two different durations of the control signal are used - 1800μs and 900μs.

For each point of the experiment, multitude of consecutive groups of vibration pulses are recorded, which correspond to the injection of the given injector. Figure 4 shows the records for one injection of the two injectors with a pressure of 25MPa, durations of the control signal - 1800μs and 900μs.

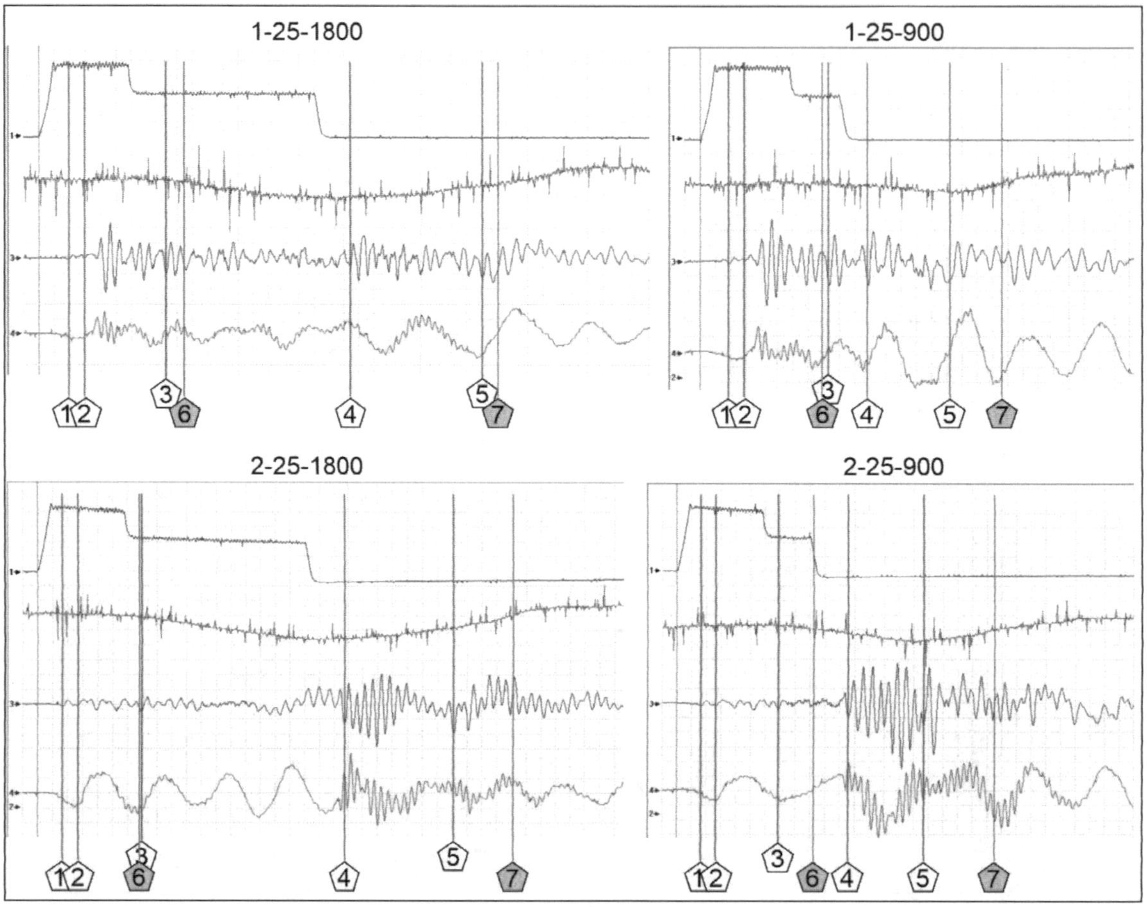

Figure 4. Recorded signals of the injectors at injection pressure 25MPa.

Above each recording is marked the injector number, the injection pressure in MPa and the duration of the control signal in μs, for example (1-25-1800). In addition to the moments explained in figure 3 and numbered from 1 to 5, figure 4 shows: the beginning of the drop in fuel pressure in front of the injector - (6); restoring the pressure to its initial level.

Figure 5 shows the records recorded at injection pressure of 80 MPa, and Figure 6 shows the records at injection pressure of 135 MPa.

It is noteworthy that after the initial deviation of the vibration signal registered with the mobile accelerometer (6), figure 1, oscillations with large amplitude and low frequency appear. They are a consequence of the used method (holding by hand) and are caused by the short-term loss of contact between the sensor and the nozzle housing. These vibrations depend on the clamping force and are largely random in nature, so they are only used to establish the initial moment of the movement of the injector associated with the beginning of lifting of the nozzle needle.

Characteristic signal variations are used to determine the moments (1) to (5) in the figures, taking into account the injection delay times for Bosch injectors shown as examples in [1].

IOP Conf. Series: Materials Science and Engineering **977** (2020) 012023 doi:10.1088/1757-899X/977/1/012023

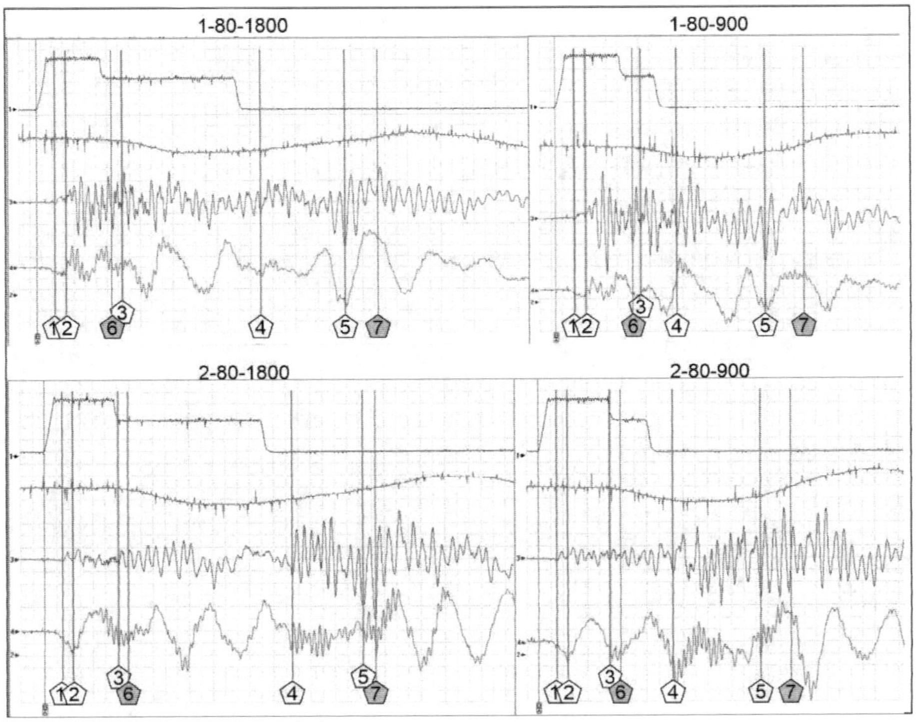

Figure 5. Recorded signals of the injectors at injection pressure 80MPa.

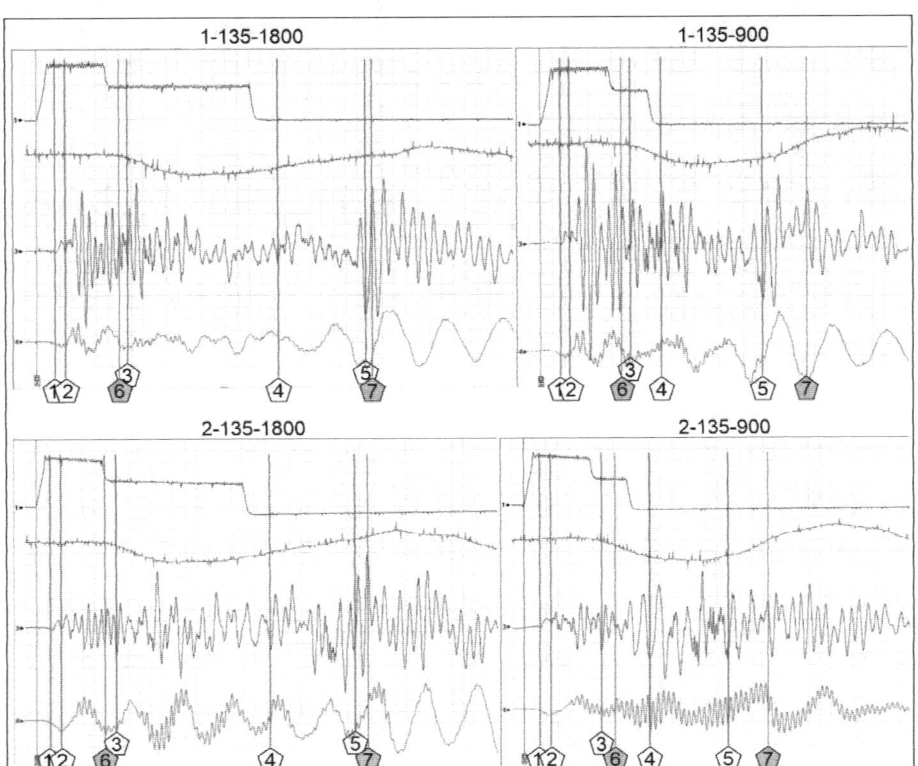

Figure 6. Recorded signals of the injectors at injection pressure 135MPa.

From the set of records for one point of the experiment, three random records were selected and the times from (1) to (7) were determined. They represent the time from the beginning of the control signal to the corresponding point. The obtained values of these times for the two injectors are averaged and are shown in table 1.

Table 1. Delay times relative to the start of the control signal.

Times (μs)			0-1	0-2	0-3	0-4	0-5	0-6	0-7
Injector 1 0445110259	25MPa	900μs	184	292	832	1084	1704	976	1988
		1800μs	184	296	828	2048	2908	940	3004
	80MPa	900μs	172	266	784	1196	1904	742	2270
		1800μs	170	264	776	2040	2778	688	3048
	135Mpa	900μs	168	248	760	1020	1864	688	2244
		1800μs	164	248	768	2036	2760	700	2828
Injector 2 0445110044	25MPa	900μs	172	272	692	1160	1668	756	2264
		1800μs	168	272	696	2076	2912	692	3212
	80MPa	900μs	156	232	599	1114	1841	678	2120
		1800μs	152	233	612	258	2636	684	2836
	135Mpa	900μs	144	236	684	1108	1792	612	2136
		1800μs	140	228	700	2044	2680	600	2880

Based on the times in table 1, determined from experiments, are calculated: the delay Δt_1 of injection relative to the beginning of the control signal; the delay of the injection end Δt_2 relative to the end of the control pulse; injection duration t_{inj}, figure 3. The results of the calculation are shown in table 2.

Table 2. Delay times and injection duration.

Delay and injection times (μs)			Δt_1	Δt_2	t_{inj}	t_p
Injector 1 0445110259	25MPa	900μs	292	804	1412	1012
		1800μs	296	1108	2612	2548
	80MPa	900μs	266	1004	1638	1528
		1800μs	264	978	2514	2360
	135Mpa	900μs	248	964	1616	1556
		1800μs	248	960	2512	2388
Injector 2 0445110044	25MPa	900μs	272	768	1396	1008
		1800μs	272	1112	2640	2520
	80MPa	900μs	232	941	1609	1442
		1800μs	233	836	2403	2152
	135Mpa	900μs	236	892	1556	1524
		1800μs	228	880	2452	2280

The last column in table 2 shows the duration t_p of the fuel pressure drop in front of the injector as a result of the fuel injection. It is obtained as the difference between the times at points (7) and (6), figure 4, figure 5 and figure 6. This time t_p can be used to assess the correctness of determining the injection duration.

3. Analysis of the Experimental Results

In order to assess the adequacy of the shown method, the influence of the injection pressure and the control pulse duration on the delay times Δt_1, Δt_2 and the injection duration t_{inj} was evaluated. The presence of a logical relationship between the specified parameters would confirm the correctness of the method. Figure 7 shows the effect of the injection pressure on the delay at different control pulse durations for the two injectors.

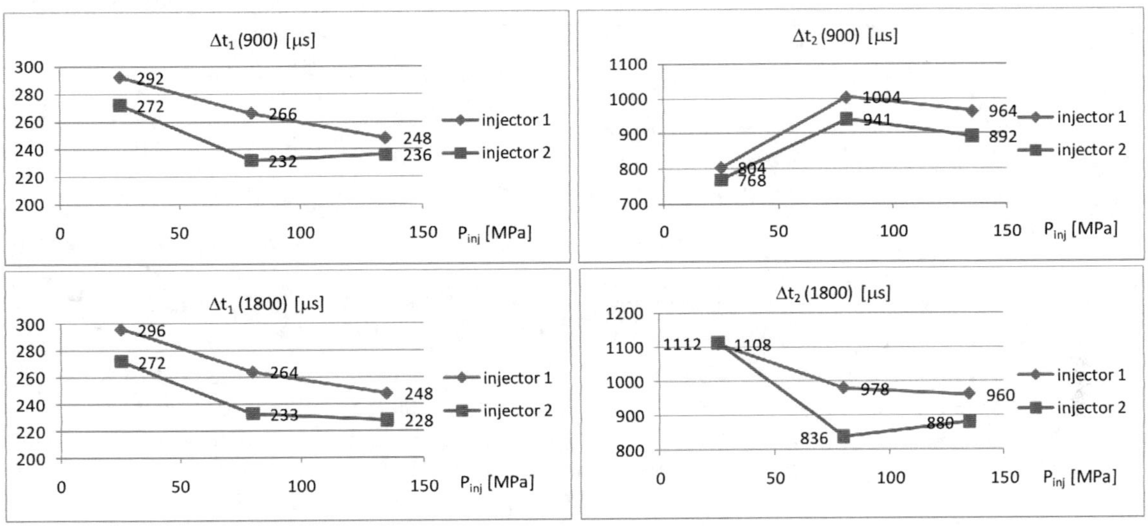

Figure 7. Delay of start and end of injection depending on pressure.

The delay of the start of injection Δt_1 relative to the beginning of the control pulse is in the range of 232-296μs.

The delay of the start of injection is not affected by the control pulse duration and decreases with increasing injection pressure by 13-16%. This is due to the increased reaction speed of the system with increasing pressure in it.

The injector with shorter length and lower mass of the control piston has less delay 20-30 μs, which is 8-12%. This is due to the lower inertia of the system.

The delay of the end of the fuel supply Δt_2 relative to the end of the control pulse is significant 800-1100 μs. Here, the effect of pressure and injection duration are mixed and further studies are needed for an accurate assessment. However, in all cases the injector with less mass of the control piston has shorter delays.

The dependences shown in figure 7 are determined by the values of the different times using the presented method. Certain influences and dependencies have repeatability and logical explanation.

Figure 8 shows how the injection duration is affected by the change in injection pressure for both injectors.

It can be seen that with different control signal durations, the injection pressure affects the injection duration differently. This is a result of the joint influence of various factors: wave phenomena in the dense medium of the fuel and the deformations of the elements of the fuel system under the influence of high pressure; combination of the action of inertial, spring and pressure forces on the movable elements in the injector. In any case, the injector with lower mass of its moving parts has a shorter injection time, which is mainly due to the shorter delay at the end of the injection.

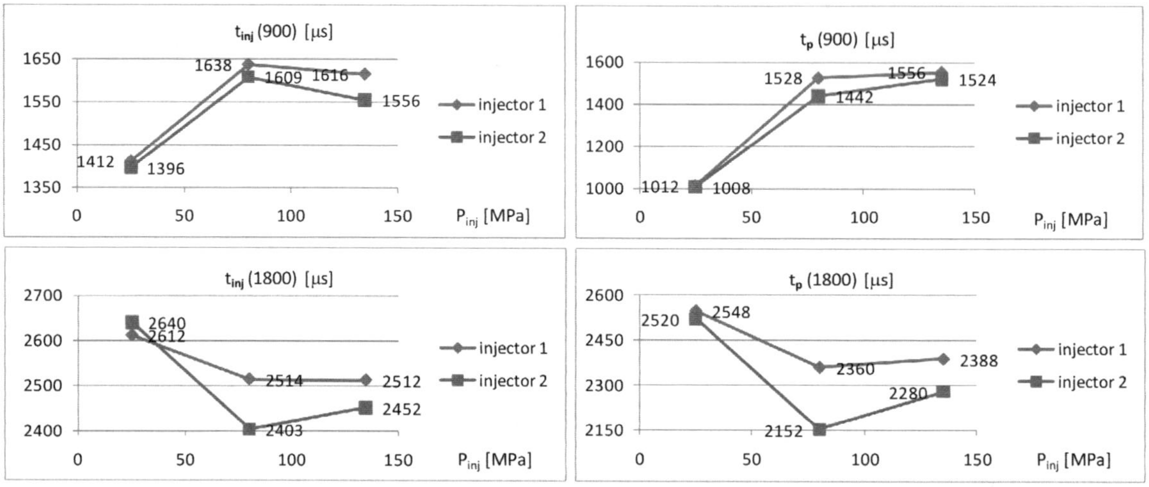

Figure 8. Injection duration depending on pressure

The interval in which there is a pressure drop is delayed in relation to the movement of the needle and the injection interval. This is completely natural, given the delayed reaction of the pressure wave in the hydraulic system. Therefore, the time values t_p of the pressure drop in front of the injector cannot be used directly to determine the injection duration t_{inj}. Nevertheless, the close values of the two times can be noted. The nature of the change of the two times is very close, figure 8, in both injectors.

4. Conclusion

The proposed method for determination of injection phases gives adequate results and can be used for research purposes. It is characterized by simplicity and ease in building the experimental setup. No design changes are required to the tested injectors.

The duration of the injection is much longer than the duration of the control pulse, the difference between them is in the range of 500-800µs and is influenced by the duration of the pulse and the injection pressure.

The determined influences of the injection pressure and the mass of the movable elements in the injector on the injection delay and duration have a logical character and confirm the non-random nature of the times determined by the presented method.

5. References

[1] Hammer J and Binder A Einspritztechnik 2010 Teil 1 IVK Universität Stuttgart

[2] Robert Bosch GmbH 2004 *Diesel-Engine Management, 4th Edition* (Stuttgart) p504

[3] Ivanov Z, Mihaylov V and Mersinkov H 2014 Optical method for measurement needle lift of electromagnetic valves for LPG/CNG injection *J. EcoVarna* **20** 415-22

[4] Christ K, Michelsburg M, Back K, Eidam A and Kiencke U 2010 Injector calibration with knock sensors for gasoline direct injection *J. Sensoren und Messsysteme* **18** (VDE VERLAG GmbH Berlin) 374

Acknowledgments

This work was completed thanks to the generous cooperation of Diesel Service Varna, which provided its equipment, benches and injectors for research.

IOP Conf. Series: Materials Science and Engineering **977** (2020) 012024 doi:10.1088/1757-899X/977/1/012024

Brake efficiency determination of light vehicles

R D Dimitrov[1,2] and D Z Ivanov[1]

[1]Department of Transport engineering and technologies, Technical University of Varna, Bulgaria

[2]E-mail: r_dimitrov@tu-varna.bg

Abstract. The present paper presents a study whose objective is to determine the brake deceleration of two classes light vehicles - class A is vehicles with weight up to 1350kg, class B is vehicles with weight up to 1000kg. The vehicles of both classes are equipped with ABS system. The measurements are made with commercial device BRAKESAFE. Measured parameters are vehicle speed, stopping distance and total time to stop. The vehicles were tested under varying conditions: with ABS system activated or deactivated; tires at different stage of wear. The obtained results are analyzed and conclusions are made about measured brake parameters under all conditions.

1. Introduction

The exact determination of brake deceleration at emergency stopping is important parameter, which is used for calculation of some values at preparation of vehicle crash expertise. In technical literature exists information about brake deceleration, but in some books that are used when preparing expertise this information is old or not accurate enough [1, 2, 3]. Also, this data in most causes is not valid for new vehicles. To improve accuracy of results it is necessary to update this database. In this study are determined brake deceleration, stopping distance and total time to stop of some of the most common light vehicles.

The value of brake deceleration is used in many formulas for determination of speed of vehicle before impact, speed of vehicle in the moment of impact, time for moving of a vehicle from point A to point B, necessary distance to stop of vehicle and etc. The exact determination of this parameter will increase accuracy of calculation and at preparation of vehicle crash expertise [4,5].

The aim of this study is to measure and determine the brake deceleration, stopping distance at emergency stopping at different vehicle speeds, different tire tread wear and with or without ABS system. For the experiments of stopping without ABS system, the tested vehicles are the same, only this system is electrically turned off.

2. Experimental study procedures

2.1. Classes vehicles and measurement device

The study was carried out at Department of Transport Engineering and Technologies at Technical University of Varna. Two classes of light vehicles are researched.

Class **A** are vehicles with weight up to 1350kg, sedan or station wagon with four and five doors, produced from 2000 till 2010 year. All of the vehicles are equipped with ABS system and use brake disc

IOP Conf. Series: Materials Science and Engineering **977** (2020) 012024 doi:10.1088/1757-899X/977/1/012024

for front and rear wheels. Class **B** are vehicles with weight up to 1000kg, sedan or hatchback with three and four doors, produced from 1999 till 2010 year. All of the vehicles are equipped with ABS system and use also brake disc for front and rear wheels.

The experiments are conducted at sunny days, dry asphalt with temperature around 34°C on a horizontal road section. Measured speed of vehicles is from 10 till 70 km/h to full stop.

The measurement device is shown in figure 1. The tester is BRAKESAFE [6], certified and produced in England.

Figure 1. Brake tester.

Parameters of the device are shown in table 1.

Table 1. Tester parameters.

№	Parameter	Value
1	Range (m/s^2)	0 ÷ 13
2	Accuracy (m/s^2)	0.2
3	Display	Graphical LCD
4	Measures force applied to brake pedal	Yes
5	Bluetooth	Wireless connectivity
6	Case	IP65
7	Weight (kg)	1
8	Approved from - Vehicle and Operator Services Agency	(VOSA)

2.2. Results

The shown results are averaged for class A from 10 tested vehicles, class B from 8 tested vehicles.

The graphs shown in figure 2 illustrate the variation of brake deceleration for both classes vehicles at speed from 20km/h, 30 km/h, 40 km/h, 50km/h and 70 km/h.

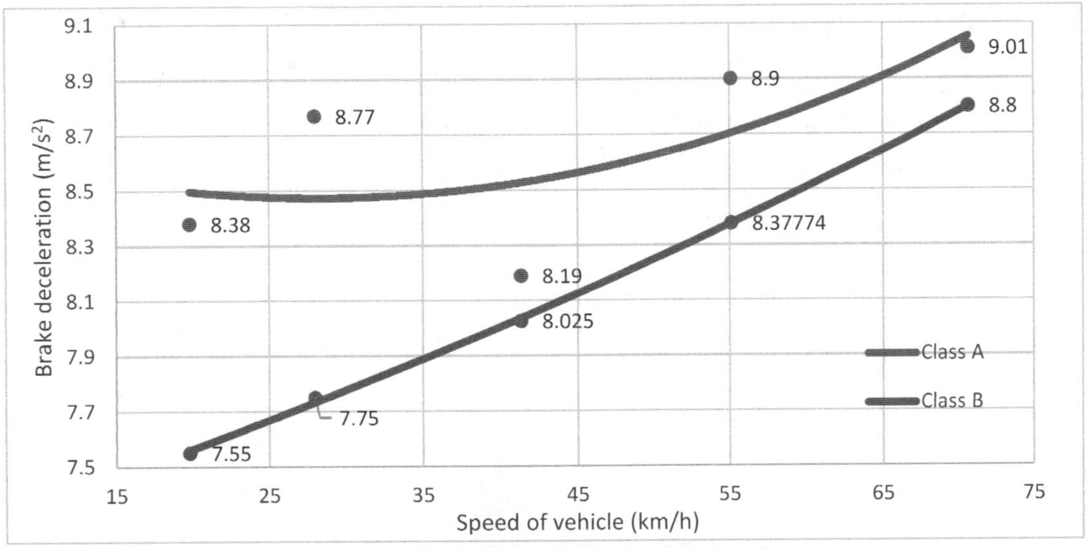

Figure 2. Variation of Brake deceleration.

At speed till 50km/h the brake deceleration have slightly lower values for vehicles of class A around 8.5 m/s^2, for vehicles of class B around 7.8 m/s^2. Over 50km/h vehicles of class A have brake deceleration around 9.0 m/s^2, vehicles of class B around 8.0 m/s^2.

Figure 3 illustrate the variation of stopping distance for both classes vehicles at speed from 20km/h, 30 km/h, 40 km/h, 50km/h and 70 km/h.

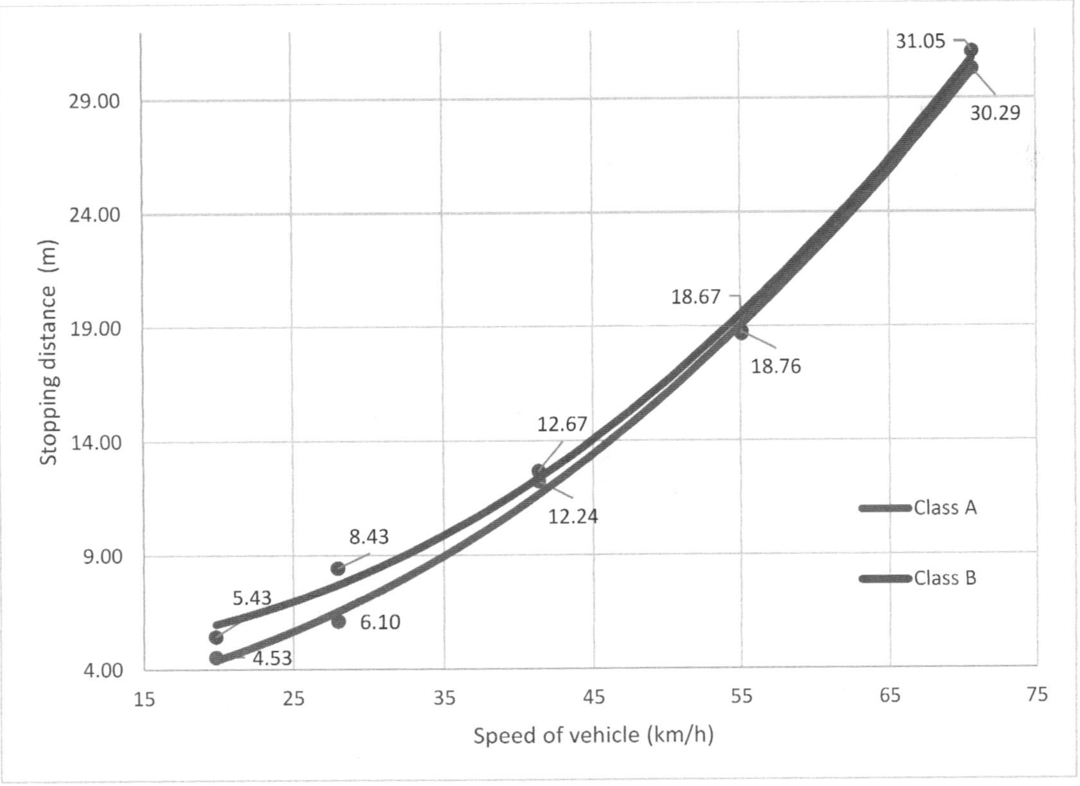

Figure 3. Variation of stopping distance.

At speed 20km/h – stopping distance is around 5m, at 50km/h – stopping distance is around 16m and at speed 70km/h stopping distance is around 30 meters.

Figure 4 illustrate the variation of total time to stop for both classes vehicles at speed from 20km/h, 30 km/h, 40 km/h, 50km/h and 70 km/h.

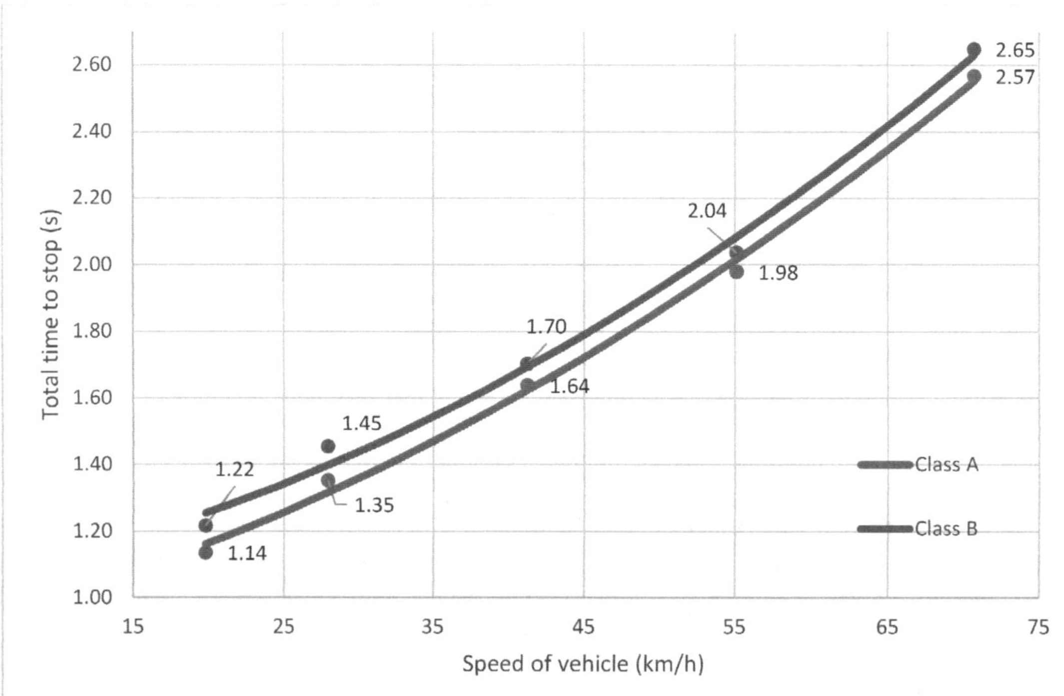

Figure 4. Variation of total time to stop.

The total time to stop is a function again of brake deceleration. At speed 20km/h – time to stop is around 1.2s, at 50km/h - time to stop is around 2.0s and at speed 70km/h – time to stop is around 2.6s.

Also is tested on the same vehicle how braking deceleration changes when the automobile stop with and without ABS system at speed 40km/h. Results are shown on figure 5. The shown values are averaged for every tested class of vehicles.

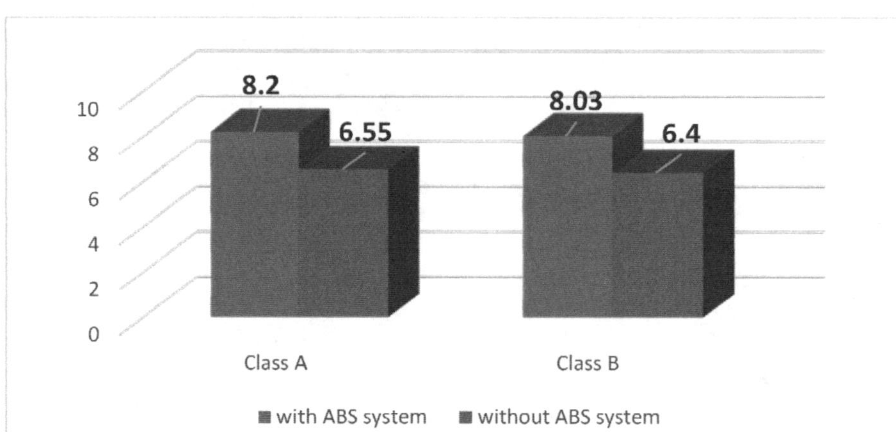

Figure 5. Braking deceleration with and without ABS system.

At the same conditions is conducted a test to determine what is the influence on braking deceleration at deferent tire tread wear at speed 40km/h. Tests are made with three type of tires with depth of protector 3mm, 6mm and 8mm. Tested vehicle is from class A, model of the tires is Continental - Conti eco contact. The results are shown in next figure.

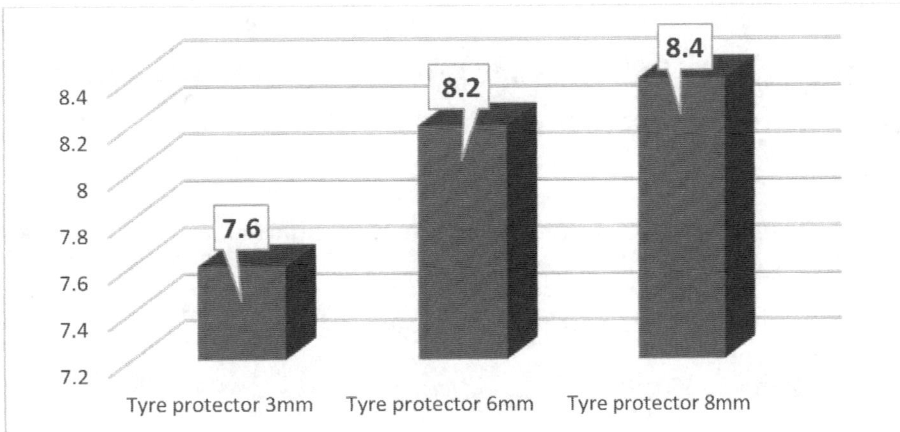

Figure 6. Braking deceleration depending from tire wear.

3. Conclusion

The analysis of the experimental results provides the basis for drawing the following conclusions:

- At speed over 50km/h the braking deceleration of vehicles from class A is with 7% higher than at speed under 50km/h;
- At speed over 50km/h the braking deceleration at vehicles class B is around 10% higher than at speed under 50km/h;
- ABS system increases braking deceleration for vehicles class A with 25%, for vehicles class B with 26%.
- The presence of an ABS system in the vehicles increases the braking deceleration (Braking efficiency) average with around 25%.
- Tire tread wear up to minimal allowable value reduces braking deceleration with 10%.

References

[1] Gillespie, T D 1992 *Fundamentals of vehicle dynamics*. Society of Automotive Engineering, Inc. 400 Commonwealth Drive Warrendale, PA 15096-001, p 250

[2] Illarionov, V A 1997 *Expert's examination of traffic accidents* (Экспертиза дорожно-транспортных происшествий), Moscow: Transport, p 255 (in Russian)

[3] Hockicko, P, Trpišová, B 2013 *Are students' conceptions about automobile braking distances correct?* In: Engineering Education Fast Forward 1973 – 2013: proceedings of the 41st SEFI annual conference, 16-20 September 2013, Leuven, Belgium. - Brussels: SEFI, ISBN 978-2-87352-004-5

[4] Kolla, E, Kohút, P, Kubjatko, T 2014 *Analysis of pedestrain body movement in pedestrian – MPV collision*, In: Automotive safety 2014: problemy bezpeczeństwa w pojazdoch samochodowych, IX international science-technical conference: Rajecké Teplice - Slovakia, 8-10 April 2014, Kielce: Wydawnictwo Politechniki Świętokrzyskiej, pp 198-208, ISBN 978-83-63792-21-3

[5] Rievaj, V, Vrábel, J, Hudák, A 2013 *Tire inflation pressure influence on a vehicle stopping distances*, In: International Journal of Traffic and Transportation Engineering Vol. 2, No 2, pp 9, ISSN 2325-0062

[6] https://turnkey-instruments.com/

IOP Conf. Series: Materials Science and Engineering **977** (2020) 012025 doi:10.1088/1757-899X/977/1/012025

A SI Engine Indicated Characteristics Determination for Gasoline and Methane Operation

R D Dimitrov[1,4], Kr Ts Bogdanov[1], M Magdziak-Tokłowicz[2], N A Andonov[1], V Z Panayotova[3]

[1]Department of Transport engineering and Technologies, Technical University of Varna, Bulgaria
[2]Department of Automotive Engineering, Wroclaw University of Science and Technologies, Poland
[3]Department of Chemistry, Medical University - Varna, Bulgaria

[4]E-mail: r_dimitrov@tu-varna.bg

Abstract. The present paper presents a study whose objective is to determine the indicated parameters of an spark-ignition engine and external mixture formation when running on methane and gasoline. The experiments were carried out using optimum adjusting parameters for engine operation on methane and gasoline fuels. The parameters measured and determined are: indicated specific fuel consumption, indicated mean effective pressure, indicated power, indicated efficiency, friction mean effective pressure and etc. over the entire engine speed range. An analysis of the obtained results is made, comparing the obtained indicated parameters at the operation of internal combustion engines with alternative fuel - methane and traditional fuel - gasoline.

1. Introduction

With the increasingly stringent environmental standards for exhaust gas set by European legislation, the use of alternative fuel sources is growing [1,3,4,5]. One option to reduce CO_2 emissions is by using alternative fuels. These alternative fuels for Spark-Ignition engines can be: LPG (liquid petroleum gas), CNG (compressed natural gas), different types of alcohols and biogas [7,8].

Methane alternative fuels, i.e. methane is of growing interest as they produce the lowest exhaust gas emissions when burnt in the engine. The disadvantages of these fuel systems are the additional large weight of the fuel tank and the lower power output upon engine retrofitting. Nevertheless, the fuel has a low cost and better environmental performance.

The indicator specific fuel consumption $\mathbf{g_i}$ gives an evaluation of the engine operating cycle economy. Indicated mean effective pressure (IMEP) $\mathbf{p_i}$ is the average pressure produced in the combustion chamber during the operating cycle. It is an expression of the theoretical, frictionless power known as indicated power $\mathbf{N_i}$. In addition to completely disregarding power lost to friction, indicated power gives no indication as to how much actual power is delivered to the propeller shaft for doing useful work. However, it is related to actual pressures that occur in the cylinder and can be used as a measure of these pressures.

IOP Conf. Series: Materials Science and Engineering **977** (2020) 012025 doi:10.1088/1757-899X/977/1/012025

Previous research has shown that when using methane as a SI engine fuel, CO_2 emissions are reduced by up to 20%, HC emissions are decreased by about two times and CO_2 emissions are reduced by an average of up to six times [10].

The greatest potential is associated with the use of methane to fuel engines. The CNG consists of about 95% methane and 5% other gases, which makes this gas a usable fuel for internal combustion engines. In addition, the exhaust gases produced when operating on this fuel have lower harmful emissions and the price of natural gas is times lower than that of gasoline [9].

There are several reasons for using methane as an alternative ICE fuel. Among these are the lower noise levels of the combustion process, the reduced exhaust emissions even with cold start, the significantly lower CO_2 emissions (about 30%), the high octane rating of methane, which makes it a suitable choice for supercharged gasoline engines as well. As compared to gasoline, methane fuels have the following disadvantages: lower density, extra energy required for their compression, gas losses during production and transportation. The majority of all gas engines are forced-ignition and intake-manifold fuel injection. Compared to gasoline, methane accounts for a significant portion of the fresh charge.

The aim of this research is to study and determine the indicated parameters of Spark-Ignition Engine when running on methane and to compare these results with the parameters of an engine running on traditional fuel - gasoline.

2. Experimental study procedures

2.1. Engine Details

The study was carried out on a gasoline engine in the laboratory of the Department of Transport Engineering and Technologies, Technical University of Varna. The experiments performed provide figures of wide-open throttle speed characteristics, with fuel-air ratio of $\alpha = 0.9$ for gasoline and $\alpha = 1.05$ for methane (these are the optimum AFR values for both types of fuel) and optimum electric spark ignition timing for each type of fuel. The characteristics were taken at crankshaft speeds of 1500 min^{-1}, 2000 min^{-1}, 3000 min^{-1}, 3500 min^{-1}, 4500 min^{-1}, and 5500 min^{-1}, with wide-open throttle.

The engine parameters are shown in Table 1.

Table 1. Tested engine parameters.

№	Parameter	Value
1	Volume [dm^3]	1.275
2	Cylinder diameter [mm]	70.61
3	Stroke [mm]	81.28
4	Compression ratio	9.75
5	Nominal power/ crankshaft speed [kW/ min^{-1}]	46/5500
6	Nominal torque/ crankshaft speed [Nm/ min^{-1}]	95/3000

2.2. Results

Table 2 indicates the principal parameters of the fuels used in the experiment.

Table 2. Parameters of the fuels used in the experiment.

Parameter	Methane	Gasoline A95
Lower heating value H_u [kJ/ kg]	45 670	44 000
Density [kg/m^3], 20°C; 760 mm Hg	0.748	720

IOP Conf. Series: Materials Science and Engineering **977** (2020) 012025 doi:10.1088/1757-899X/977/1/012025

Molecular mass [kg/mol]	16.77	107
Combustion rate [sm/s]	33	43
Combustion air required for complete combustion of 1 kg of fuel [kg]	16.20	14.7
Octane number - MM	130	95
Flash point [°C]	540	744
Mass composition H/ C [%]	25/ 75	14.5/ 85.5

The graphs shown in Figures 1 to 7 show the experimental results. Fig. 1 illustrate the variation of Indicated specific fuel consumption over the entire engine speed range with optimum control parameters when engine is running on gasoline and methane.

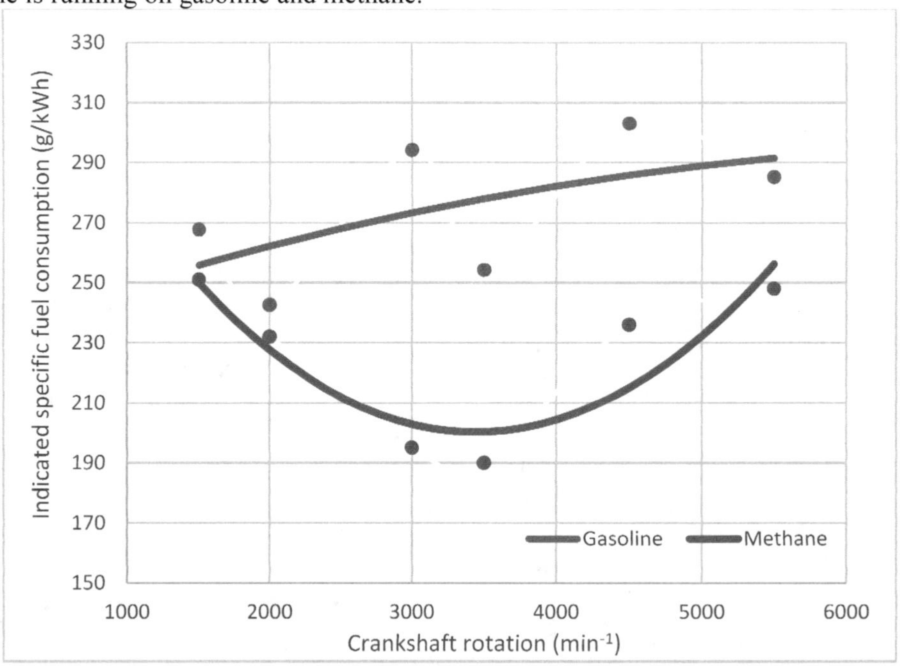

Figure 1. Variation of ISFC.

When the engine operates on methane, the minimum value of Indicated specific fuel consumption is 190 g/kWh at 3500 min^{-1}, with gasoline, the minimum value of Indicated specific fuel consumption is 254 g/kWh at the same crankshaft rotation. In terms of comparison between the two types of fuel, the Indicated specific fuel consumption marks an average decrease of approximately 33 % when the engine works on methane.

Figure 2 show the variation of Indicated mean effective pressure. When engine work on methane, the values of IMEP are lower because of air-fuel mixtures is lean and have higher heat losses for heating a larger cyclic quantity of air.

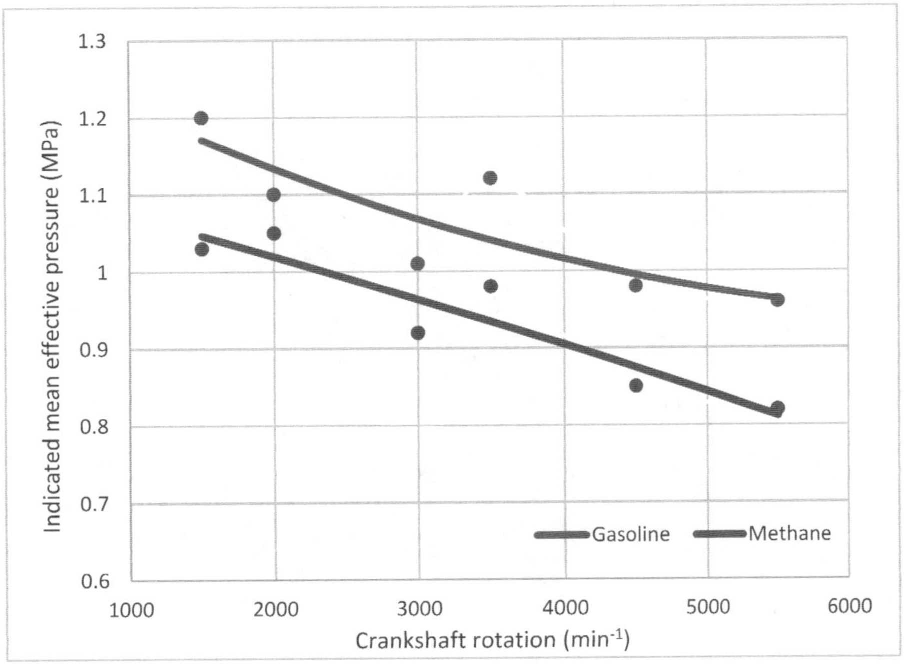

Figure 2. Variation of IMEP.

Fig. 3 show the variations of Indicated power. The higher values are obtained at gasoline fuel, because of the indicated diagram have bigger area and the effective parameters are better.

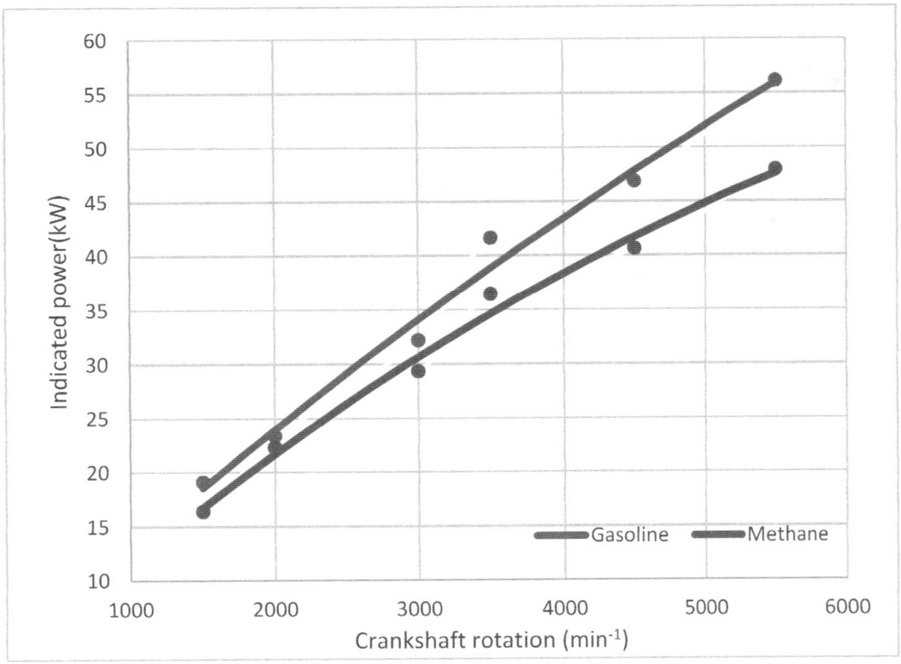

Figure 3. Variation of Indicated power.

Figure 4 show the variation of friction power. At both fuels the friction power is almost equal, because of the mechanical losses in the engine do not change.

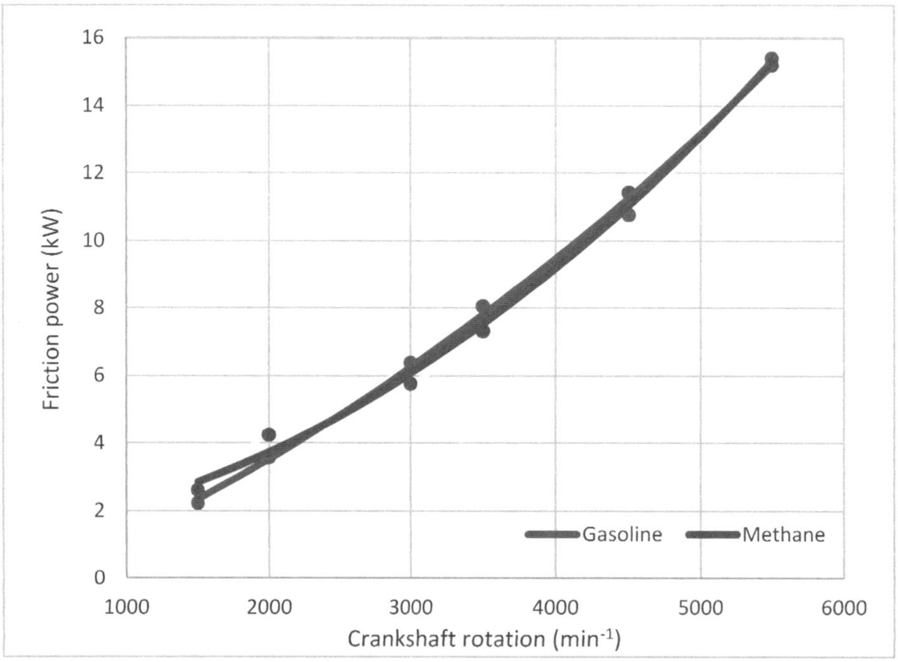

Figure 4. Variation of Friction power.

The variation of Friction mean effective pressure is shown on figure 5. The lower values are for gasoline, because of lower heat losses at burning process.

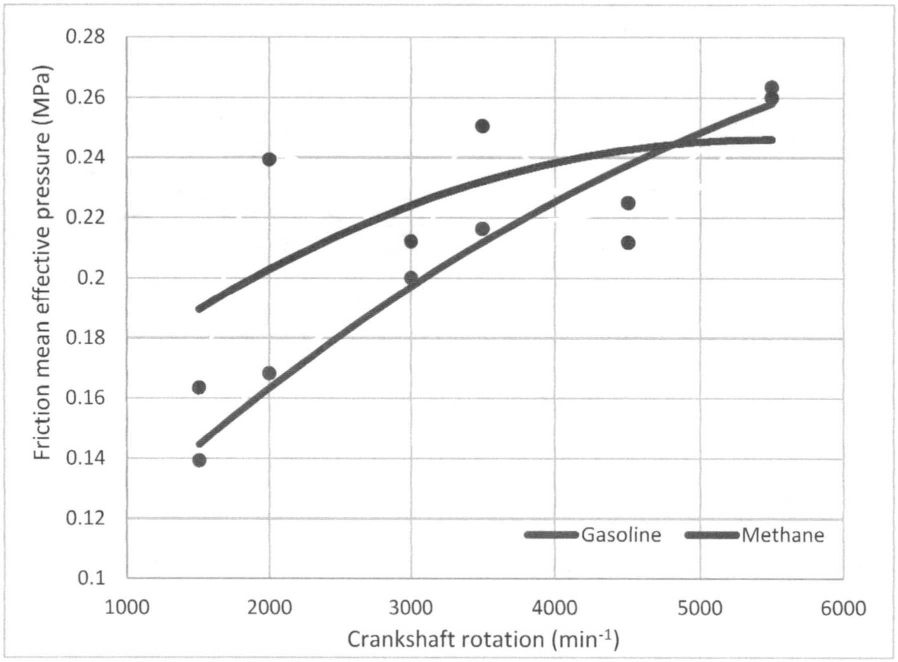

Figure 5. Variation of Friction mean effective pressure.

Figure 6 and 7 shows the variation respectively of Mechanical efficiency and Indicated efficiency. The indicated efficiency at methane fuel is around 20% better than gasoline fuel. This shows that the combustion process when engine work with methane is more perfectly.

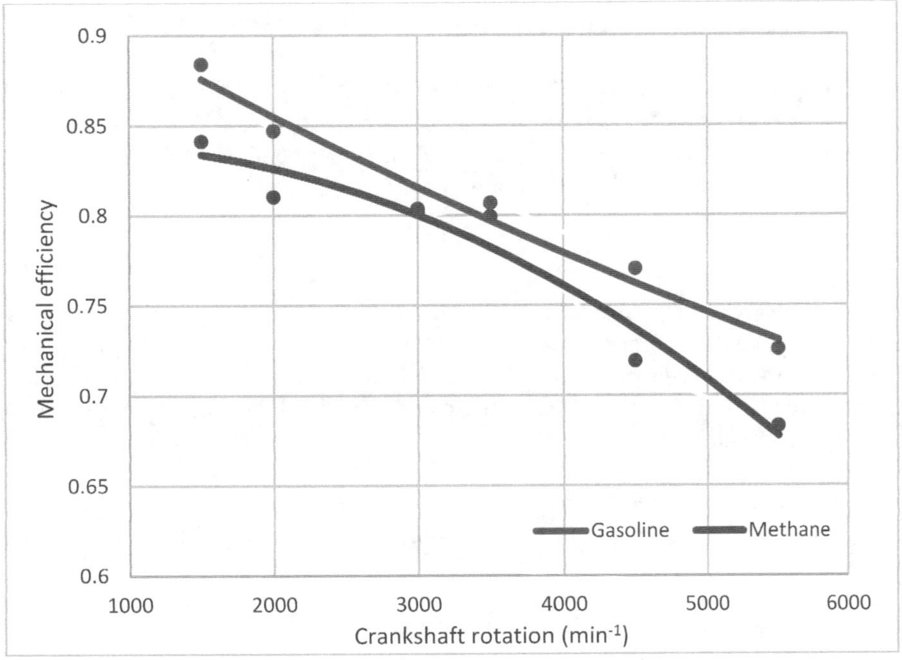

Figure 6. Variation of Mechanical efficiency.

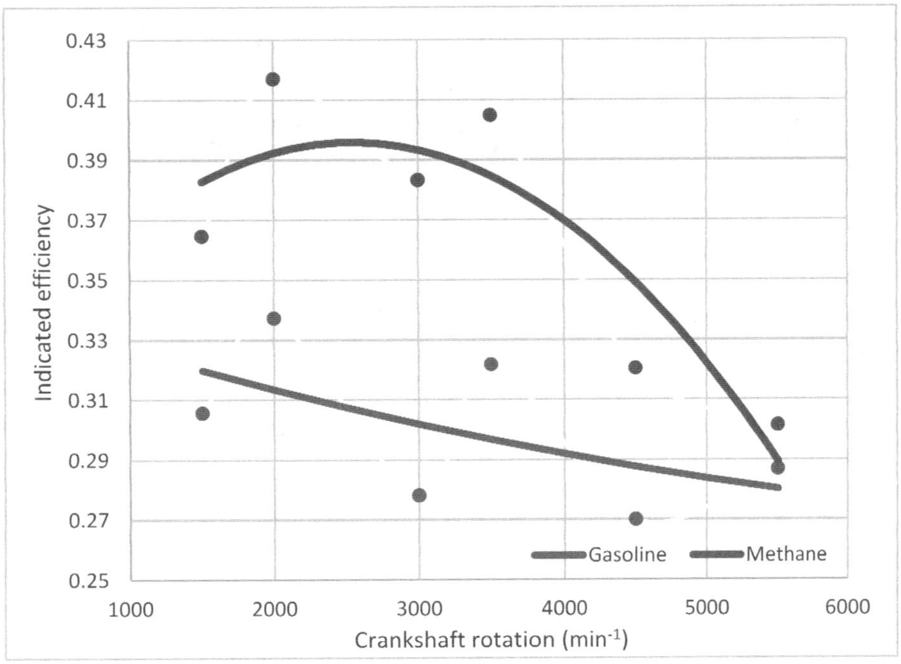

Figure 7. Variation of Indicated efficiency.

3. Conclusion

The analysis of the experiment results provides the basis for drawing the following conclusions:

- The indicated efficiency at methane fuel is around 20% better than gasoline fuel because of that the combustion process when engine work with methane is more perfectly;
- Indicated specific fuel consumption marks an average decrease of approximately 33 % when the engine works on methane.
- When engine work on methane, the values of IMEP are lower because of air-fuel mixtures is lean and have higher heat losses for heating a larger cyclic quantity of air.
- The higher values of indicated power are obtained at gasoline fuel, because of the indicated diagram have bigger area and the effective parameters are better.

References

[1] Petrakides S, Butcher D, Pezouvanis A and Chen R 2018 On the combustion of premixed gasoline-natural gas dual fuel blends in an optical SI engine *J. Power Technologies* **98** 387–95
[2] Heywood J 1988 *Internal Combustion Engine Fundamentals* (McGraw Hill) p 960
[3] Wattanavichien K and Azetsu A 2004 Studies of visualized diesohol combustion phenomena in IDI engine (CNG and Alternative Fuels, Oxygenated Fuels) *Proc. Int. Symp. diagnostics and modeling of combustion in internal combustion engines 2004.6* 423–30
[4] Iliev S 2018 Comparison of Ethanol and Methanol Blending with Gasoline Using Engine Simulation *Biofuels - Challenges and opportunities* ed Mansour Al Qubeissi (IntechOpen) chapter 9
[5] Iliev S 2018 A Comparison of Ethanol, Methanol and Butanol Blending with Gasoline and Relationship with Engine Performances and Emissions *Proc. of the 29th DAAAM Int. Symposium (Vienna, Austria)* ed B Katalinic (DAAAM International) pp 0505-0514
[6] Hu E, Huang Z, Liu B, Zhehg J and Gu X 2009 Experimental study on combusting characteristics of a spark ignition engine with natural gas – hydrogen blends combining with EGR *Int. J. of Hydrogen Energy* **34**(2) (Elsevier) pp 1035–44
[7] Dimitrov E, Tashev A and Pantchev S 2015 Methodology for experimental investigation of the influence of gaseous fuels on the performance of the diesel engine operating on gas-diesel cycle *Bultrans-2015 Proceedings* pp 103-7 (In Bulgarian)
[8] Serrano L, Matos de Carvalho P, Pires N, Pereira N and Gameiro da Silva M 2019 Comparing sustainable alternatives for energy source for the automotive sector *Energy for Sustainability Int. Conf. (Turin, Italy)* Conference Paper
[9] Gou M, Detuncq B, Guernier C and St-Germain P 1990 Performance of a single cylinder engine fuelled by a mixture of natural gas and gasoline *SAE Technical Paper*
[10] Pielecha J, Merkisz J, Kurtyka K. Comparison of Vehicle Exhaust Emissions with Gasoline Engines of Different Ecological Classes in Road Tests: Research Methods and Solutions to Current Transport Problems pp 330-340, Poznan University of Technology Poznan, Poland

IOP Conf. Series: Materials Science and Engineering **977** (2020) 012026 doi:10.1088/1757-899X/977/1/012026

Characteristics of vacuum solenoid valves in case of various malfunctions

Stoyan Stoyanov[1], Veselin Mihaylov[1,2] and Zdravko Ivanov[1]

[1]Technical University of Varna, Department "Transport Engineering and Technologies", 9010 Varna, Bulgaria

[2] E-mail: v_mihaylov@tu-varna.bg

Abstract. Vacuum solenoid valves are used to control actuators, in most cases connected with the emission system. In general, the diagnosis of modern cars is difficult due to the large number of elements and the fact that usually they gradually change their characteristics but remain operational. This is especially important for elements without feedback. The purpose of this research is to determine characteristics of faulty valves, which could be useful in diagnostics process. The results show change of valve characteristics at some of the most common problems during long term operation of the vehicle.

1. Introduction

Vacuum solenoid valves, also known as Electro-pneumatic pressure converters, Electric switch over valves or Boost control solenoids are used to control many systems of engines by controlling actuators - for example, variable geometry of the turbocharger, EGR valves, various bypass or throttle valves etc. [1-3].

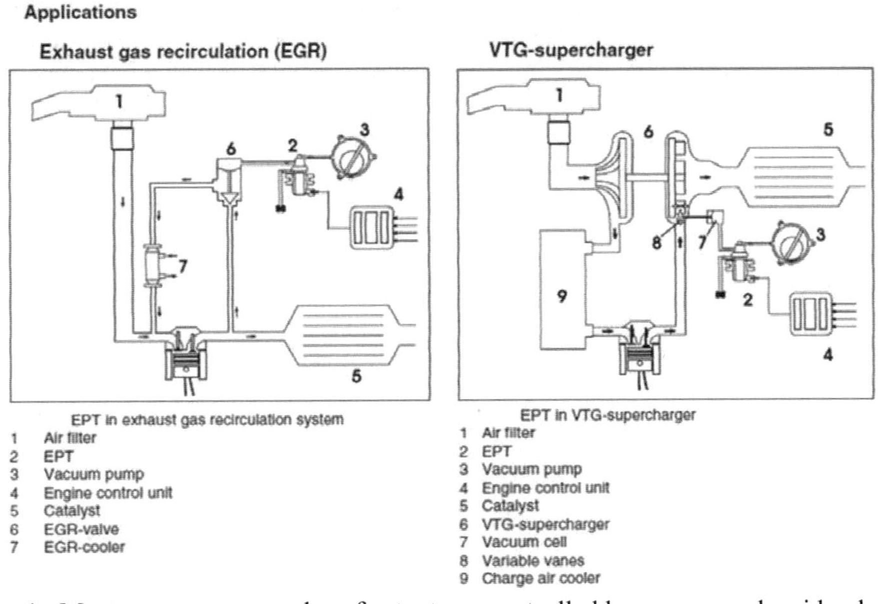

Figure 1. Most common examples of actuators, controlled by vacuum solenoid valves [3]

During the operation of the vehicle, these valves inevitably change their characteristics or are damaged, which leads to incorrect operation of the engine and ultimately to an increase in emissions of harmful components into the exhaust gases. Modern electronic engine control units (ECU) are equipped with a function for self-diagnosis and fault detection in the systems they operate. In the event of a fault in any of these systems, a trouble code is registered in ECU memory, which can be read by a diagnostic and test tools [4,5]. The car repair shops of the respective car brand have information systems and a database in which the technicians can find detailed information about the nature of the trouble code and the ways to repair the breakdown. Apart from that, there are many sites on the Internet with information about the meaning of these fault codes - for example https://www.obd-codes.com, and many others. When the nature of the fault is related to the control of an electrical circuit, a corresponding fault code is recorded in the ECU's memory which clearly directs the technician to the problem. For example, codes P0243, P0245 and P0246 which mean - "Turbo pressure regulation electro valve control - Open circuit, Short circuit to earth or Short circuit to positive ", indicates the corresponding problem in the electrical circuit of the turbocharger vacuum solenoid valve. Similar codes also exist for defective EGR vacuum solenoid valve and other engine systems [3]. There are also faults in which one or more elements of a system changes their characteristics in such a way that their electrical parameters remain within some normal limits, which prevents the registration of a fault, but the system still does not work properly. There are algorithms for detecting such faults based on the readings of the various sensors in the engine management system, but the detection of a particular element of the whole system is difficult due to the fact that the discrepancy may be due to each element of the system separately. As an example, OBD trouble code P0234 - „Turbocharger pressure signal: Pressure too high in relation to the reference value " is stored in ECU memory when there is a difference between the turbocharging pressure reference value and the turbocharging pressure measured. Other possible trouble codes that are registered with the failure of the turbocharger vacuum solenoid valve are: P0238 - „Turbocharger pressure signal - Short circuit to positive, or open circuit" which is recorded in case when measured value is too high and fault code P0100 - Flowmeter signal - Air flow not plausible, is recorded when there is a negative difference of 200 to 300 mg/stroke between the measured air flow and the air flow reference value. In all these cases, the ECU switches to an emergency engine control strategy and cutting of the EGR valve and EGR butterfly in all operating modes, cutting of the "exhaust heat recovery" function, limitation of engine torque and change to "Limp Home" mode [3,6,7]. The possible causes of these trouble codes according to the information systems of the authorized workshops for the given car brand can be turbocharging pressure sensor, variable geometry of the turbocharger jammed in the "maximum turbocharging" position, faulty turbocharger, vacuum problem, air flow sensor, EGR valve, electrical harness, any leak of air etc. Additional measurements and checks are required to detect the specific faulty element in the system according to the manufacturer's instructions and technical experience. Figure 2 shows a method for checking the condition of the valve provided by its manufacturer by which several of the possible faults are checked [3].

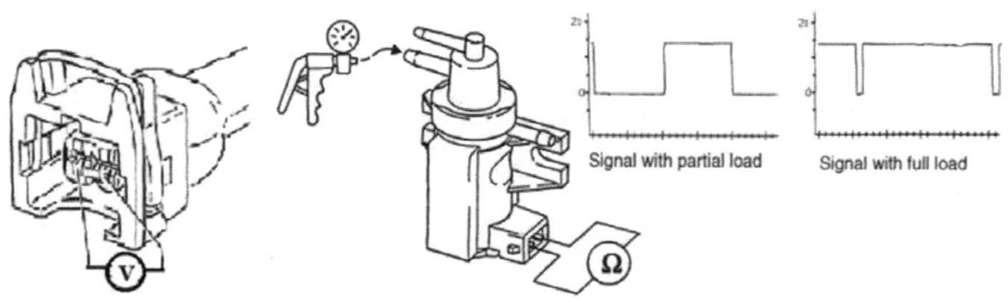

Figure 2. Test points according to the vacuum solenoid valve manufacturer [3]

There is a lot of research in the scientific literature related to the pneumatic control of various devices in automobile engines [1,2]. In most of these studies, it is either a matter of mathematical models and a detailed description of the operation of these valves or of the characteristics of new valves [8].

The purpose of this study is to determine the changes in the characteristics of vacuum solenoid valves after they have been replaced in car workshops due to a failure in the system they operate. For this purpose, several damaged valves were tested, and their characteristics were determined and compared with the characteristics of a new valve.

2. Experimental set-up

To determine the static characteristics of faulty valves, the following test stand has been used. The stand enables to be explored the set pressure at different duty cycle of PWM signal. The stand is shown on figure 3. Constant adjustable pressure can be maintained at the inlet of the tested valve by means of the vacuum pump, the pressure regulator and the tank. The valve outlet is connected to an actuator chamber. For monitoring and recording of pressures, the inlet and outlet are connected to a 2-chanel MAP sensor. The control signal with variable pulse width (PWM) is supplied by a DAC device. The control signal is fed to the tested valve through an amplifier designed to control electromagnetic devices with a built-in flyback diode. The values from the pressure sensors and the control signal are recorded by a digital oscilloscope [8].

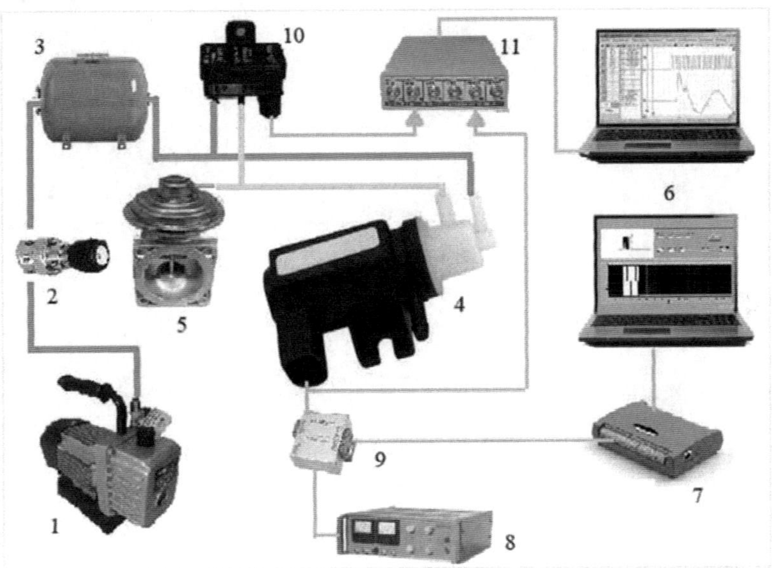

Figure 3. Experimental test stand:
1 – vacuum pump; 2 – pressure regulator; 3 – vacuum tank; 4 – vacuum solenoid valve; 5 – actuator; 6 – laptops; 7 – DAC for control signal generation; 8 – power supply; 9 - amplifier; 10 – 2-channel absolute pressure sensor; 11 – DAQ device.

3. Results and discussions

For current study were used electro pneumatic valves, made by Pierburg (figure 3, pos.4). The tested valves have been replaced in car repair shops due to a malfunction in the system of which they are a part. The valve is controlled by signal with constant frequency of 250 Hz and varying duty cycle. The characteristics pressure vs duty are determined by increasing the duty cycle from 0% to 100 % and back in steps of 10 %. In addition to electrical and other mechanical malfunctions that are possible, the study identified the following faults in the vacuum solenoid valves (figure 4): - clogging of the air filter (pos.1), micro deformations of the valve plate (pos.2), partial rupture of the membrane (pos.3).

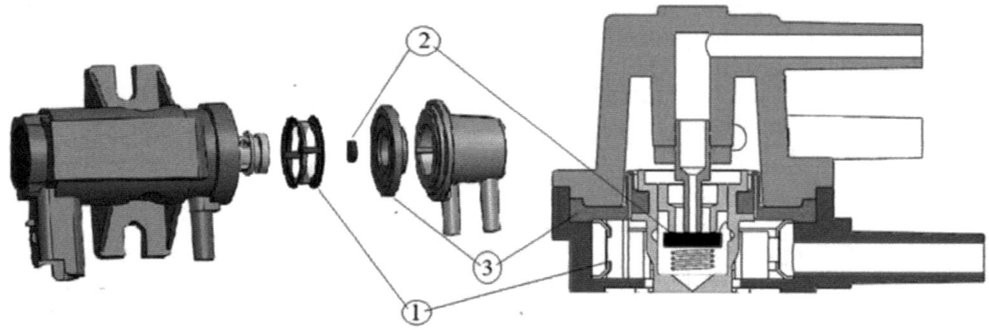

Figure 4. Vacuum solenoid valve construction and elements prone to cause malfunction
1 – air filter; 2 – valve plate; 3 – rubber membrane.

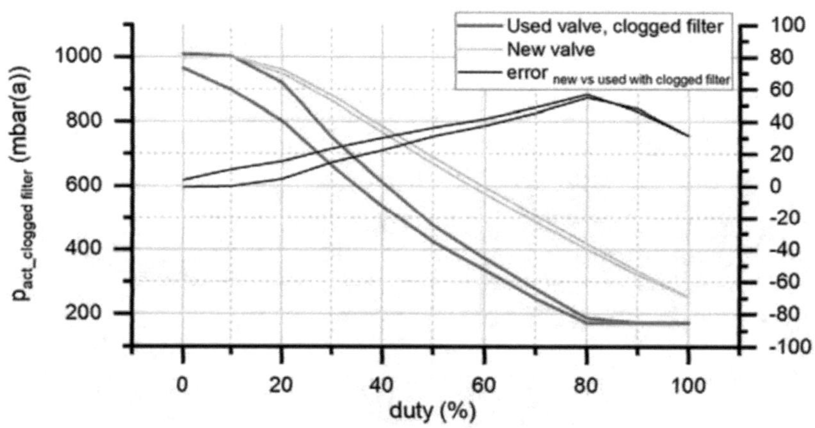

Figure 5. Regulated pressure vs duty cycle of new valve and used valve with clogged filter

One of the most common problems of these valves is clogging of the filter which is venting the valve to the atmosphere. On figure 5 with green colour is shown a new valve and with red a faulty valve (on this and the following graphs). The characteristic of regulated pressure vs duty cycle differs considerably from the reference one and the faulty valve is providing more vacuum in the whole range, with difference reaching up to 50 %.

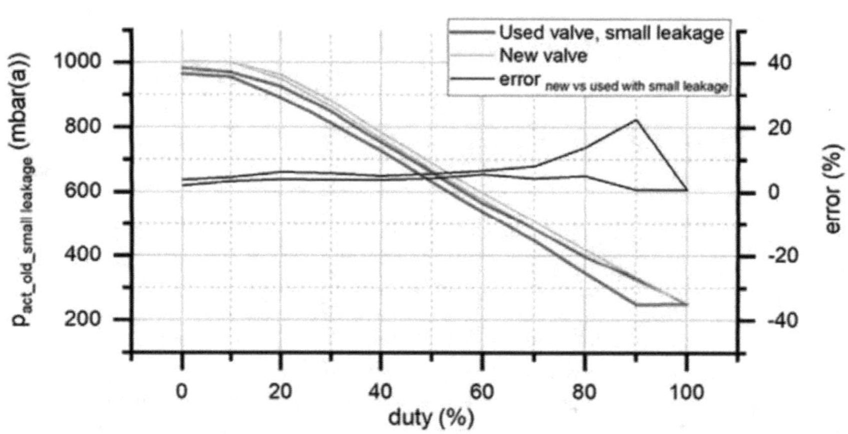

Figure 6. Regulated pressure vs duty cycle of new valve and used valve after prolonged operation

On figure 6 and figure 7 are shown examples of gradual change of characteristic of the valve during prolonged operation, where the problem is micro deformations of the rubber valve plate and this leading to the valve plate (figure 4, pos. 2) not sealing well. Usually the valve is supplying more vacuum to the actuator. In one of the cases the vacuum valve has wider hysteresis than usual at high duty ratios (figure 6).

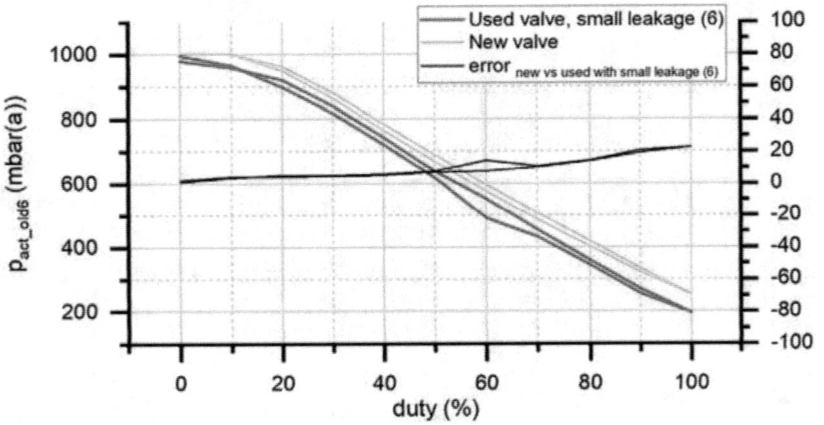

Figure 7. Regulated pressure vs duty cycle of new valve and used valve after prolonged operation (another example)

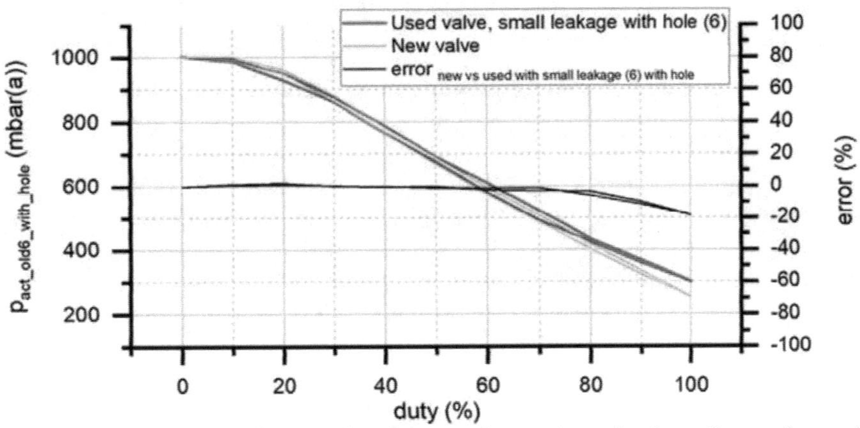

Figure 8. Regulated pressure vs duty cycle of new valve and used valve after prolonged operation with partial rupture of the rubber membrane

Figure 8 illustrates the behaviour of a pneumatic valve with a small partial rupture of the rubber membrane. Here the problems occur at upper medium and high duty ratios. It should be noted that these valves can deteriorate very fast from some point on.

Figure 9 shows the effect of a faulty power stage of the part of ECU, which is controlling the Vacuum solenoid valve. There is a flyback diode, which is allowing the current to decrease slowly when the power transistor is switched off during PWM. When this flyback diode is burned out, the tendency is to supply much less vacuum at the whole practical working range (10-90 %).

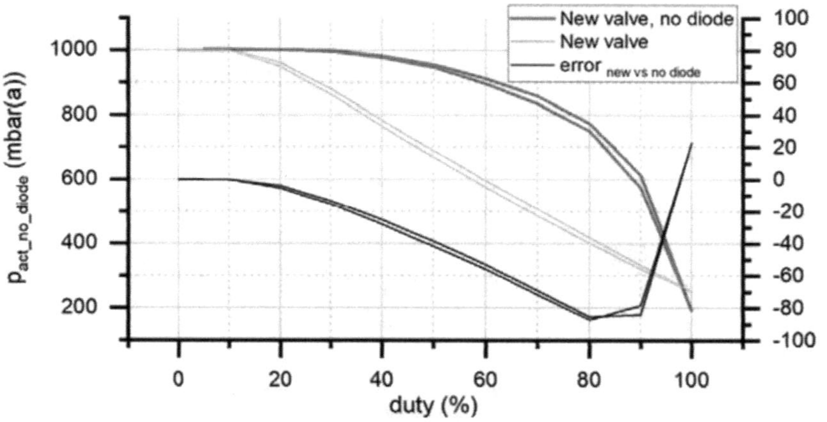

Figure 9. Regulated pressure vs duty cycle of new valve, with faulty ECU power stage (burned out diode)

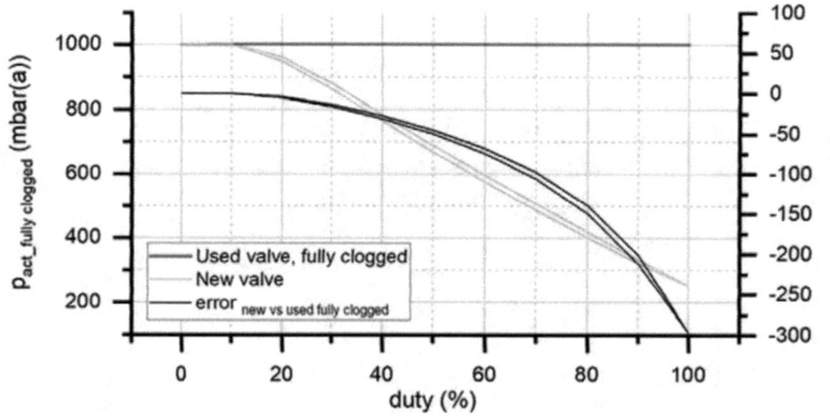

Figure 10. Regulated pressure vs duty cycle of new valve and used totally inoperative valve

On the last figure is shown a valve which resistance is in the recommended tolerances by the service documentation, but mechanically is totally inoperative.

4. Conclusion

Experimental research has been done and characteristics of several faulty electromagnetic pneumatic valves are determined. These characteristics can help in the diagnostics process of valves, due to the lack of feedback signal.

From this experimental research can be concluded that:

1. As the valve ages, as well as when the venting to the atmosphere is clogged, the tendency of change is to increase the regulated vacuum.
2. Damage (puncture) to the rubber membrane leads to decrease of regulated vacuum.
3. A flyback diode is essential for correct operation of the valve. In case of malfunction in it the deviation from demand value can reach up to 80 %.
4. For accurate diagnostics of this type of valves it is advised to use specialized equipment and additional testing than that proposed by OEM service recommendations.

Acknowledgment
This research was supported by the Bulgarian Ministry of Education and Science through National Programme "Young scientists and postdoctoral students".

References

[1] Mehmood A 2012 Modeling, simulation and robust control of an electro-pneumatic actuator for a variable geometry turbocharger. *Computers and Society [cs.CY]. Universite de Technologie de Belfort-Montbeliard NNT BELF 0188*

[2] Thomasson A 2014 Modeling and control of actuators and co-surge in turbocharged engines *Linköping studies in science and technology* Dissertations No 1590

[3] Pierburg Service information, Electropneumatic pressure transducer – EPT Product summary *si 0065 GB MSI-PG 09.02*

[4] Shinde B, Kore S, Thipse S 2016 Comparative Study Of On Board Diagnostics Systems-EOBD, OBD-I, OBD-II, IOBD-I and IOBD-II *IRJET Volume: 03 Issue: 03 e-ISSN: 2395 – 0056 p-ISSN: 2395-0072*

[5] Moniaga J, Manalu S, Hadipurnawan D and Sahidi F 2018 Diagnostics vehicle's condition using obd-ii and raspberry pi technology: study literature *IOP Conf. Series: Journal of Physics: Conf. Series* ***978*** *012011* doi :10.1088/1742-6596/978/1/012011

[6] Oluwaseyi M, Sunday A 2020 Specifications and Analysis of Digitized Diagnostics of Automobiles: A Case Study of on Board Diagnostic (OBD II) I*nternational Journal of Engineering Research & Technology Vol. 9 Issue 01* ISSN: 2278-0181

[7] Posada F, German J 2016 Review of LVD OBD requirements under the European, Korean and Californian emission programs *International Council on Clean Transportation*

[8] Stoyanov S and Mihaylov V 2020 A Study on Characteristics of vacuum solenoid valves *IOP Conf. Ser.: Mater. Sci. Eng.* accepted

IOP Conf. Series: Materials Science and Engineering **977** (2020) 012027 doi:10.1088/1757-899X/977/1/012027

Changes in fuel supply parameters of Common Rail nozzle after major service intervals

Trifon Uzuntonev[1,2] and Sergey Belchev[1]

[1]Department of Transport Engineering and Technologies, Technical University - city of Varna, 1 Studentska Str., city of Varna, Bulgaria

[2]E-mail: uzuntonev_trifon@abv.bg

Abstract In a number of countries of the European Union the average age of vehicle fleet is over 15 years, and in the entire European Union vehicles are on average 11.1 years old. This means that the environmental parameters are largely determined by the technical condition of the current vehicle fleet and in particular by the condition of fuel systems operated over extended periods of time. The article examines the changes in the behaviour of electromagnetic nozzles of Common Rail fuel system as a result of irreversible wear and deformation, which in turn affects the hydraulic characteristics and controllable fuel supply parameters. The influence of the changes in the electromagnetic nozzle's adjustable geometric clearances on the injected fuel and the dephasing of this process in relation to the moment of transmission of the control pulse are examined. An analysis of the impact of these phenomena on the work process development in terms of deterioration of the vehicle's environmental characteristics is made.

1. Introduction

In a number of developed European countries, the average age of the vehicle fleet is 11.1 years according to data taken from European Automobile Manufacturers' Association (ACEA). The top three positions are held by Luxembourg (6.4 years), followed by Austria (8.2 years) and Ireland (8.2 years). This stands in contrast to countries like Poland, Romania, Bulgaria, Lithuania, Estonia, etc. In Bulgaria, the average age of vehicle fleet is over 18 years. These important statistics determine the approaches of the individual governments to the introduction of restrictive measures to reduce ambient air pollution. Apparently, the introduction of restrictions on the movement of vehicles that do not meet certain eco-standards in certain urban areas for countries listed in the second half of this list is discriminatory. In this case it is more important to pay attention to the changes in the environmental performance of vehicles after major service intervals, the reasons for its deterioration, and the measures needed to bring them into service.

The prolonged operation of electromagnetic nozzles of Common Rail battery fuel system leads to irreversible wear and deformation, which in turn affects the hydraulic characteristics and controllable fuel supply parameters. They in turn lead to negative impact on the work process organisation, with all the accompanying negative consequences.

The aim of the presented work is to determine the level of influence of wear and deformation in electromagnetic nozzles CRI1 on the adjustable geometric clearances AH (Ankerhub) and RLS (Restluftspalt), fig.1. The changes in these clearances may lead to any possible changes in the amount of injected fuel and nozzle speed, which is expressed as extension of the response time RT (Response

Time out). These deviations from the standard adjustments cause deterioration of the power and economic performance of the engine, and of its environmental performance.

2. Prerequisites and Basic Concepts

The quantity of injected fuel and its time of injection into the engine cylinder in the Common Rail electromagnetic nozzles is controlled by an electronic pulse of a certain nature and duration. It is controlled by a card inserted in the computer and remains unchanged under the same operating conditions during the entire operation of the vehicle. The injected fuel quantity is a function of both the nature of this pulse and the physical phenomena occurring in the structure of the nozzle itself, figure 1.

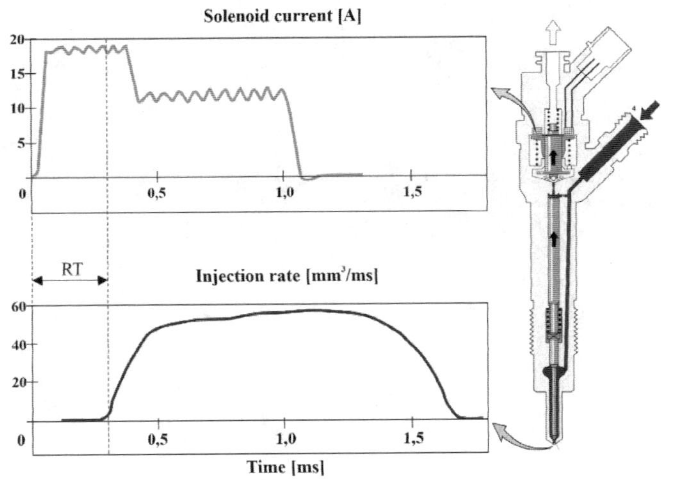

Figure 1. Fuel supply control of electromagnetic nozzle

Despite the construction features of the different generations of Common Rail electromagnetic nozzles, they all follow similar principle of operation and have common adjustable geometric clearances. CRI1 nozzle of the company Bosch 0 445 110 076, designed for Puegeot 2.0 HDI, is subject to analysis in the article. It largely corresponds to the idea of the article - establishing changes in its work after major service intervals comparable to the average statistical age of the vehicle fleet. Figure 2 shows a detailed section of the electromagnetic part of the nozzle with the most important geometric clearances AH and RLS. They are regulated during the initial production of the nozzle and their values largely determine the basic fuel supply parameters. Ball valve lift AH determines the chocking of fuel as it flows out of the nozzle control chamber. The increased AH lift leads to increased portion of injected fuel. The occurrence of preconditions for more intensive wear and tear in this group should be taken into account as a negative consequence of such an increase. The strength of interaction between the armature and electromagnet is determined by the size of residual air gap RLS and obeys Coulomb's law:

$$F = k \times (q_1 \times q_2) / r^2 \text{, where} \tag{1}$$

q_1 and q_2- are fixed point electric charges
R is the distance between charges, in this case the residual air gap RLS
K is a coefficient depending on the measurement system

In this dependence, the distance r = RLS is an adjustable value and has a second-degree influence on the magnitude of interaction force. In the actual structure, this force overcomes the pre-adjusted spring tension and determines in this sense the system speed. This means that, for example, the

IOP Conf. Series: Materials Science and Engineering **977** (2020) 012027 doi:10.1088/1757-899X/977/1/012027

increased value of adjustable geometric parameter RLS and the maintained duration of control pulse will result in delayed opening of ball valve and change in the time of fuel supply, figure 2.

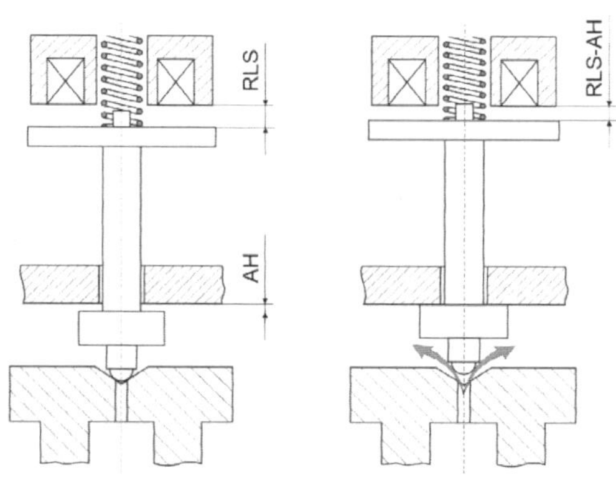

Figure 2. Adjustable geometric parameters of the electromagnetic nozzle

During prolonged operation of an electromagnetic nozzle, numerous closures of the control ball valve are made in its conical seat. This causes deformations in this structural element and "sinking" of the ball, figure 3.

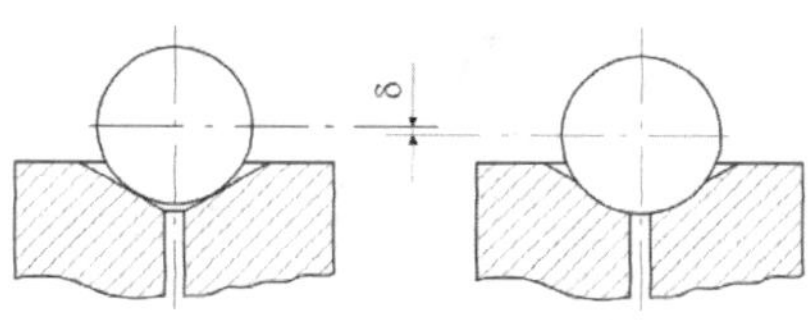

Figure 3. "Sinking" of the ball in the conical seat

This deteriorates the sealing condition, since the contact is now made on the surface and not on the line, as it was at the beginning of operation. The emphasis in this case is placed on the fact that the ball "sinking" automatically affects the values of the most important geometric clearances AH and RLS. In this case, the two parameters increase their values. The accuracy of their adjustment is up to 2 microns, and the actions are performed in a specific environment. This implies a noticeable impact of relatively small deformations on the main fuel supply parameters, such as injected fuel quantity and actual beginning of fuel injection. The simultaneous increase of the two adjustable parameters AH and RLS causes different effect on injected fuel quantity. Higher AH values suggest fuel increase, but at the same time, higher RLS values cause a delay in the response time of RT nozzle. As a result of the wide-ranging effect on fuel supply process due to the increased geometrical parameters AH and RLS, the injected fuel quantity changes insignificantly. The main fuel supply parameters, being the beginning of fuel supply process and its differential characteristics, will be negatively affected. This means that injected fuel quantity during continuous operation of the nozzle must be objectively measured after checking and adjusting the main adjustable geometric parameters that change during operation.

IOP Conf. Series: Materials Science and Engineering **977** (2020) 012027 doi:10.1088/1757-899X/977/1/012027

3. Experimental Studies

Standard CMX 3000 test bench for Common Rail nozzles was used to establish quantitative relationship between the level of wear and tear in the valve group of the examined electromagnetic nozzle and the basic fuel supply parameters. The accuracy of measurement of injected fuel quantity is 0.1 mm^3. The bench has an additional option to measure the response time RT of the electromagnetic nozzle. For this purpose, the fuel is injected into additional measuring chamber before being led to the flow meter. Piezoelectric sensor, which detects the time of pressure increase as a result of fuel injection, is installed in the chamber. This is considered to be the beginning of fuel supply. Despite certain conditions mainly related to the fact that the fuel is fed into a laboratory chamber rather than into the engine cylinder, the present case concerns comparative studies. Therefore, factors that are not taken into account have the same disturbing effect in all experiments. The RT parameter is read in μs and is displayed on the bench.

The different levels of wear and tear in the nozzle valve group are simulated by setting AH and RLS values in advance with the help of standard adjusting washers. Based on a preliminary analysis of the physical wear and tear process, the clearances were planned to be increased at regular intervals.

The quantities of injected fuel Q and the response time of RT nozzle were measured. The physical meaning of RT parameter is the delay of actual fuel injection, relative to the time of supply of the control pulse, figure 1. The aim of the experiment is to establish the quantitative dependence of this parameter on the time-varying air gap RLS and valve lift AH. To perform the experiment, standard programme set in the bench database for implementation of characteristic modes for the specific nozzle was used.

Tables 1, 2 and 3 show the results of the experiment with variation of RLS parameter at AH = 0.052mm = const. The value of AH parameter corresponds to the standard initial setting.

Table 1. Experiment results AH=0.052mm RLS= 0.100 mm

	$p_{Common\ Rail}$ bar	t μs	RT μs	Q mm^3
TEST 0	1350	1000	345	51.1
TEST 1	800	500	378	16.7
TEST 2	250	600	458	4.9
TEST 3	800	160	372	2.3

Table 2. Experiment results AH=0.052mm RLS=0.080 mm

	$p_{Common\ Rail}$ bar	t μs	RT μs	Q mm^3
TEST 0	1350	1000	335	51.7
TEST 1	800	500	372	17.0
TEST 2	250	600	448	5.1
TEST 3	800	160	365	2.4

Table 3. Experiment results AH=0.052mm RLS= 0.060 mm

	$p_{Common\ Rail}$ bar	t μs	RT μs	Q mm^3
TEST 0	1350	1000	323	52.3
TEST 1	800	500	356	18.6
TEST 2	250	600	430	6.2
TEST 3	800	160	348	3.0

4. Analysis of the Experimental Results

The attached tables containing simulations of different operating modes (combinations of pressures in the fuel battery and duration of the control pulse) set in the test bench database show reduction of the nozzle's response time RT at increased pressures in the fuel battery $p_{Common\ Rail}$. This is normal, given the fact that the pressure in the nozzle's control chamber is proportional to that in the fuel battery and differs slightly due to the chocking in the inlet. The higher pressure helps control valve to open faster. Therefore, in all experiments where RLS has different values, the response time for TEST0 ($p_{Common\ Rail}$=1350 bar t =1000μs) is the smallest. The analysis made shows that the smaller air gap RLS contributes to the greater speed of the nozzle, which is expressed in reduction of the response time RT. This applies to all operating modes for which the tests have been conducted, figure 4.

Simultaneously with the experiments, tests were conducted to establish the influence of the different values of control valve AH lift on the injected fuel quantity. In this case, the relationship is unequivocally defined. As the valve lift increases, with the air gap being maintained, fuel quantity increases as a result of the faster pressure drop in the control chamber. This applies to all operating modes involved in the tests. The increased lift AH in TEST3, which is identified as a pre-portion of fuel before the main fuel portion, has the most negative effect. In absolute terms, this quantity is smaller and even any insignificant change in the increase leads to the possibility of implementing a work process with great advance of release and with all resulting negative consequences for the environmental characteristics and power and economic indicators of the engine.

The article does not present the results of these experiments due to the known influence of AH on the quantitative characteristics of injected fuel quantity.

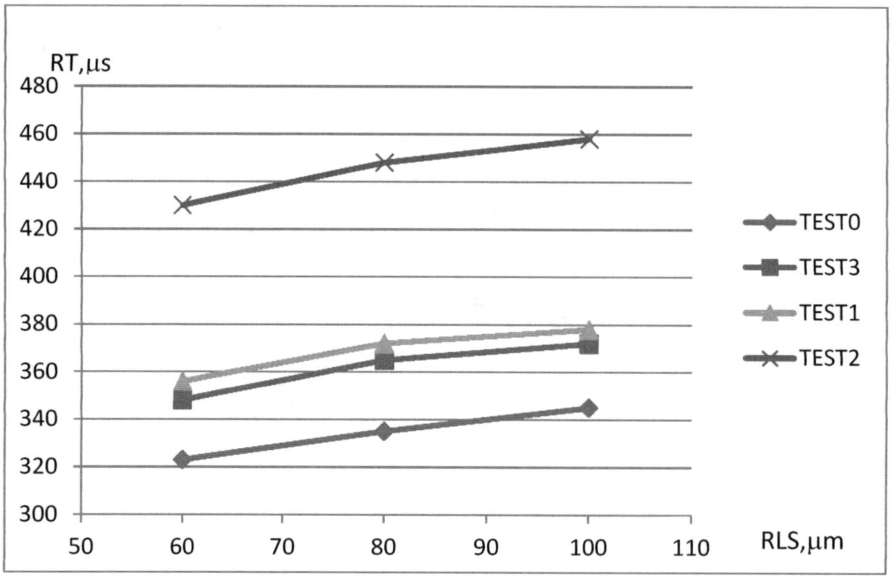

Figure 4. Relationship between RLS and RT in different operating modes

5. Conclusion

The prolonged operation of electromagnetic nozzles Common Rail leads to wear in the control valve movable elements. This leads to deterioration of its sealing capabilities and changes in important adjustable clearances. They, in turn, have negative impact on the injected fuel quantity and change the actual injection process beginning. Deteriorated organisation of the work process affects negatively the power and economic performance of the engine, and its environmental performance. The article analyses the mechanism of this process and makes a quantitative assessment of the relationship between the level of wear and tear and the level of deterioration of the nozzle's hydraulic characteristics.

In the presence of standard Common Rail nozzle test bench, deviations in the electromagnetic nozzle's operation can be determined with sufficient accuracy. The accumulation of sufficient database allows for assessment of the RT values based on the level of wear and tear and deviation from the nozzle's normal operation.

If the control valve's sealing is preserved within the permissible limits, it is possible to restore the initial adjustments of the nozzle and to maintain its operability for a further period.

6. References

[1] Robert Bosch GmbH 2004 *Diesel-Engine Management, 4th Edition* (Stuttgart) p504

[2] Xiong J and Gu H 2017 An Intelligent Dual-Voltage Driving Method and Circuit For a Common Rail Injector for Heavy-Duty Diesel Engines *J. IEEE Access* **5** 27681-27689

[3] Dongiovanni C and Coppo M 2012 Accurate Modelling of an Injector for Common Rail Systems *Fuel Injection* Siano D (Rijeka: InTech) p262

[4] J.Hammer, A.Binder Einspritztechnik Teil 1 IVK Universität Stuttgart

[5] Klaus Krieger Diesel-Einspritztechnik für Pkw-Motoren Überblick über Verfahren und Ergebnisse, MTZ Motortechnische Zeitschrift 60 (1999)

[6] Rolf Leonhard, Johann Warga, Thomas Pauer, Msrkus Rückle, Matthias Schnell Magnetventil Common-Rail-Injektor mit 1800 bar MTZ 02/2010 Jahrgang 71

[7] Rolf Leonhard, Johann Warga Common Rail System von Bosch mit 2000 bar Eispritzdruck für Pkw MTZ 10/2008 Jahrgang 69

IOP Conf. Series: Materials Science and Engineering **977** (2020) 012028 doi:10.1088/1757-899X/977/1/012028

Study the impact of fuel and exhaust gas recirculation on HCCI combustion

Svetoslav Mihalkov and Plamen Punov[1]

[1]Department of combustion engine, automobiles and transport, Technical University of Sofia, 8, Kl. Ohridski blvd., 1756, Sofia, Bulgaria

[1] E-mail: plamen_punov@tu-sofia.bg

Abstract. This paper presents a numerical study of fuel/fuel blends impact on rate of heat release at different exhaust gas recirculation rate in case of homogeneous charge compression ignition (HCCI) combustion. A direct injection engine is used for simulation which also allows mixture preparation in intake manifold leading to premixed combustion. The numerical analysis was conducted by means of an engine model developed in advanced simulation software AVL Boost. The combustion model is based on skeletal reaction mechanism of C7H16(n-heptane) that uses 26 species and 66 reactions. Additionally, to the main fuel, methane and hydrogen was added. Thus, the influence of fuel blends was evaluated at EGR rate within the range of 0% to 40%.

1. Introduction

Conventional spark-ignition (SI) and compression-ignition (CI) engines offer high thermal efficiency in conversation the fuel energy. However, the combustion process in both engines leads to high NOx and soot formatted in the cylinder due to high local temperature and non-homogeneous charge especially in diesel engines. As a result, complex after treatment system needs to be implemented in passenger cars that leads to higher vehicle's cost as well as to restriction in the optimal engine management. Moreover, CO_2 emissions from ICEs need to be further reduced according to the new European regulation for passenger cars that was imposed in 2020. The new CO_2 reference limit is 95 g/km. In short term, replacing the gasoline by compressed natural gas (CNG) in SI engines could reduce CO_2 emissions by 25% in NEDC [1]. This study also reported around 50% reduction of NOx while CO and CH slightly increase. However, the emissions were measured after a three-way catalytic converter (TWC). Thus, TWC is still needed to respect Euro 6 limits. Methane in mixture with CO_2 in a form of biogas could be successfully used in stationary SI engines where biogas is residual product [2]. However, increased concentration of CO_2 significantly reduces the engine output power.

In order to reduce in-cylinder NO_x and soot more complex approach is needed. For that, low temperature combustion (LTC) of homogeneous mixture need to be implemented [3]. Pachiannan et al. in [4], presented a comprehensive review of performance and emission characteristics of LTC when different fuel was used. LTC combustion can be achieved by different approaches such as: homogeneous charge compression ignition (HCCI), premixed charge compression ignition (PCCI), partially premixed charge compression ignition (pPCCI), reactivity controlled compression ignition (RCCI) and etc. The main differences between HCCI, PCCI and pPCCI is the duration of ignition delay period. HCCI is recognised by long ignition delay while pPCCI offers ignition delay similar to conventional CI engines [5]. Operating range in terms of local temperature/air-excess ratio of HCCI for NOx and soot reduction

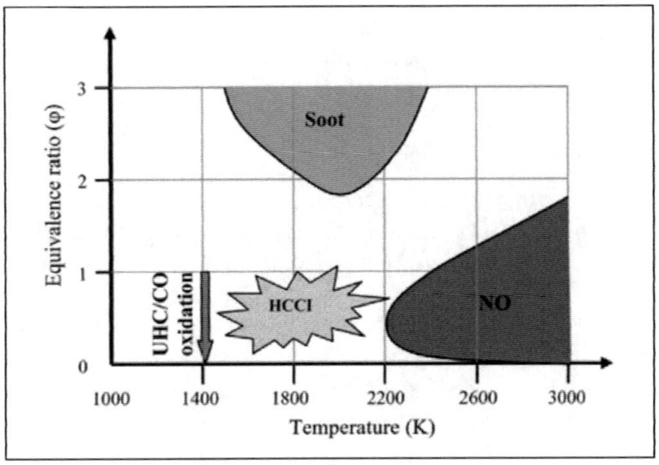

Figure 1. Operating mode of HCCI combustion on local φ-T diagram [8]

is shown in Figure 1. RCCI is a combustion process where usually two fuels with different reactivity are used [6,7]. LTC is very sensitive to numbers of factors such as: fuel reactivity, injection timing, injector characteristics, injection pressure, piston bowl geometry, intake charge temperature, exhaust gas recirculation (EGR), turbocharging, compression ratio and etc. [8–10]. The influence of fuel/fuel blends and EGR rate on combustion characteristics was widely studied in the literature [11–15]. The fuel has an impact on HCCI combustion by their reactivity. The impact of ethanol/diesel and ethanol/biodiesel blends on emissions was experimentally studied in [13]. It was reported that an average 56.5% of diesel fuel can be replaced by ethanol with 70% NOx and 61% smoke opacity reduction. Calam et al. [16] compared combustion characteristics, engine performance and emissions on several fuels to n-heptane operating in HCCI combustion. It was reported that octane number has significant impact on HCCI combustion as CA50 was more controllable with high octane fuels while E25 is more suitable for HCCI combustion with minimum pressure rise.

The combustion process in ICEs is often studied numerically by means of 1D-0D simulation approach [17]. It allows to investigate the combustion characteristics using different blends in both SI and CI engines without complex CFD simulation. However, HCCI combustion is usually two or three stage combustion with low temperature phase (LTF), intermediate phase (IF) and high temperature phase (HTF) [4,18] that cannot be expressed with conventional diesel combustion mechanism. Phenomenological HCCI combustion models are based on detailed chemistry reactions of the fuel in 0D cylinder model. The reaction mechanism on n-heptane and blends was studied in [19–21]. Zeuch et al. [21] proposed a comprehensive skeletal reaction mechanism of n-heptane, while Tsurushima in [20] presented a skeletal reaction mechanism with 33 species and 38 reactions.

Thus, the paper aims to study numerically the impact of fuel/fuel blends on the rate of heat release, pressure rise, engine performance and pollutant formation at different exhaust gas recirculation rate. The main fuel was n -heptane know as primary reference fuel (PRF) while in the blends the n-heptane is mixed with methane and hydrogen.

2. HCCI combustion mathematical background

HCCI combustion can be divided in three phases, such called: LTF, IF and HTF. The first phase depends on the characteristics of the charge flow; the intermediate phase depends on the kinetics of the chemical behavior of the fluid, and the last phase is influenced by chemical and turbulent mixing conditions.

In order to obtain the rate of heat release in 0D cylinder model following correlation can be used:

$$\frac{dQ_F}{d\alpha} = \sum_{i=1}^{nSpcGas} u_i . MW_i . \dot{\omega}_i \tag{1}$$

where: $nSpcGas$ - number of species in the gas phase u_i - species internal energy [J/kgK]; MW_i - species molecular weight [kg/kmole]; $\dot{\omega}_i$ - species reaction rate [kmole/m^{3}s]

The species mass fractions are calculated as follows:

$$\rho \frac{d\omega_i}{d\alpha} = MW_i . \dot{\omega}_i \tag{2}$$

where: ρ - mixture density [kg/m^3] and ω_i - species mass fraction [-].

The reaction rate of each species $\dot{\omega}_i$ is calculated based on a specified set of chemical reactions that describe the auto-ignition process. In this study a reduced skeletal reaction mechanism on C7H16 (n-heptane) proposed by Barroso [22] was used. This reduced mechanism consists of 26 species and 66 reactions as it based on skeletal mechanism of Newson that involves 67 species and 254 reactions. The analysis reported by Barroso, concerning heat release rate parameters CA10 and CA50 for reduced mechanism revealed maximum deviation of 2°CA degrees compared to the initial mechanism.

The thermodynamic properties of each species is estimated as follows:

$$\frac{c_p}{R} = a_1 + a_2 T + a_3 T^2 + a_4 T^3 + a_5 T^4 \tag{3}$$

$$\frac{H}{RT} = a_1 + \frac{a_2}{2} T + \frac{a_3}{3} T^2 + \frac{a_4}{4} T^3 + \frac{a_5}{5} T^4 + \frac{a_6}{T} \tag{4}$$

$$\frac{S}{R} = a_1 lnT + a_2 T + \frac{a_3}{2} T^2 + \frac{a_4}{3} T^3 + \frac{a_5}{4} T^4 + a_7 \tag{5}$$

This skeletal reaction mechanism cannot be used to estimate the NOx and soot formation, thus, it was used to predict heat release rate and NO for different fuel blends and EGR rate.

3. Simulation model

The engine that was numerically studied is 2.0 liter four cylinders' direct injection engine, developed for passenger cars. The maximum output power when operating with diesel in conventional compression-ignition mode is 101 kW at 4000 rpm as the maximum torque is 320 Nm at 2000 rpm. The engine is equipped with variable geometry turbocharger. The boost pressure is limited to 1.4 bar. Common rail direct injection fuel system of the engine is delivered by Delphi. The maximum injection pressure is limited to 1600 bar. The engine is equipped with EGR and post treatment system including catalytic converter and diesel particulate filter (DPF). The cylinder head is equipped with four valves per cylinder. The main geometrical parameters of the engine are listed in Table 1.

Table 1. Main engine parameters

Type of engine	Direct injection
Number of cylinders	4
Displacement	2 L
Cylinder bore	85 mm
Cylinder stroke	88 mm
Compression ratio	17,6
Valves per cylinder	4

This engine was considered suitable to be adapted for HCCI combustion using diesel fuel or fuel blends as it offers direct injection in the cylinder while port injection of high octane or gaseous fuel could be further implemented. The engine model was developed in advanced simulation software AVL Boost (Figure 2). The model is based on 0D cylinder modeling considering uniform thermodynamics parameters in the combustion chamber and 1D unsteady flow modeling into intake and exhaust pipes. The main engine data such as: engine type, operating parameters, friction losses and firing order were

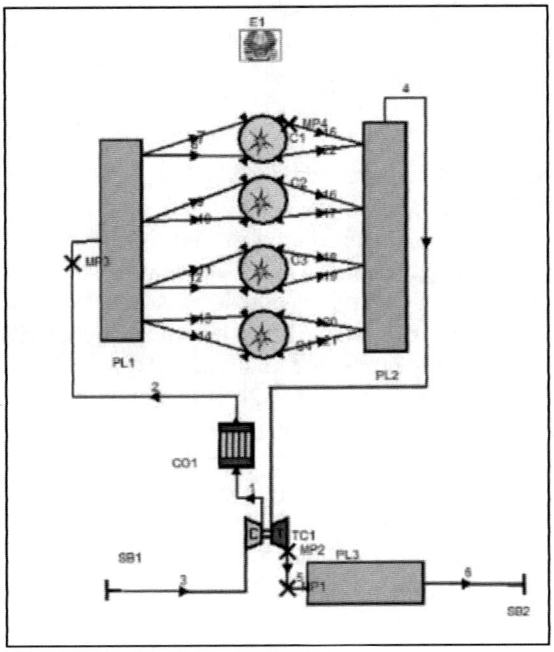

Figure 2. Engine simulation model, built in advanced simulation software AVL Boost

defined in the Engine element - E1. Cylinder geometry was imposed in elements C1 to C4. The single zone HCCI combustion was chosen which required definition of general species transport and detailed reaction mechanism including reaction coefficients. In the element named "Cylinder", gas to cylinder wall heat transfer was defined as well as the valves lift curves and valves discharge coefficients. The intake and exhaust geometry was presented by pipes and plenums (PL1 and PL2). Moreover, an air intake intercooler was placed after the compressor - CO1. The turbocompressor model (TC1) used simplified modeling with constant pressure ratio and efficiency on the compressor side and equivalent turbine discharge coefficient as well as constant overall efficiency.

Table 2. Simulation constraints

Parameter	Value		
Engine speed	2500 rpm		
Injected fuel	1.5e-005 kg/cycle		
Boost pressure	0.3 bar		
EGR	0%	20%	40%
Fuel mixture	100% C7H16		
	90% C7H16 + 10%CH4		
	80% C7H16 + 20%CH4		
	95% C7H16 + 5%H2		

In order to establish the EGR rate a simplified approach was used. There was no physical connection between exhaust and intake manifold while the EGR rate was determined by defining the gas composition in intake manifold.

The simulations were carried out at constant engine speed, constant total injected fuel mass and constant intake pressure. In imposed limitations the combustion process, engine performance and NO formation were studied with four fuels at several EGR rate. All limitations and variable simulation parameters are listed in Table 2.

4. Results and discussion

4.1. Study the heat release rate

The heat release rate (HRR) was studied numerically when engine is fueled by: pure n-heptane (C7H16), mixture 90% n-heptane + 10% methane, mixture 80% n-heptane + 20% methane and mixture 95% n-heptane + 5% hydrogen. The results are presented in Figure 3. When no EGR is applied the start of combustion was found to be too early - 25°CA, BTDC. The same value was observed for pure n-heptane and other fuel blends. Increasing the EGR to 20% and 40% leads to late start of combustion as it accounts to 15°CA, BTDC when n-heptane was used. However, when higher EGR rate (40%) is applied the combustion duration is longer for n-heptane due to slight increase in HRR during the LTF. As a result, the maximum HRR for EGR of 40% was at 5°CA, ATDC providing the maximum IMEP. Adding methane and hydrogen in fuel blends reduced the effective EGR rate. In the cases when 10% methane and 5% hydrogen was added to n-heptane the engine IMEP is close to zero when EGR was 40%. Moreover, increasing the methane to 20% in fuel blends leads to effective combustion only without EGR. The maximum HRR is similar for all cases without EGR while 20% EGR leads to reduced maximum HRR value. Here, it was observed the effective combustion with 40% EGR only for pure n-heptane as it significantly reduced maximum HRR to 140 J/deg. When fuel blends were studied, it was observed that the maximum of HRR was close to TDC or after TDC which means higher IMEP and engine thermal efficiency. Using 20% methane in the fuel blend leads to lower intensity of LTF and higher HRR during the HTF as a result maximum HRR occurs at 3°CA, ATDC. However, this fuel blend limited effective EGR range below 20%.

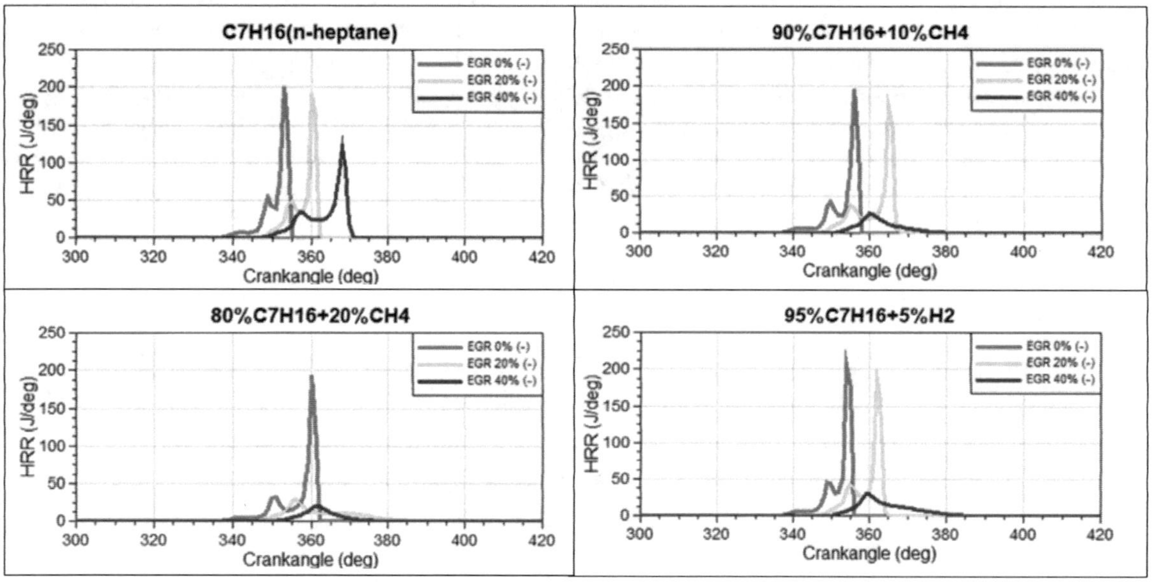

Figure 3. HRR obtained with different fuel/fuel blends and EGR rate

IOP Conf. Series: Materials Science and Engineering **977** (2020) 012028 doi:10.1088/1757-899X/977/1/012028

4.2. Study the in-cylinder pressure

In-cylinder pressure was also studied in order to analyze maximum pressure rise and to evaluate the engine performance. Thus, the impact of fuel/fuel blends and EGR rate within the range of 0% to 40% on in-cylinder pressure was investigated. The results are presented in Figure 4. The maximum in-cylinder pressure was observed for all studied fuel blends to be without EGR. Despite of the fact that start of combustion and CA50 were different for each fuel blend, the maximum pressure was estimated to be within the range of 115 bar to 120 bar. The pressure maximum angular position was affected by fuel blends as it was far before TDC for pure n-heptane while it reached TDC when 20% methane was added. In case of pure n-heptane the maximum pressure had the same value even when 20% EGR was applied. However, the position of maximum pressure was close to TDC. Increasing the EGR to 40% for n-heptane leads to lower in-cylinder pressure as the maximum value was estimated to be 92 bar at 10°CA ATDC. Appling higher EGR rate (40%) for fuel blends leads to very low integral heat release, thus very low IMEP was estimated.

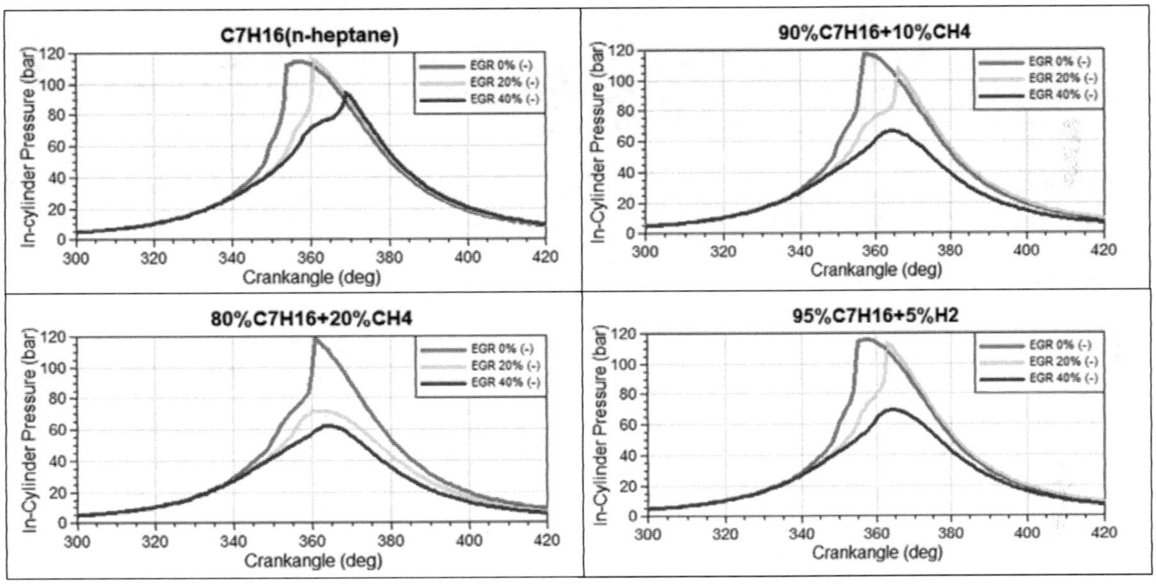

Figure 4. In-cylinder pressure with different fuel/fuel blends and EGR rate

4.3. Engine performance and NO analysis

In HCCI combustion the HRR and in-cylinder pressure curve are very sensitive to fuel reactivity and in-cylinder conditions. Thus, the engine performance is also affected while the start of combustion and CA50 are sensitive to fuel/fuel blends and EGR rate. Estimated engine output performance, maximum pressure rise and NO fraction are summarized in Table 3.

Table 3. Numerical results

Fuel mixture	EGR [%]	Engine torque [Nm]	Engine power [kW]	Pressure rise [bar/deg]	Mass fraction NO [-]
C7H16	0%	30.9	8.1	27.8	5.08e-006
	20%	46.9	12.3	26.8	1.08e-006

	40%	56.3	14.7	7.2	6.94e-008
90%C7H16 10%CH4	0%	36.5	9.6	21.5	4.68e-006
	20%	53.8	14.1	19.8	5.01e-007
80%C7H16 20%CH4	0%	43.9	11.5	22.9	3.22e-006
	20%	14.1	3.7	4.2	5.42e-011
95%C7H16 5%H2	0%	33.4	8.7	32.3	4.88e-006
	20%	50.1	13.1	24.7	8.54e-007

5. Conclusion

The impact of fuel/fuel blends and EGR rate on combustion, performance and NO formation in HCCI combustion was numerically studied. Four fuels were proposed: pure n-heptane (C7H16), 90% n-heptane + 10% methane, 80% n-heptane + 20% methane and 95% n-heptane + 5% hydrogen. For each fuel/fuel blend EGR rate was varied within the range of 0% to 40%. The engine speed was 2500 rpm, total fuel mass was 1.5e-005 kg/cycle and the boost pressure was 0.3 bar. The results revealed that EGR rate has significant impact on start of combustion and CA50 of HRR. The higher range of EGR can be applied when pure n-heptane is used. Adding methane to the n-heptane reduced maximum HRR and retarded start of combustion. It had positive effect to IMEP and engine performance. However, higher concentration of methane limited the effective EGR rate. The engine could not operate with 40% EGR when the fuel blend consists 10% and 20% methane. Moreover, methane concentration more than 10% also reduced EGR rate below 20%. The small quantity of hydrogen has similar effect to 10% methane in the fuel blend. It provided similar engine performance but higher maximum pressure rise with EGR within the range of 0% to 20%. The simulation revealed higher engine performance with pure n-heptane and 40% EGR. Here, the engine torque accounted to 56.3 Nm and output power was 14.7 kW. This case offered lower maximum pressure rise of 7.2 bar/deg and lower NO fraction of 6.94e-008. Similar engine performance was observed with fuel blend of 90% n-heptane and 10% methane. However, the maximum pressure rise is much higher as well as the NO fraction.

Acknowledgment

The study was funded by project BG05M2OP001-1.001-0008 "National Centre for Mechatronics and Clean Technologies" Operational Programme Executive Agency, Ministry of Education and Science, Bulgaria.

The numerical simulations in the study were done by means of the advanced simulation package AVL AST. We gratefully acknowledge the AVL company for providing us with the opportunity to use AVL simulation products for numerical studies at the Faculty of Transport of the Technical University of Sofia.

References

[1] Bielaczyc P, Woodburn J and Szczotka A 2014 An assessment of regulated emissions and CO2 emissions from a European light-duty CNG-fueled vehicle in the context of Euro 6 emissions regulations Appl. Energy 117 134–41

[2] Dimitrov R, Ivanov Z, Zlateva P and Mihaylov V 2019 Optimization of biogas composition in experimental studies E3S Web Conf. 112

[3] Krishnamoorthi M, Malayalamurthi R, He Z and Kandasamy S 2019 A review on low temperature combustion engines: Performance, combustion and emission characteristics

Renew. Sustain. Energy Rev. 116 109404

[4] Pachiannan T, Zhong W, Rajkumar S, He Z, Leng X and Wang Q 2019 A literature review of fuel effects on performance and emission characteristics of low-temperature combustion strategies Appl. Energy 251 113380

[5] Gan S, Ng H K and Pang K M 2011 Homogeneous Charge Compression Ignition (HCCI) combustion: Implementation and effects on pollutants in direct injection diesel engines Appl. Energy 88 559–67

[6] Reitz R D and Duraisamy G 2015 Review of high efficiency and clean reactivity controlled compression ignition (RCCI) combustion in internal combustion engines Prog. Energy Combust. Sci. 46 12–71

[7] Li J, Yang W and Zhou D 2017 Review on the management of RCCI engines Renew. Sustain. Energy Rev. 69 65–79

[8] Bendu H and Murugan S 2014 Homogeneous charge compression ignition (HCCI) combustion: Mixture preparation and control strategies in diesel engines Renew. Sustain. Energy Rev. 38 732–46

[9] Thangaraja J and Kannan C 2016 Effect of exhaust gas recirculation on advanced diesel combustion and alternate fuels - A review Appl. Energy 180 169–84

[10] Mathivanan K, J. M. Mallikarjuna and Ramesh A 2016 Influence of multiple fuel injection strategies on performance and combustion characteristics of a diesel fuelled HCCI engine – An experimental investigation Exp. Therm. Fluid Sci. 77 337–46

[11] Lü X C, Chen W and Huang Z 2005 A fundamental study on the control of the HCCI combustion and emissions by fuel design concept combined with controllable EGR. Part 1. the basic characteristics of HCCI combustion Fuel 84 1074–83

[12] Lü X C, Chen W and Huang Z 2005 A fundamental study on the control of the HCCI combustion and emissions by fuel design concept combined with controllable EGR. Part 2. Effect of operating conditions and EGR on HCCI combustion Fuel 84 1084–92

[13] Gawale G R and Naga Srinivasulu G 2020 Experimental investigation of ethanol/diesel and ethanol/biodiesel on dual fuel mode HCCI engine for different engine load conditions Fuel 263 116725

[14] Putrasari Y, Jamsran N and Lim O 2017 An investigation on the DME HCCI autoignition under EGR and boosted operation Fuel 200 447–57

[15] Aydoğan B 2020 An experimental examination of the effects of n-hexane and n-heptane fuel blends on combustion, performance and emissions characteristics in a HCCI engine Energy 192 116600

[16] Calam A, Aydoğan B and Halis S 2020 The comparison of combustion, engine performance and emission characteristics of ethanol, methanol, fusel oil, butanol, isopropanol and naphtha with n-heptane blends on HCCI engine Fuel 266 117071

[17] Iliev S 2015 A comparison of ethanol and methanol blending with gasoline using a 1-D engine model Procedia Engineering vol 100 (Elsevier Ltd) pp 1013–22

[18] Blomberg C K, Wright Y M and Boulouchos K 2018 A phenomenological HCCI combustion model in 0D and 3D-CFD Fuel 226 365–80

[19] Xu H, Yao C and Xu G 2012 Chemical kinetic mechanism and a skeletal model for oxidation of n-heptane/methanol fuel blends Fuel 93 625–31

[20] Tsurushima T 2009 A new skeletal PRF kinetic model for HCCI combustion Proc. Combust. Inst. 32 II 2835–41

[21] Zeuch T, Moréac G, Ahmed S S and Mauss F 2008 A comprehensive skeletal mechanism for the oxidation of n-heptane generated by chemistry-guided reduction Combust. Flame 155 651–74

[22] Raya B 2006 Chemical kinetic mechanism reduction, multizone and 3D-CRFD modelling of homgeneous charge compression ignition engines

AUTHOR INDEX